"Current and Projected Technologies for Food Protection—Recommendations and Implementation"
Proceedings of the 1986 Conference for Food Protection

CHARLES W. FELIX

Food Protection Technology

LEWIS PUBLISHERS, INC.

Library of Congress Cataloging-in-Publication Data

Food protection technology.

Papers presented at the Third Conference for Food Protection, Ann Arbor, Mich., Aug. 17–20, 1986.
Includes bibliographies and index.
1. Food contamination—Congresses. 2. Food—Microbiology—Congresses. I. Felix, Charles W. II. Conference for Food Protection (3rd : 1986 : Ann Arbor, Mich.)
TX511.F678 1987 664′.07 86-27745
ISBN 0-87371-047-9

Second Printing 1988

LEWIS PUBLISHERS, INC.
121 South Main Street, P.O. Drawer 519, Chelsea, Michigan 48118

PRINTED IN THE UNITED STATES OF AMERICA

PREFACE

The Conference for Food Protection was conceived in the late 1960s as the brainchild of environmental health officials responsible for food safety in federal, state, and local jurisdictions, and their counterparts in the food industry. They decided that much would be gained by having the leaders in food protection in the United States come together for a few days of earnest discussion about the major food protection problems of the day, with a view to recommending solutions to those problems.

The first Conference for Food Protection was held in Denver, Colorado, in the spring of 1971. Attendance was by invitation only. About 400 food protection specialists participated. Their discussions resulted in more than 100 recommendations, which ranged from a sanguine call for the formation of a single federal food protection agency to the introduction of a new method of sanitation control now known as Hazard Analysis Critical Control Point (HACCP) procedure.

The Denver Conference was financed by the U.S. Food and Drug Administration under a contract with the American Public Health Association. In the years following, APHA made a valiant effort to secure funding for a follow-up conference, but to no avail. Federal funds for a reprise of the undertaking were no longer available. At last, APHA sent an invitation to like-minded professional and trade associations to join it in a Study Committee for a Conference for Food Protection. More than 70 organizations, including federal agencies and national food companies, banded together to promote the concept of a conference that would be self-supporting.

Under that multiple sponsorship, the Second National Conference for Food Protection was held in Washington, DC, in May 1984. Again more than 400 interested parties attended. They included not only federal, state, and local health officials, but also industry executives, academicians, and consumer representatives. Their deliberations resulted in more than 150 recommendations for the improvement of food safety to the year 1990.

The very first recommendation to come out of the 1984 Conference was the decision to establish a continuing organization, national and international in scope, that would see to it that a food protection conference would be held every two years. The Conference was incorporated in 1985 and, with the assistance of the National Sanitation Foundation, which adopted the Conference as a management project, a third meeting was scheduled for the summer of 1986.

The 1986 Conference met in Ann Arbor, Michigan, on August 17–20, 1986. Over 200 Conference members convened to consider the theme "Current and Projected Technologies for Food Protection—Recommendations and Implementation." As in the 1984 Conference, they were divided into seven working committees representing the major technical perspectives of the participants. These committees, operating units of the Conference organization, include Toxicology; Microbiology; Good Manufacturing Practices and Quality Control; Standards and Regulations; Education and Training; New Foods, Food Processing, and Packaging; and the Conference Program Committee, which is responsible for making recommendations regarding the structure and policy of the Conference as an organization.

The 60 recommendations of the 1986 Conference were the product of the following procedure: The 25-member executive committee decided among themselves the subjects that would be discussed in each committee under the agreed-upon theme. They invited technical experts to develop "white papers" on the selected subjects. The white papers were reviewed by appointed reviewers who represented the spectrum of Conference membership: government, industry, consumers, and academia. The papers were distributed to registrants before the meeting. At the meeting, participants reviewed the main points of the papers, including suggested recommendations, and entered into discussion.

The recommendations finally agreed upon in each committee, and a few minority reports of dissenting views, were reviewed by the Conference's five councils. The councils correspond to the principal interest groups represented at the meeting: government, industry, consumers, academia, and professional associations. Their viewpoints on the appropriateness of the technical recommendations were then conveyed to the membership assembled in plenary session for voting. A vote was then taken on each of the recommendations in turn, following full discussion of the issues as desired.

What you will find in these pages are the white papers which formed the basis for discussion at the Conference, and the recommendations which came out of those discussions. In between the two, there were hours and hours of serious, sometimes heated, exchange on the issues, as people of good will and sound science sought to resolve the serious problems of food safety presented to them.

This book should appeal to anyone who has a responsibility for food safety or an interest in it. That will include regulators at every level of government, food industry executives, professors and students in schools of food science and public health, and very many consumers whose interest in food protection has been aroused by the ever more frequent incidents of food contamination and foodborne illness that have been reported in the media in recent years.

This volume will also be of keen interest to members of the Conference for Food Protection. It is the official proceedings of their third Conference meeting and a continuation of the dialogue they entered into when they joined the Conference as individual or organizational members. We hope it will attract those who read it to enter into that dialogue by becoming, themselves, members of the Conference for Food Protection, committing themselves and their organizations to the goal of the Conference as enunciated in its constitution:

"To foster public health by promoting safety in the production, processing, packaging, sale and service of foods; and to provide a continuing forum for the exchange of information among all national and international parties interested in food safety."

To all who share that ambition, I recommend the Conference for Food Protection and this publication of its 1986 proceedings.

Charles W. Felix
Chairman
1986 Conference for Food Protection

TABLE OF CONTENTS

EDUCATION AND TRAINING

NEW FOODS, PROCESSING, AND PACKAGING

INTRODUCTION: A New Manifesto for Food Safety

Sanford A. Miller

Director, Center for Food Safety and Applied Nutrition
U.S. Food and Drug Administration

The Conference for Food Protection has made tremendous progress since its first meeting in 1971. It is heartening to know that an ongoing organization, with a constitution and by-laws, has now been established to provide structure and continuity. The Center for Food Safety and Applied Nutrition at the Food and Drug Administration is proud to have been involved from the beginning and to continue as a supporting member of this newly incorporated Conference. All signs indicate that this organization will play an ever-growing role in identifying and expressing future issues of food safety policy. Its special approach of bringing together the best among industry, government, academia, and consumers is to be applauded. Today, food protection decisions carry important social and economic implications. Now, more than ever, food safety issues are complex and far-reaching in scope. To address them constructively and efficiently, we need to recognize the multiplicity of their components and to develop ways to bring them together. This Conference is a primary step in this direction.

In this introduction, I intend first to present my views of the scientific and policy environment of this Conference. Second, I will articulate what I consider the primary goal of food safety policies in today's environment. Third, I'll present some strategic concepts that Conference members should keep in mind as they review and formulate recommendations. Finally, I want to mention the societal and intellectual underpinnings for acting upon the Conference's recommendations.

ENVIRONMENT

The environment in which we consider food protection today can be described in two words: change and complexity. But enmeshed with these descriptors are a number of realizations that present a richer, more complete picture of the modern context for food protection.

Multidimensional Problems

Most prominent among these is our increasing recognition of the multidimensional nature of food protection issues. No longer do our problems appear univariable. Consider, for example, just this minor litany of questions we face: What is the role of diet in immune response? How does nutritional deprivation affect reproductive function and fetal and perinatal development? What is the bioavailability of nutrients in fabricated foods? What impact will new food substances have on nutritional balance and toxicology? What are the optimal nutritional needs of laboratory animals and how do animal diets affect toxicological test endpoints? What are the beneficial consequences of some food constituents? What are valid health messages in food labeling? What methods can we develop to provide reliable, routine quantitative and qualitative analyses of food constituents and contaminants? What factors cause foodborne microbiological virulence? How can microbiological detection programs be improved? The list could go on and on. The questions are not simple. Rather, they deal with complicated processing technologies, complex issues of chronic and functional toxicities, and the mysterious web of human biology. What *is* clear is that addressing them will require our brightest scientific minds working at the frontiers of chemistry, microbiology, nutrition, toxicology, genetics, food technology, and a host of other disciplines and specialties.

Nutrient Patterns

A second factor shaping today's environment is our growing realization that nutrient patterns are critical in the disease process. Our dietary state, as determined by these patterns, provides a basic physiologic state that influences our responses to challenges from microorganisms, chemical and environmental stresses, and degenerative disease processes. Conversely, we are beginning to recognize that microbiological load, chemical load, and other factors can modify human nutritional requirements. The National Research Council dramatically recognized this point in its report "Diet, Nutrition and Health," where it emphasized that "From the standpoint of public education and public health . . . it is considerably less important to identify isolated compounds that cause or protect against certain cancers than it is to identify dietary patterns that enhance or minimize overall risk."

Today, virtually every tolerance or action level we set, and every recommendation concerning the safety of an additive we make, is affected by our knowledge or ignorance of the role of diet in the toxicity of individual substances. With our recognition of this fact, we need to formulate better questions. Today we ask, "Is a compound safe?" and "How many bacteria can I allow?" We need to broaden these questions, to ask: Safe for whom? In what kinds of diets? Under what conditions? And when? We need a greater focus on the "It depends . . ." aspect of our questions, always remembering the importance of nutrient patterns.

Health Promotion through Food

Yet another characteristic of today's food policy environment centers on the mounting belief that food is more than sustenance—that food can play a significant role in enhancing the quality of our lives. Modern attitudes assume that disease is not part of our destiny, but that heart attacks, cancer, and other devastating debilities can be controlled. And diet is viewed as a vital part of that control system. In this environment, more is expected—in data, knowledge, and understanding—from us all, whether we work in government, industry, or universities. Food issues now encompass not only *protection* from hazards, but also *promotion* of health, via better diet.

Complex and Changing Technology

Although there are certainly many others, I'd like to mention one additional factor that I view as extraordinarily important in defining our environment. I refer to the ever-increasing complexity of our food production, processing, and distribution system. Because of concentration within the food industry, a problem in a single plant today can affect millions of people of broadly diversified physiologic states. Just a few months ago, for example, a maker of frozen Italian-style meat, cheese, and pasta items voluntarily recalled its products because of possible *Salmonella* contamination. These foods were packaged in a variety of sizes from 2½-pound aluminum pans to 23-pound plastic bags for commercial use. The firm involved had 1000 distributors and its products went to caterers, restaurants, universities, hospitals, and similar institutions.

Certainly, one of the greatest challenges posed by this environment of new and rapidly changing food technologies is developing appropriate monitoring and regulatory strategies for dealing with them. Traditional technologies in which control concepts have long been well known have been replaced with new technologies interacting in new ways, making it more difficult to identify critical safety points.

Perhaps nowhere is this situation better illustrated than in the milk business. In the past, we've concentrated on pasteurization. Now, however, the economics of milk production and distribution increase the possibility of postpasteurization contamination. In one milk incident that occurred between March and April 1985, more than 16,000 culture-confirmed cases of illness from *Salmonella typhimurium* occurred in northern Illinois and the surrounding states. The food vehicle was 2% low-fat milk produced by a dairy that supplied about 40% of the milk consumed in the Chicago area. Estimates indicate that as many as 200,000 individuals may have been affected. The Illinois Department of Health reported the outbreak directly caused the deaths of two people and was a contributing factor in the deaths of several other individuals who had underlying illness and succumbed to infection.

Our concerns are increasing as new and more frequent incidents of food microbial contamination are becoming apparent. In part, the problems may seem greater because we are more alert to their possibility. But we can't ignore the fact that certain susceptible populations may now be exposed to microbial hazards they might otherwise not have known. And new technologies—such as computer-operated processing—are challenging our vigilance and our resourcefulness. The changes in our food system have been essential for its effective, efficient, and smooth operation, but they have created a new world for formulators of protection measures and policies. And I have not mentioned the constantly changing, often political environment, or the impact on the gross national product of food safety decisions.

GOAL OF MODERN FOOD POLICY

Given this new world, a world of multidimensional problems, new and complex technologies, and changing social, political, and economic attitudes; a world in which nutritional patterns become important; and a world in which we must deal with new sets of questions, what should be the goal of food protection policies? For me, that goal is straightforward. We must provide the American people with a food supply that meets *all* of their expectations. More complex are those expectations themselves. Certainly, Americans expect a safe food supply: a food supply that won't interfere with their health. But beyond that, they also want a food supply that will promote and improve their health. Moreover, they want food that is abundant, pleasing, varied, and economical. Acceptance of this ambitious goal means we must have food policies that encourage innovation and at the same time assure that innovation is nonthreatening.

STRATEGIC CONCEPTS

How do we accomplish that goal? This Conference has assumed responsibility—at least in part—for suggesting one set of options, one set of pathways to achieving that goal. In the past, it has undertaken this responsibility diligently and admirably through extensive and carefully contemplated recommendations. Now, however, as members review past recommendations and consider new ones, I'd like to challenge them to take the process a step farther—to make recommendations with certain key strategic concepts in mind.

Problem vs Discipline Orientation

Primary among these is the importance of becoming oriented to problems rather than disciplines. We must embrace our issues in the broadest sense of the word. And we must provide the education, the resources, and the environment for doing so to those who deal with food issues. In a world of multidimensional issues permeating a society of rapidly changing technologies, it is not sufficient for us to ask only about toxicology, or only about chemistry, or solely about any other single discipline. When we talk about coronary artery disease and fats, for example, we should also talk about the kinds of fats, the technology of fats, lifestyle, economics, and a host of other considerations.

Need for Priorities

A second key concept I challenge Conference members to keep in mind is the need for priorities. As each year goes by, those of us concerned with food protection are experiencing the growing pressures and the mounting frustrations which arise from the necessity of satisfying competing priorities. Take, for example, the tension now existing between the need for vigorous compliance activities and the need for a strong, stable science base. *Better* enforcement requires the very *best* in science. How do we adequately satisfy these needs? I encourage Conference members to think about this situation and to consider the importance of allocating resources as well as identifying activities. I also encourage them to think about priorities with respect to the types of food hazards. As I've noted, a resurgence of interest regarding microbial issues is under way. The direct and indirect costs of foodborne illnesses could and do amount to billions of dollars annually. But more tragic are the fatal or chronic effects that may follow these episodes, effects that often affect susceptible populations such as the elderly and the newborn. These facts must be weighed against other food protection concerns. For example, at the same time concern about microbial hazards is increasing, the public and the Congress continue to pressure regulatory agencies to increase their efforts with respect to pesticides and food chemicals, even though Center for Food Safety review of data on nearly 1600 food chemicals is offering encouraging indications about the safety of those substances. How do we factor in these perspectives to our food policies?

We cannot abandon *any* issues of food safety, but we must think carefully about those which are most important to us—those which carry the biggest public health impact. We must arrange our priorities accordingly. Conference members considering their recommendations should ponder how they relate to each other, how they affect various populations, and how they may influence public health. Priority setting is a soul-wrenching process which we all must face. Priorities will be set whether we do it directly, or whether we allow them to be established through apathy or neglect. I ask that this reality be kept in mind.

Integrated Policies

Food policies arise on a variety of fronts, and their sphere of influence is far-reaching. It's critical that this fact not be forgotten, and that efforts be made, where possible, to integrate food policies. To some extent, this is already happening. We are seeing efforts at international harmonization of standards through activities such as the Codex Alimentarius. Moreover, other meetings and groups already have recognized the importance of integrated policies. Notable among these has been the Administration's Working Group on Agricultural Research Needs, which acknowledged the need for improved linkage between human nutrition requirements and agricultural production. I sincerely hope that this Conference will continue this emerging trend of looking at policies in an integrated fashion.

INSTITUTIONAL AND INTELLECTUAL STRUCTURES FOR ACCOMPLISHING RECOMMENDATIONS

In my view, even if recommendations clearly establish priorities, integrate policies, and arise from a well-defined philosophy, the mechanisms through which they will be implemented must be addressed. If we are to approach problems in a broad-based fashion, we need the human and institutional resources to support such an approach.

At FDA, we have been moving in the direction of providing these resources. We began in 1979 by developing our Research Plan. Through its universal approach, this plan identified research needs within the context of the needs of the Center for Food Safety and Applied Nutrition. In this process, competing and overlapping needs also became more apparent. Next, we initiated new research approaches, such as our integrated nutrition-toxicology program. This program focuses the agency's resources into understanding the role that diet and nutrition play in the development and progression of toxicity. And third, we have altered our philosophical approach by moving from a set of descriptive activities to a set of activities that ask "Why?" Today, for example, microbiological efforts now aim to go beyond identifying hazards to also understanding their mechanisms and their interactions with other areas, such as nutrition and toxicology.

But the FDA is just one organization. The tasks that the recommendations of this Conference will imply are for all of society. I can't emphasize too strongly that we need to think about the settings in which we can achieve the broad-based education our leaders and scientists in food policies must have. And I can't emphasize too strongly that we must also consider the settings in which they will work—settings that offer the inspiration and insights that arise from the stimulation of collegiality and the freedom of independence. What institutions can best serve these needs? How can we best inform our colleagues of our conclusions, our goals, and our needs?

Finally, we must think about constituencies. We are entering a new era of nutrition and interest in health messages and health claims. What will be these messages? How will they be told? Who will provide them and who will decide? Agencies? The Congress? Consumer pressures? Economic interests? These are profound questions just beginning to emerge. The answers could dramatically influence the health and attitudes of our citizens. Conference members should think about these issues as well.

CONCLUSION

We are fortunate that our nation enjoys active citizens and an open government. We are free to ask, to challenge, to question, and to debate what we do and what affects us. Out of this system has come the safest, most abundant food supply in the world. But equally important, out of this system has come an aversion to complacency and a commitment to the ideal that we can improve; that we can grow; and that we can be better. This Conference embodies that commitment, and I congratulate its members on the constructive role it has taken in the arena of food policy. To paraphrase Charles Felix and Martha Rhodes in the preface to the report of the last Conference: This is the beginning, not the end, of a dialogue between regulators, industry representatives, academics, and consumer advisers. We in government have learned that assuring food protection is a continuous exercise, an evolving process of new situations and new solutions.

The issues are difficult and the task is not easy. Persons involved in food protection must both respond and initiate. They must remember that presenting questions without contemplating answers can be an empty effort, and that pursuing answers to poorly conceived questions can lead to misdirected efforts. The Conference for Food Protection is a unique means for addressing today's complex and critical issues of food policy. For this reason, its role is vital. Oliver Wendell Holmes once said, "Life is action and passion. A man must share the action and passion of his times or be thought not to have lived at all." In its own way, this is what this Conference is all about.

Conference for
Food Protection
P.O. Box 1468
Ann Arbor, Michigan 48106
Executive Office
Telephone 703/777-7448

Here, by Committee, are the people principally involved in developing the Conference agenda:

TOXICOLOGY
Officers
Gary Flamm, Food and Drug Administration, Chairman
Robert Neal, Chem. Industry Inst. of Toxicology, Co-Chairman
Sheldon Murphy, University of Washington, Secretary
Consultants
Melvin Andersen, US Air Force
Frank Cordle, FDA
Michael Gough, Environ Corporation
Ronald Hart, Natl. Ctr. for Toxicology Research
Robert Moolenaar, Dow Chemical Co.
Joseph Rodricks, Environ Corporation
Alan Rulis, FDA
Robert Scheuplein, FDA
Michael Taylor, Esq., King & Spaulding

MICROBIOLOGY
Officers
Frank Bryan, Consultant, Chairman
Carl Vanderzant, Texas A&M U. Co-Chairman
William Brown, ABC Research Corporation, Secretary
Consultants
Douglas Archer, FDA
J.S. Bailey, Russel Research Ctr.
Jane Ballard, Washington State Health Dept.
Charles Bartleson, Washington State Health Dept.
Paul Blake, Centers for Disease Control
Dean Cliver, Food Research Inst.
Donald Corlett, Del Monte Research Center
Nelson Cox, Russel Research Center
Michael Doyle, Food Research Inst.
Daniel Fung, Kansas State University
John Guzewich, New York State Health Dept.
Paul Hartman, Iowa State U.
James Jay, Wayne State U.
John Kobayashi, Washington State Health Dept.
Anthony Sharpe, Health & Welfare Canada
John Silliker, Silliker Laboratories
Ewen Todd, Health & Welfare Canada

STANDARDS & REGULATIONS
Officers
Edward C. Heffron, Michigan Dept. of Agriculture, Chairman
Irving Bell, Coca Cola Co., Co-Chairman
Ruth G. Desmond, Federation of Homemakers, Secretary
Consultants
John Adams, Natl. Milk Producers Federation

James Brown, National Restaurant Association
Robert Davis, Philadelphia Dept. of Health
Sherwin Gardner, Grocery Manufacturers of America
Maurice Guerrette, N.Y. Dept. of Ag. & Markets
Mary Heslin, Conn. Dept. of Consumer Protection
Carroll Johnson, Friendly Ice Cream Corp.
George Kupfer, Milwaukee Health Dept.
Thomas Masso, Minn. Dept. of Agriculture
Allen Mallthys, Natl. Food Processors Assn.
John McCutcheon, USDA
Frank McLaughlin, University of MD
Robert Reeves, Inst. of Shortening and Vegetable Oils
Darrell Schwalm, FDA
Robert Sielken, Texas A&M University
Joel Simpson, Dobbs International Services
Merrill Thompson, Chadwell & Kayser

GOOD MANUFACTURING PRACTICES AND QUALITY CONTROL

Officers

Howard E. Bauman, Pillsbury, Chairman
Philip E. Nelson, Purdue University, Co-Chairman
Earl Montgomery, U.S. Dept. of Agriculture, Secretary

Consultants

Dane Bernard, Natl. Food Processors Assn.
Roland Blanchett, Pillsbury Company
James Chambers, Purdue University
Cleve Denny, Natl. Food Processors Assn.
Roger Dickerson, Jr., FDA
Irving Flug, University of Minnesota
Thomas Holzinger, Borden
Ralph Johnston, USDA
Joseph Marcy, Rampart Pkg.
Elmer Marth, University of Wisconsin
Joseph Mayo, Pillsbury Company
Thomas Mulvaney, FDA
Donald Pusch, Pillsbury Company
Dale Seiberling, Private Consultant

EDUCATION & TRAINING

Officers

David McSwane, Indiana University, Chairman
Harry Haverland, Private Consultant, Co-Chairman
John C. Chrisman, Creative Dining, Secretary

Consultants

Rosella Bannister, Michigan Consumer Education Center
James Brown, National Restaurant Assn.
Evelyn DeNike, FDA
Lilyan Goossens, FDA
Esther Shapiro, City of Detroit

NEW FOODS, PROCESSING, AND PACKAGING

Officers

Irwin A. Taub, U.S. Army Natick Labs, Chairman
Daniel F. Farkas, U. of Delaware, Co-Chairman
Donald F. Emery, General Mills, Secretary

Consultants

Dane Bernard, Natl. Food Processors Assn.
Gregory Cramer, FDA
Peter Elias, Intl. Food Irradiation Project
Dennis Heldman, Campbell Soup Company
Marcus Karel, Mass. Institute of Technology
Robert Keown, University of Delaware
Daryl Lund, University of Wisconsin
Gerad McCowin, FDA
Nanette Newell, Genetic Engineering Consultant
Darrell Wood, Food Research Institute

CONFERENCE PROGRAM

Officers

Charles W. Felix, Private Consultant, Chairman
Martha E. Rhodes, FL Dept. of Agriculture, Co-Chairman
Lillian Mohr, Florida State University, Secretary

Consultants

Richard Dawson, U.N./Food & Agriculture Organization
Earl Dietz, National Sanitation Foundation
Fernando Guevedo, U.N./Pan American Health Org.
John Moore, Environmental Protection Agency
Howard Pippin, Health & Human Services/FDA
Judith Segal, U.S. Department of Agriculture/FSIS

TOXICOLOGY

1. An Historical Perspective on FDA's Use of Risk Assessment

Robert J. Scheuplein

Food and Drug Administration

Risk assessment can have very different meanings depending on the viewpoint of the author and the context of his application. Even from the narrow standpoint of the regulation of the safety of foods, FDA's practice of risk assessment, and its flip side, safety assurance, has undergone some significant changes through the years. Some have argued that these changes are so profound that they amount, in effect, to a conceptually different judgment—away from a comfortable assurance of absolute safety toward an assurance of some very small yet distinctly uncomfortable level of risk. If we do not believe that the food supply is less safe now than in former years—and we do not—how is it that the public perception of foodborne risk is now more pervasive? Part of the answer is that we have become the victims of both the advances in our technology and of the increased sophistication of our concerns. But the fine print, the inherent uncertainty in safety evaluation, has always been with us. During the hearings and the congressional committee work on the Food Additives Amendment of 1958, this uncertainty was given official status by the inclusion of the following "Concept of Safety" in the House Committee Report accompanying the Bill.[1]

> The concept of safety used in this legislation involves the question of whether a substance is hazardous to the health of man or animal.
>
> Safety requires proof of a reasonable certainty that no harm will result from the proposed use of an additive. *It does not—and cannot—require proof beyond any possible doubt that no harm will result under any conceivable circumstance.*
>
> This was emphasized particularly by the scientific panel which testified before the subcommittee. *The scientists pointed out that it is impossible in the present state of scientific knowledge to establish with complete certainty the absolute harmlessness of any chemical substance.*
>
> In determining the "safety" of an additive, scientists must take into

consideration the cumulative effect of such additive in the diet of man or animals over their respective life spans together with any chemically or pharmacologically related substances in such diet. Thus, the safety of a given additive involves informed judgments based on educated estimates by scientists and experts of the anticipated ingestion of an additive by man and animals under likely patterns of use.

Reasonable certainty determined in this fashion that an additive will be safe will protect the public health from harm and will permit sound progress in food technology.

The legislation adopts this concept of safety by requiring the Secretary to consider in addition to information with regard to the specific additive in question, among others, the following relevant factors: (1) the probability of consumption of the additive and of any substance formed in or on food because of the use of such additive; (2) the cumulative effect of such additive in the diet of man or animals, taking into account any chemically or pharmacologically related substances in such diet; and (3) safety factors which qualified experts consider appropriate for the use of animal experimentation data.

In determining the safety of an additive, the Secretary would have to consider not only the food to which the additive is directly added, but also other foods derived from such foods. For example, in evaluating the safety of an additive for poultry feed, the Secretary would have to consider any residues that might appear in eggs produced by the poultry.

Similarly, in determining the safety of additive-treated cattle feed, account would have to be taken of residues of the additive in the milk or edible flesh of the animal.

Since the scientific investigation and other relevant data to be taken into consideration by the Secretary include information with respect to possible cancer-causing characteristics of a proposed additive, the public will be protected from possible harm on this account.

The italic phrases particularly would seem even today to provide both sufficient awareness of the true nature of safety assurance as a judgmental art and also provide sufficient flexibility to deal with it. But the entire signal from Congress was ambiguous. While the concept of safety was sequestered safely in the Committee report, the Delaney Clause, with which it was wholly inconsistent, was established firmly in the statute and its implicit guarantee of absolute safety established firmly in the public mind.

The analytical process described in the Committee's Concept of Safety has in recent years come to be called "risk assessment," as it has also become more complex and arguably more quantitative. It has also become increasingly formalized as more of the decisions of Federal agencies regulating chronic health hazards have become more bitterly controversial. In 1983, the National Research Council under an FDA funded contract published its report on "Risk Assessment in the Federal Government: Managing the Process."[2] As the various health regulatory agencies developed procedures for risk assessment, the procedures themselves became a focus of criticism by scientists, industry representatives and public interest groups. The National Academy's objectives were to analyze the basis for the criticism and to recommend means of improving the reliability and credibility of the process.

It may be useful, in order to gain more perspective on the roots of the controversy, to take a look at the historical development of risk assessment as practiced for approximately 80 years by the FDA—primarily by the "Foods" part of the Agency.

THE CRUCIAL ROLE OF ANIMAL STUDIES

Although Dr. Harvey Wiley's early experiments on food additives were conducted on human volunteers—the famed "Poison Squad"—it soon became clear that studies with animals, not humans, would have to provide the scientific basis for FDA's food safety evaluations. Before the House Committee on Interstate and Foreign Commerce in 1906, Dr. Wiley noted that determining the physiological effects of chemicals on human subjects was not as easy or as straightforward as with animals. Concerning the use of borax as a preservative, the following discussion took place between Congressman James R. Mann and Dr. Wiley:[3]

> *Mr. Mann*:
> Does your report show that in your opinion the use of borax has a deleterious effect upon the organs of the body?
>
> *Dr. Wiley*:
> Of course, you understand, Mr. Mann, the tests that we have made are not the same as those made upon animals fed for pharmacological experiments because after a given time the animals are killed and their organs are examined, and the changes in the cells are studied by the microscope. We were precluded from doing that.
>
> *Mr. Mann*:
> Is that your conclusion?
>
> *Dr. Wiley*:
> My conclusion is that the cells must have been injured, but I had no

demonstration of it, because I would not kill the young men and examine the kidneys.

This colloquy between the Congressman and Dr. Wiley compresses into a brief moment a still accurate picture of the purpose and the need for animal studies, and the difficulties the FDA faces when explaining or interpreting such data. For 52 years between the Food and Drugs Act of 1906 and the Food Additives Amendment of 1958, the legal burden of proof that a food substance was unsafe fell on the government. A suspect substance was invariably tested in FDA (or USDA) laboratories in order to develop the scientific evidence that it might be injurious to consumers. In June 1950, the House created a select committee to investigate the use of chemicals in food products. The Committee Report, issued in June 1952, concluded that the law was inadequate to insure protection of the public health, and recommended new legislation to require premarket safety testing for chemicals.[4] In July 1958, the committee reported out a bill which was eventually signed into law as the Food Additives Amendment of 1958. This bill shifted the burden of proving the safety of food additives to the manufacturer.

The use of animal data as the experimental basis for FDA's risk assessment of food components continues to the present day. In their chapter in the 1955 revision of FDA's "Procedures for the Appraisal of the Toxicity of Chemicals in Food, Drugs and Cosmetics," Dr. Arnold J. Lehman and Dr. Geoffrey Woodard began their discussion on the necessity for animal studies with the following quotation:[5]

> People are rather unpredictable and don't always die when they are supposed to, and don't always recover when they should. All in all, we must depend heavily on laboratory experimentation for sound and controllable basic principles. — A. L. Tatum

And they continued:

> there have been statements, in many ways by many people, to the effect that in the final analysis it is the experience with man that really counts. However, this experience with man may be painful, perhaps even tragic, in its consequences, and to a great extent avoidable. To a large extent, the distress accompanying experience in man can be materially reduced if full advantage is taken of the usefulness of animal toxicity experimentation. . . . Some of the pertinent advantages that derive from animal experimentation are:
>
> (1) Relatively large numbers of animals can be used under conditions in which the majority of variables inherent in biological experimentation can be controlled. Environmental factors, such as tempera-

ture, humidity and housing, can be kept constant. A diet designed at will can be held constant. A strain of known genetic background and of a predictable incidence of spontaneous lesions may be chosen.

(2) At any given selected time during animal experiments, the animals can be sacrificed for exhaustive or specific histological examinations. Damage not detected by the usual whole animal organ function tests can often be found by the pathologist.

(3) The life span of most laboratory animals is such that several years of equivalent experience in man can be telescoped into a few weeks or months of animal experimentation, and a whole lifetime into two rat years. Abundant evidence that this can be done exists in the field of chemical carcinogenesis. Here the effects in man of ten to 30 years' exposure to a carcinogen have been duplicated in 12 to 30 months in laboratory animals.

(4) Moral considerations prevent the pushing up of dosages in clinical experimentation. In the experimental laboratory, comparable data can be developed without jeopardy to human life. Doses of a chemical can be pushed to the point of actual damage. Thus, in many cases it is possible to discover in a relatively short time, by larger dosages, an injury that a smaller amount of the chemical may only infrequently produce in man.

(5) In the vast majority of biochemical systems occurring in nature, we find a remarkable similarity, whether the system is in a yeast cell, a mouse liver, or a man. The differences are usually minor and infrequent, and have been overemphasized. It is important, of course, to be aware of the existence of these differences. In fact, when such differences in biochemical systems with species have been found and have become understood, they have become important research tools for the biologist. Most striking evidence for this point is found in the entire field of antibiotic therapy. Here selective toxicity with respect to species differentiation has been exploited with happy results for one of the species—man.

It would be foolhardy to pretend that all possible effects in man can be predicted from results of animal experimentation. Some of the areas that have been troublesome are analgesic, sedative, stimulant or depressant effects on the higher brain centers: blood and bone marrow disturbances: sensitizers: and chemotherapeutic activity.

Of course, systematic epidemiological investigations — including fortuitous observations from occupational exposures—provide in almost all cases an irrefutable basis for assessment of risk. Such observations of actual human harm were pivotal in the gradual strengthening of FDA's authority through the years, particularly in the drug area. They are, however, clearly deficient as a feasible procedure for the detection of foodborne chronic risks. And with the change in the law in 1958 requiring preclearance of regulated additives, animal investigations of food ingredients became essential.[7]

All of the early toxicological investigations conducted by FDA during the early years and up until the 1930s were acute or subchronic studies mainly on drugs. Animals predict acute effects in other animals and in humans fairly well, or it was learned quickly enough where they did not and predictions of human response were subsequently scaled accordingly. The general relation that man is approximately 10 times as sensitive as the rat on an mg/kg basis derives from such early work on drugs.[6] Until the thirties and forties, slight extensions of these studies were used for evaluating the safety of food substances also. For example, in 1925, the USDA authorized the direct addition of nitrite to cure meat on the basis of acute studies and until 1972 the ADI of saccharin of 1.0 g/d was based essentially on acute studies.

ADVENT OF CHRONIC STUDIES AND THE INCREASING COMPLEXITY OF TOXICOLOGICAL ISSUES

In the late twenties the increasing use of insecticide sprays lead to concerns over chronic ingestion of lead and arsenic in the form of lead and calcium arsenate from residues left on fruits and vegetables. In January 1927, an advisory committee recommended that the U.S. Department of Agriculture undertake the study of the chronic intoxication of both lead and arsenic. Funds eventually became available, and in 1935 this task was assigned to the newly established Division of Pharmacology in the Food and Drug Administration. Despite a large scientific literature that contained more than 10,000 references on either lead or arsenic toxicity, no study was found that definitely established in a single species of animal just how little arsenic or lead would cause toxicity when ingested over long periods of time.[8] A long-term study in animals was needed to answer this question, and studies in dogs and rats were begun. Basic work in toxicology, particularly in nutritional toxicology in the thirties and forties, was also laying the essential groundwork supporting the need for chronic studies.

The colors used in foods, drugs and cosmetics are an interesting example of this trend in approach to toxicology. The FD&C Act of June 25, 1938 contained the requirement that only certified coal-tar colors may be used for foods, drugs, and cosmetics. At the time, there were some 300 individual colors in use. On the basis of acute oral studies and 60-day subacute feeding studies in rats, and acute intraperitoneal studies in mice, as well as guinea pig sensitization tests, 27 *External D&C Colors*, 49 *D&C Colors*, but only 17 *FD&C* colors were certified. Only colors with chronic studies were deemed safe enough to permit lifetime, daily ingestion by both adults and children.

By 1943, toxicologists were beginning to appreciate the chronic study as an indispensable means of obtaining information on the toxicity of such diverse substances as ergot, heavy metals, e.g., lead, arsenic, cadmium, and selenium, and glycols as well as the colors. Chronic studies offered a means of observing three types of potential effects that are impossible to detect in an acute study: (1) a delayed effect—one that can be induced by a single administered dose but that requires time to develop, e.g., cancer; (2) an effect brought about by accumulated physical damage, e.g., heavy metal toxicity, and (3) an effect produced by substances that can be stored in tissue on continual administration, e.g., lipid-soluble substances; organic mercury. Chronic studies were considerably less than a lifetime in those days; their length has gradually increased over the years.

However, such studies were used only infrequently prior to the advent of technological aids which were rapidly introduced into food processing after WW II. They received additional impetus in the 50s with concern over the use of pesticides and the enactment of the Pesticides Residues Amendment of 1954 which required conformity with a tolerance established by FDA.

The 1955 FDA monograph cited above[5] contained the following statement:

> A food substance should show no chronic toxicity in animals in an amount equivalent to 100 times the proposed human use level, i.e., a safety factor of 100 should be present.

This monograph and the previous one in 1949[9] were official recognition that the approach to the appraisal of the safety of foods was somewhat different from that of a drug and that the emphasis on chronic studies required further guidance in their design, conduct, and interpretation. With any chronic study one can expect the pathologic changes to be less clearly related to treatment than acute studies because (1) they are occurring more slowly, usually at lower doses, and cannot be checked and replicated in short time period, and (2) because they may be confounded with other chronic degenerative changes. Since the tests were longer, they involved a considerable investment of time and resources; it made sense to ensure that the data that industry was relying on to support its product and was submitting to FDA was designed appropriately. But other questions were also asked. What is the most appropriate toxicological end point? For example, with lead, is it the effect on growth rate, the effect on reproduction, paralysis from the various forms of encephalopathy, blood changes, enzymatic changes, or tumor formation? Subtle incipient toxicological signs whose detection depended as much or more on the sensitivity of the methods as on the biological significance of the effects were being observed. Toxicologists were beginning to be able to detect biochemical and physiological changes whose toxicological significance was uncertain.

FOCUS ON CARCINOGENICITY

When carcinogenicity was the end point, the problems of study design, conduct, and interpretation mounted. The FDA monograph in 1949 made no mention of carcinogenicity studies; they were subsumed in the section on chronic studies. The 1955 monograph has a separate section on carcinogenicity and contains the statement:

> . . . positive results in these animal tests can be taken as creating a suspicion that the chemical under study may be carcinogenic for man, but do not prove it to be so.

Although it is not definitely known whether all animal carcinogens will also produce cancer in humans, the converse is true. With the possible exception of arsenicals, every chemical known to produce cancer in man will likewise do so in one or more species of animal. It has been argued that these observations provide evidence for the qualitative validity of animal carcinogenicity studies.

The 1955 monograph also recognized the difficulty in detecting weak carcinogens, the advisability of testing in two species of rodents, the care needed to select genetically appropriate animals, and the importance of histologic evaluation of tumor-bearing animals.

Four years later the 1959 monograph[6] contained a separate section entitled "Carcinogenicity Screening" and added: dietary factors, the proper number of animals, the grading of malignancy and others to the growing number of scientific issues to be considered.

The cancer studies produced special problems of their own.

1. They exacerbated the familiar question of human relevance. The genetic component of carcinogenesis makes it of paramount importance to select a strain of animal that is not only readily available, generally healthy and disease free, but also one that at the same time exhibits good sensitivity to carcinogens. But there is no certain way to do this.

2. The desire to test for weak carcinogens requires a large number of experimental animals to compensate for deaths unrelated to treatment. Unfortunately information increases only in proportion to the square root of the number of animals and frequently the number of animals used represents a compromise between what we want to know and what we can afford to find out. Even with the use of 50 or more animals per group, the significance of the results may be borderline.
3. The use of high test doses (the MTD issue) to induce a certain yield in tumors in a manageable number of animals may lead to shifts in metabolic pathways and target organs that can obscure the interpretation of the study, as well as the prediction of human risk at lower levels.
4. And last, there is the problem of obtaining statistical significance. The confounding which can produce false negative and false positive responses can be more frequent when the studies are designed to magnify a weak effect to within observable limits so that the results can be extrapolated 2–4 orders of magnitude.

All of these technical problems were controversial issues in our evaluation of saccharin, aspartame, cyclamate, nitrites, and FD&C Red #2.

THE SAFETY DECISION – INTERACTION BETWEEN SCIENCE AND THE LAW

The 1958 Food Additives Amendment established Sec. 409 of the Act and subsection 409(c)3 directs FDA not to approve a food additive petition: . . . if a fair evaluation of the data

> A. fails to establish that the proposed use of the food additive, under the conditions of use . . . will be safe . . .

Two points warrant explanation: (1) First, by requiring that the data *establish* safety, Congress clearly placed the burden of proving safety on the petitioner. This is important because if data fall in the grey area and there is a genuine uncertainty about safety, the petitioner loses. The placement of the burden of proof on the petitioner is a fundamentally important factor, and it even has a bearing on how scientific data are interpreted. The FDA scientist who accepts an obligation to protect the public can take a skeptical view of a seemingly minor irregularity in the data. The petitioner or his scientific staff who after all have firsthand knowledge of the study, and also some degree of enthusiasm for the product, sometimes cannot understand what appears to them a highly conservative and wildly improbable interpretation of their data. But this tension is inevitable, given the mission of FDA and the placement of the legal burden of proof. There are also more subtle problems. Scientific concerns and suspicions over what may be lurking behind irregularities and deficiencies in data tend to arise from the leading edge of our scientific imagination where the influence of new and often unproven scientific developments is greatest. In contrast, the resolution of these uncertainties tends to lag behind in a skeptical demand for adequate information to answer the questions that have been raised. The question of when enough is enough, is one of the tasks of our senior level scientists who must try to assure that a sensible balance is maintained between careful scrutiny and reasonable assurance. (2) The second essential point is the meaning of *safe*. It is interpreted by FDA and the courts as meaning that . . . *there is a reasonable certainty in the minds of competent scientists that the substance is not harmful under the intended conditions of use*. Congress, in predicating this standard on its Concept of Safety, was mindful of expert testimony emphasizing the impossibility of proving the absolute harmlessness of any chemical. However, with the simultaneous promulgation of the Delaney Clause, it seems to many people now as it did to FDA for many years that Congress excepted carcinogenic food additives or constituents in food additives from the inherent limitations of the scientific method. A Paracelsus may assert that the dose makes the poison, but the Delaney Clause requires that a noncarcinogen must be a noncarcinogen at all tested doses.

Another vital point about these standards is that they are so phrased to require proof of a negative. The fault is not with the law. Hazard can be measured, safety cannot. As long as carcinogenic effects are excluded, this definition did not and does not give FDA great difficulty. Safety was established by demonstrating an effect level, finding a threshold or no-effect level (NOEL) at a lower dose and then "backing off" by a factor of 100 to locate the "safe dose." As long as scientific opinion agreed that toxicity effects in animals were thresholdable and reasonably akin to those in humans, this could be considered an acceptable demonstration of "a reasonable certainty of no harm."

The situation is fundamentally different for carcinogens. How do you determine that a chemical is not carcinogenic in an animal study? It is generally held that carcinogen bioassays cannot be used to determine carcinogen thresholds. Depending on your view of cancer mechanisms, such thresholds either don't exist in principle, or they do, but at doses too low to be observed experimentally with a reasonable number of animals. So the traditional approach of finding a NOEL and backing off to lower doses won't work. Of course, if there is a positive effect, the carcinogens governed by the Delaney Clause are banned anyway. For the carcinogens that are not covered by the Delaney Clause, some risk at all doses must be assumed and questions of risk acceptability and risk assessment arise.

LIMITED SENSITIVITY OF WHOLE ANIMAL STUDIES

The measurement of both toxic and carcinogenic responses by means of animal studies are governed by the same statistical limitations. The observed outcome of no tumors among 100 animals in an idealized case with no background incidence only provides assurance, at 99% probability level, that the true risk is under 4.5%. This is so because the experiment with 100 animals is statistically too insensitive to detect at that level of confidence an incidence smaller than 4.5% *if it were to occur in the study*. Actual animal experiments with background incidences in the control groups and nonrandom differences in animal breeding, selection, husbandry, and in other confounding variables are even less sensitive. Similarly the observations of noeffect in ordinary toxicology studies are equally insensitive to the existence of signs of toxicity below the sensitivity of the method. Why then has the safety factor approach, which is predicated in the belief of a true experimental no-effect level, been widely accepted for classical toxicology and generally rejected for carcinogens? The answer is that toxicologists believe in true thresholds for the vast majority of substances with the possible exception of carcinogens. In addition, they believe that the distribution of individual thresholds in animals and humans are sufficiently narrow and the distribution averages sufficiently similar in animals and humans that a safety factor of 100 adequately encompasses the spread of responses.* To the statistician's question, "How can you be certain you are failing to detect an adverse response below the sensitivity of the study?" the toxicologist has generally replied, "I have selected the doses (by previous studies) to give an easily measurable effect in the animals, e.g., an ED_{50}, and then I have reduced the dose until the effect is unobserved. Many other similar studies including those in which such animals have lived for a lifetime with no ill effects following such no-effect doses convinces me that my experiment truly determines a threshold."

*Skin sensitization and a few other idiosyncratic responses to chemicals constitute exceptions to this general rule. Here it is likely that the distribution of individual thresholds is extraordinarily wide.

Unfortunately, this confidence in the existence of a measurable population threshold and in its narrow distribution is not nearly so well placed with carcinogenesis. It was apparent from the work of investigators as early as the 1940s that exposure to low levels of some carcinogens for an extended period approaching major portions of an animal's lifetime could be far more hazardous than predicted solely on the basis of total dose.[10-12] (These studies, however, were conducted for the most part on relatively potent carcinogens, e.g., the nitrosamines and the carcinogenic polynuclear hydrocarbons and not on the weaker carcinogens — a distinction that may yet prove significant.) The discovery that a single exposure to an initiator could produce a latent effect that was essentially permanent and might be manifested even if treatment with the promoter were delayed for up to half the lifetime of the animal,[13, 14] and the discovery that small persistent doses can be more effective than single large doses in producing tumors[10] were new and surprising findings to toxicologists. Such observations emphasized the potential importance of low doses and suggested that for carcinogens the experimental no-effect level observed with a limited number of animals might truly reflect the test's statistical insensitivity. These observations supported the concerns of scientists and regulators during the 1958 and 1960 Congressional hearings and helped prompt Secretary Flemming's testimony in support of the Delaney Clause.[15]

> The preponderance of scientific evidence clearly dictates our position: Our advocacy of the anti-cancer proviso in the proposed color additives amendment is based on the simple fact that no one knows how to set a safe tolerance for substances in human foods when those substances are known to cause cancer when added to the diet of animals. I should like to underline again one statement of Dr. (G. Burroughs) Mider's review of the role of certain chemical and physical agents in relation to cancer. It is this:
>
> > "*No one at this time can tell how much or how little of a carcinogen would be required to produce cancer in any human being, or how long it would take the cancer to develop.*"
>
> This is why we have no hesitancy in advocating the inclusion of the anti-cancer clause.
>
> Unless and (until there) is a sound scientific basis for the establishment of tolerances for carcinogens, I believe the Government has a duty to make clear — in law as well as in administrative policy — that it will do

everything possible to put persons in a position where they will not unnecessarily be adding residues of carcinogens to the diet.

There are currently two extreme classes of models to account for the increase in the carcinogenesis response rate with dose, both of which preclude the possibility of a measurable population threshold. One is premised in the existence of individual thresholds or tolerances that are so widely distributed that some animals (or humans) will react even at the smallest dose. This was the premise behind the original model of Mantel and Bryan.[16] The other extreme class of models is premised on an essentially uniform distribution of test animals, i.e., uniform in their inherent responsiveness, and a stochastic model for carcinogenesis in which the probability of response is an increasing function of dose, from the smallest dose to the highest. Virtually all current mathematical models, multistage, multihit, logistic . . . fall into this latter category although the more recent ones tend to be a blend of the two extremes.[17]

THE INCREASING ROLE OF PEER REVIEW

The concerns over the sensitivity of carcinogen studies led to efforts to obtain consensus as to the proper design, conduct, and interpretation of a carcinogen bioassay. A test with too few animals would be suspect on its face and a test with too many animals would be unmanageable, as well as too profligate of time and resources. The idea was to devise the best test possible, describe the protocols for its conduct and the criteria for its interpretation, and then hold everyone accountable to the rules. Bonafide negative studies are those that are negative under these rules, and the problem of inadequate sensitivity is resolved to legal satisfaction, if not to scientific certainty.

The increasing role and reliance on study protocols, GLPs, interpretive guidelines, and peer reviews is too familiar to emphasize here. What may not be so apparent is that their increasing role represents an admission that science is less able to reach strictly objective decisions. A supposedly straightforward scientific evaluation process now requires more of the skills usually expected of accountants, negotiators, and, of course, attorneys. Because of the inadequacy of our basic scientific information, we can only assess the evidence of carcinogenicity by negotiated debate, a process less capable of revealing nature's truths than gaining consensus and otherwise preparing the position against legal or societal challenges.

The first systematic effort by FDA to develop a consensus view on the appropriate design, conduct, and interpretation of cancer bioassays for food additives and pesticides began in 1969. FDA established an Advisory Committee on Protocols for Safety Evaluation with a special Panel on Carcinogenesis. The Committee consisted entirely of scientists from outside the FDA, but included Dr. Leo Friedman, who, in the interim became Director of the Division of Pharmacology and Toxicology, Bureau of Science, at FDA. Their comprehensive report was published in 1971.[7] Part of it grappled with the problems posed by the statistical insensitivity of carcinogen bioassays. It was recognized that the upper confidence limits on negative tests were uncomfortably large. Even with as many as 1000 test animals and using only 90% confidence limits, the upper limit yielded by a negative experiment is 2–3 cancers per 1000 test animals. "No one would wish to introduce an agent into a human population for which no more could be said than it would probably produce no more than 2 tumors per 1000."[7] It was clear that, despite the difficulties that are introduced, the only practicable basis for estimating the safe dose was by extrapolation downward from results obtained at elevated doses well above the use level. It was recommended that testing should be done in two species of rodents at doses and under experimental conditions likely to yield maximum tumor incidence. This would mean the use of doses several orders of magnitude above the actual use levels. The Committee endorsed, as a prudent model, a form of linear extrapolation quite similar to the Gaylor-Kodell procedure[18] sometimes used by CFSAN today. And following Mantel and Bryan,[16] the Committee referred to a defined level of "virtual safety" of 10^{-8}.

For agents not judged carcinogenic, the Committee recommended the use of a 100-fold safety factor.[7] The noncarcinogenic chemicals are evaluated on the basis of toxic effects other than carcinogenicity. While recognizing that ". . . no absolute guarantee of the safety of such a use level could be made," the Committee was unwilling to recommend using linear extrapolation from an upper confidence limit to the use level for substances for which there was no positive evidence of carcinogenicity. Such a policy of applying the linear model to chemicals found negative in adequate bioassays could lead to unrealistic regulatory decisions. For example, a chemical yielding no tumors in 100 mice at 1.0% in the diet would be limited to 0.22 ppm in the human diet extrapolating from the upper 99% confidence limit of the experimental result, assuming no tumors in the controls and using no special interspecies "scale-up" factor:

$$[d_{use} = 10^{-6}/\mathrm{UCL} \times d_{exp} = 10^{-6}/0.045 \times 0.01 = 0.22 \times 10^{-6}]$$

Were the linear model applied to all chemicals in this way, it could result in the elimination of many of them on the grounds of potential carcinogenic risk, even though there was no experimental evidence of their carcinogenicity. The Committee's recommendations and FDA's preference to use separate procedures: high to low dose extrapolation for adequately tested carcinogenic compounds and appropriate safety factors for others, have persisted to the present day. The current procedure is not strictly valid scientifically for there can be carcinogenic compounds which do not exhibit a statistically significant result because the experiment was too small. The protocols are a compromise. The fundamental problem is that one cannot assure risk levels in the range of 10^{-6}–10^{-8} based upon a few hundred test animals if significant usage levels are desired.

While endorsing testing at high doses and high-to-low dose extrapolation as a pragmatic solution to the problem of test insensitivity, the FDA Advisory Committee was aware the solution itself was troubling. They emphasized the extreme unreliability of extrapolations outside the observable experimental range and warned against placing ". . . excessive reliance on mathematical sleight of hand, particularly when the dose-response curves are largely empirical descriptions, lacking any theoretical physical or chemical basis. . . ." Later groups were to emphasize the problems of interpretation and human relevance that are exacerbated by the use of high doses—the MTD (maximum tolerated dose) issue—and this is still a controversial area.[19,20]

The linear model is, in the vast majority of cases, the most conservative of the extrapolation models and its predicted risk at low doses serves as an upper bound to the risks predicted by the other models. It was for this reason, i.e., the conservative nature of the extrapolation, that it was suggested by the Committee. They felt it was the practical equivalent of applying the Delaney Clause to all bonafide carcinogens.

APPLICATION OF RISK ASSESSMENT TO FDA REGULATION

The first official application of extrapolation-based risk assessment was the animal drug regulations in July 1973 which authorized how the "methods for examination" called for by the DES proviso were to be conducted. In 1962, Congress, at the urging of the Secretary of HEW, altered the flat prohibitions against added carcinogens embodied in the 1958 food additives and 1960 color additives Delaney Clauses (Sections 409(c)(3)(A) and 706(b)(5)(B)). The Secretary was permitted to approve carcinogenic food and color additives intended for use as feed additives for animals if two conditions were met:

1. The additive did not adversely affect the target animal.
2. "No residues" of the additive were found in the animal by

methods of examination approved by FDA through regulations.

In 1968, new animal drugs were made subject to the same "no residue" exception. There was to be a great deal of internal difficulty in implementation of this provision. (It wouldn't be promulgated by FDA in final form for 23 years.) The problem was that one couldn't interpret "no residue" literally, since there would always be some residue below the sensitivity of the approved method. For a long time, continuing into 1977, the FDA believed it was obliged to insist on the use of increasingly more sensitive methods as they were developed. This resulted in a continually decreasing "no residue" level as analytical methods improved and a "moving target" for industry to try to comply with. It was clearly an unsatisfactory regulatory procedure and FDA came under great pressure to change it. The 1973 proposed regulation was the first attempt to do so, and it relied on the original Mantel Bryan extrapolation[16] (despite the Advisory Committee's recommendation) and suggested a virtually safe risk in the range of 10^{-8}.[21]

The DES proviso raised two basic questions: (1) How could an acceptably sensitive method be determined in each specific case, when each drug was different, and (2) Having once selected a method, should its sensitivity be keyed to new refinements in chemical methods of detection. The Agency answered the first question in the first version of SOM the same way mathematicians sometimes answer theirs, by a transformation of variables. The intent and meaning of "no residue" was transformed into "no risk" and this was factored into dose and potency. This permitted a rational selection of a method that was specifically tailored to the toxicological activity of the drug and the characteristics of its use and yet had generic application. To the second question: Should the sensitivity of the method increase as refinements in analytical detection occur? The Agency said yes from 1962 to 1977 – and so the quest for a "receding zero" began. If you think that this was an obviously poor decision, I ask you to recall the temper of the times. Just two years earlier, Congress had reaffirmed the Delaney Clause. In the Hearings prior to the 1960 Color Additive Amendments, Congress listened over and over again to testimony like the following:

> Unless and until there is a sound scientific basis for the establishment of tolerances for carcinogens, I believe the Government has a duty to make clear – in law as well as in administrative policy – that it will do everything possible to put persons in a position they will not unnecessarily be adding residues of carcinogens to the diet. – Secretary Flemming (1960)

FDA was then and is now committed to reducing as much as possible the levels of carcinogens in the food we eat. So it was not unreasonable for FDA to argue that it was bound by the law and a commitment to public health to use its discretion to sanction only those methods that reduced the risk as much as analytical science would allow. The problem with this approach was not the desirability of the result but with the feasibility of the requirement. Analytical methods developed so rapidly and so profoundly that it became arguable whether additional improvements provided greater safety or merely resulted in the costly reduction of levels that were already insignificant. Congress certainly did not anticipate in 1960 that trace carcinogenic constituents would be detectable in virtually every food or color additive by 1979.

It took four years for the Agency to respond to the comments of the July 1973 proposal and draft the more comprehensive 1977 version of SOM.[22] The "virtually safe" level was now estimated using the modified Mantel and Bryan extrapolation from an upper 99% confidence limit to a maximum lifetime risk of 10^{-6}.[23] This rule, although it was remanded to the Agency on the basis of a procedural error, fixed for some time to come the ideas of a defined level of "virtual safety" at 10^{-6}, a prescribed level of statistical assurance in the range of 95–99%, and a conservative rule for extrapolation. This latter conservatism increased with the publication of a revised SOM in March of 1979 that now required a linear extrapolation from the 99% confidence limit and a maximum lifetime risk of 10^{-6}.[24] The route was a bit roundabout, but the March 1979 SOM had arrived back at an extrapolation procedure virtually identical to the 1971 Advisory Committee's recommendations.

As the SOM document passed through three versions, its aims became more comprehensive and its size increased. The 1973 SOM required 4½ pages in the *Federal Register*, the 1977 SOM, 25 pages, and the 1979 SOM, 44 pages. This expansion was in part the result of a then developing trend toward more openness and public participation in regulatory decisions. The force of events, including a strong consumerist movement and the lack of public acceptance of several agency safety decisions, was directing governmental agencies to broaden the base of their public participation. The Agency felt that public and judicial acceptance of regulatory decisions could best be secured by articulating their basis in greater detail and proposing any requirements in the form of intelligible written rules, by assuring public participation in their development and by holding itself accountable to these rules. The 1979 SOM reflected this policy, and it constituted the set of FDA's official principles and procedures governing the entire scientific-regulatory evaluation of potentially carcinogenic drugs and additives used in food-producing animals. It represented a detailed written elaboration of the 1984 House of Representatives Committee's "Concept of Safety" applied to a specific category of additives, and updated to include extrapolation-based risk assessment.

CHANGING PERCEPTIONS OF THE CARCINOGENIC HAZARD—THE SEARCH FOR GREATER REGULATORY FLEXIBILITY

The Delaney Clause provides, essentially, that no food additive is permissible in food (in any amount, however slight), if the substance in question (at any dose, however great) causes cancer in man or animals.[25] During the 1960s and 1970s, FDA found some limited flexibility in this tightly drawn provision. The quality of the cancer study had to meet FDA's scientific standards, i.e., the design and conduct of the study was measured against established scientific opinion of what constituted a credible study. When a study failed to pass muster on these grounds, the substance couldn't be found to "cause cancer . . . in animals" and the Delaney Clause did not apply. However, this particular source of flexibility was sometimes a "catch 22," because of the requirement for a demonstration of safety in the general safety provisions of Sec. 409. If such substances raised a reasonable suspicion that the additive might be carcinogenic despite the equivocal nature of the studies, they could be banned or disallowed anyway on the basis of failing to meet the "reasonable certainty of no harm" standard.

Diethylpyrocarbonate, a food additive approved for use in beverages, and mercaptoimidazoline, a food additive used in the production of rubber sealing gaskets for use in contact with food, were withdrawn from use under the general safety provisions of the Act, Sec. 409(c), on the basis of scientific studies inadequate for withdrawal under the Delaney Clause.[26,27]

A slightly more exploitable area of flexibility developed from the concept of secondary carcinogenesis. It seemed to FDA officials that if a substance could be shown to produce cancer in animals only as a consequence of demonstrable toxicity and only at high doses, and that if a reliable threshold for the toxicity could be demonstrated, then it could be argued that low dose levels were not carcinogenic and the Delaney Clause was inapplicable. Selenium, a naturally occurring essential nutrient and hepatocarcinogen at high doses only, was approved as a supplement to animal feeds on this basis.[28] The essence of this argument was that the intent of the Delaney Clause was to proscribe the addition of carcinogens at any level because there was no way of establishing a safe level. But in the case of secondary carcinogens, one could establish safe tolerances, indeed levels having no risk of carcinogenesis whatsoever, and therefore the premise supporting the application of the Delaney Clause to carcinogens generically did not pertain to secondary carcinogens. The legal basis of this argument may be debatable, but it is a good example of FDA's effort to try to make scientific sense of the law within its delegated authority. Similar arguments were made for Melengestrol acetate residues in cattle[29] and estradiol benzoate residues[30] in steers, heifers, and calves. Part of the argument in the case of estradiol was the consideration that the amount of added chemical constituted a small fraction of the amount present naturally as a result of normal metabolism and also a small fraction of the amount practically unavoidable in the normal diet.

The DES proviso, as already discussed, provided a legal niche on which to attach a risk assessment procedure. No such legal support existed, or then seemed to exist, in the direct food additive provisions of the statute. There developed, however, a steadily increasing concern over the increased likelihood that many substances would be eliminated unnecessarily because they contained traces of carcinogenic impurities. Aniline, for example, is a weak carcinogen and is also the original starting reagent for most colors. It is probably technically impossible and certainly prohibitively expensive to eliminate every last trace of aniline or its carcinogenic derivatives from all colors. Similar arguments could be made for important solvents and polymers used in the processing and packaging of food. In response to these concerns, the Agency took another hard look at the Delaney Clause and discovered still more flexibility. The Delaney Clause provides ". . . that no additive shall be deemed to be safe if it [i.e., the additive] is found to induce cancer when ingested by man or animal, or if it [i.e., the additive] is found, after tests which are appropriate for the evaluation of the safety of food additives, to induce cancer in man or animal. . . ." Under a strictly literal interpretation, the Clause applies when an additive as a whole is found to induce cancer and not otherwise. Under such a reading of the provision, the Delaney Clause would not apply to a carcinogenic constituent of an additive absent a finding that the additive as a whole induces cancer. Upon this legal construction, the "trace constituents policy" or "impurities policy" was developed. Under the general safety provisions, however, FDA claimed broad authority to evaluate fully the potential carcinogenicity of both the additive and its impurities, and under these provisions elected to use extrapolation-based risk assessment to control the risk from any carcinogenic impurities that were present. So, for the third time in its history, FDA chose an interpretation of the Delaney Clause that further narrowed its scope and effect. The new policy was first applied to D&C Green No. 6, a color containing low levels of p-toluidine, and it has since been upheld in the courts.[31]

RISK ASSESSMENT BY FDA—CURRENT PRACTICE

FDA's long history with safety evaluation has taught it to be wary of unsupportable claims of precision in an endeavor that, while science-based, is still very subjective. Nonetheless, the need to act promptly forces FDA to make safety decisions in a climate of scientific uncertainty. It has, therefore, adopted extrapolation-based risk assessment despite its recognition that it is not at present a fully adequate procedure.

There is little question that some form of risk assessment needs to be used, now and even more so in the future. The reason for this is simple, we cannot ban every carcinogen we can detect and, if we could, it wouldn't be helpful. We must make distinctions between acceptable and unacceptable risks, or, if you will, between the presence of known poisons at safe levels and the presence of poisons at threatening levels. Saccharin is simply not regarded the threat that aflatoxin is gram for gram, despite our inability to determine the precise ratio of their potencies in humans at any dose and despite our inability to really prove that either are human carcinogens. Similarly, despite our inability to demonstrate a human dose-response curve for all but a handful of carcinogens, we are not willing to disregard our dose-response principle, nor the fact that there is a roughly 8-fold order of magnitude difference in consumption between a macronutrient like sugar and a minor contaminant at a part per billion. Therefore, given the calculus of risk estimation, i.e., its dependence on potency and exposure, we need to make an attempt at quantification or semiquantification, despite the fact that the rules of the risk assessment calculus are still rudimentary.

In fact, the use of mathematical models that go much beyond these statements does not at present seem to me and to many others to be scientifically justifiable. At FDA, we have used models for the purpose of quantitative risk assessment sparingly, with caution and reluctance, but we do use them to help us justify and explain the kinds of decisions I referred to. Our policy is to use the simplest procedure that makes maximum use of the data and relevant toxicological knowledge, incorporates scientific developments in the risk assessment area when they are regarded as credible by the scientific community and, most importantly, describes the assumptions and limitations of the assessment. Extrapolation-based risk assessment is not at present a fully adequate procedure, but with new developments in defining effective target organ doses and applying pharmacokinetic modeling techniques, it is improving. Our regulatory task is hopeless without some means of distinguishing between large and small risks. This has been recognized by all the health and safety agencies that deal with low level risks from chemical carcinogens, and the OSTP Carcinogen Document is the latest attempt to make the various agency procedures as uniform and consistent as scientifically possible.[32]

To my mind, the presentation of a risk figure by FDA is more of a restatement of our toxicological-exposure judgments in summary form. I hesitate to call it the result of a calculation since that tends to obscure the fact that many choices and assumptions are inherent in the mathematical model, e.g., the treatment of background incidence, or are made after the model has been chosen, e.g., the use of MLE point estimates or upper confidence levels. I believe our decisions to permit low level carcinogens are actually motivated by an unprovable, but to me, more compelling synthesis: that given the observed toxicity of the chemical in experimental animals, its dose-response in these studies, the anticipated low exposure, and the overall judgments on the comparative toxicity, physiology, etc. . . . between humans and the animals used, that the risk to humans is probably negligible. A more precise statement is generally not possible until more basic information is available on carcinogenic mechanisms at a range of doses in animals and humans. As we use it, a quantitative risk figure is regarded as an integration of this judgment into a single, unambiguous (but currently oversimplified) statement of a worst-case risk. It is an assessment that is intended to illustrate the approximate dimensions of the risk and not to forecast the actual consequences.

In the case of saccharin, we used the extrapolated risk figure primarily as a rhetorical device to suggest that the animal studies bore some relation to human risk. The media's amusement over "800 cans of diet soda" calculated as the human dosage equivalent to the animal studies tended to disparage completely the significance of saccharin's animal carcinogenicity. In the case of lead acetate, it was used to demonstrate that even the conservative bias in our risk calculations did not produce a worst-case risk large enough to worry about. When we are quite certain we are in the range of negligible cancer risks, we use extrapolation-based risk assessment to "draw a line in the sand," because such fixed limits are often useful and sometimes even essential for rational regulation.

Of course, it is possible to uncouple the mathematical modeling of the risk from the biological judgment and to produce a model-driven risk assessment. But to do this, you need to have greater faith in the current extrapolation techniques than we have at FDA. The current extrapolation models are not well based on biological mechanism and rely on very few measured quantities, usually only dose rate and incidence. Sound biological judgment includes far more complex information and combines it with expert intuition and experience. It is a subjective judgment, but an informed one, and it includes far more information.

What the statute demands from us is a reliable safety decision, not a formal and inherently unstable method of justification or classification. It is instructive to look at risk assessment in this way, for in effect it is a means of classification. It produces a classification in which the number of categories is infinite and where each is assigned a risk figure. Our food additive provisions are currently more compatible with two categories: those shown to be safe and those not. The British have five categories for their foodstuffs which are assigned on the basis of advisory committee reviews. And IARC has four categories for their carcinogens. Both the IARC and the British systems attempt to judge the strength of the overall experimental evidence as we in FDA do; incidentally, neither system currently employs extrapolation-based risk assessment.

Have we come closer to the "limits of scientific certainty," as so often claimed? Is this what fundamentally underlies so many of the recent disputes between the agencies, the industry, and the courts? If so, what precisely is the nature of the uncertainty? Our methods for detecting substances in food and in the environment are dramatically better than they were. Analytical chemists are detecting fewer nanograms and picograms of a greater variety of chemicals in food, air, water, and soil. Surely the degree of certainty regarding our potential exposure to carcinogens has vastly increased. Similarly, our toxicological methods, while improving at something less than breakneck pace, have also gradually improved. Surely, the certainty with which we can assure the safety of our food supply has increased in some small proportion to advances in our knowledge of cancer mechanisms, the use of gene-tox tests, and the efforts directed to the refinement of the bioassay and its interpretation. How does scientific uncertainty increase when scientific knowledge expands, scientific methods improve, and, aided by dramatic advances in technology, more reliable scientific information is produced? Generally, I'd suggest that our increased scientific uncertainty is produced by:

1. The advance of science itself
2. The misfit between chemical detection and toxicological evaluation
3. The static nature of the statutes

The first part of the answer is familiar to every scientist and it is that science itself is destabilizing. Every time science opens a door to explore a territory, it discovers more doors within it. Science can only answer the questions that are put to it. If we remain uncertain, it must be that our questions have also grown more sophisticated and have more than kept pace with advances in science and technology. Second, to a large degree, there is a misfit between the relative ease of detecting a substance, even in nanogram amounts, and the difficulty of assessing its long-term health effects. The former yields to advances in electronics, miniaturization and their adaptation and distribution in the form of analytical instrumentation. The latter requires an even more precise knowledge of the biology of humans and their animal surrogates. The vast improvement in our methods of analytical detection have exposed carcinogens in the food supply at levels and commensurate risks too low for our carcinogen bioassays to evaluate with comparable precision. We are capable now, more than before, of asking scientific questions that we can't answer.

Finally some of our statutes themselves, based as they are on the science of simpler times, contribute not to our uncertainty, but to our difficulty in dealing with it. The Delaney Clause is, of course, the best example. By promising no carcinogenic additives in food at a time when more sensitive methods are finding more carcinogens at lower levels in food and in additives, the law is in effect automatically being tightened.

REFERENCES

1. H.R. Rept. No. 2284. 1958. 85th Cong. 2d sess.
2. National Research Council. March 1, 1983. Risk Assessment in the Federal Government: Managing the Process. National Academy Press, Washington, DC, 191.
3. U.S. Congress. Feb. 13, 1906. House Committee on Interstate and Foreign Commerce. Hearings on the Pure-Food Bills, 282–283.
4. H.R. Rept. No. 2356. 1952. 82d Cong. 2d sess.
5. Lehman, A.J., et. al. 1955. Procedures for the appraisal of the toxicity of chemicals in foods, drugs and cosmetics. *Food, Drug, Cosmetic Law Journal* 10: 679–748.
6. Staff of the Division of Pharmacology, Food and Drug Administration, Dept. of HEW. 1975. Appraisal of the Safety of Chemicals in Foods, Drugs and Cosmetics, Third Printing, Published 1959 by The Association of Food and Drug Officials of the United States. 1.
7. Food and Drug Administration Advisory Committee on Protocols for Safety Evaluation: Panel on Carcinogenesis Report on Cancer Testing on the Safety Evaluation of Food Additives and Pesticides. 1971. *Toxicol. and Appl. Pharmacol.* 20: 419–438.
8. Calvery, H.O., Lang, E.P., and Morris, H.J. 1938. The chronic effects on dogs of feeding diets containing lead acetate, lead arsenate and arsenic trioxide in varying concentrations, *J. Pharmacol. and Exp. Ther.* 64 (No. 4): 364–387.
9. Lehman, A.J., Lang, E.P., Woodard, G., Draize, J.H., Fitzhugh, O.G.,

and Nelson, A.A. 1949. Procedures for the appraisal of the toxicity of chemicals in foods. *Food, Drug, Cosmetic Law Quarterly* 412–434.

10. Druckrey, H. 1943. Quantitative Grundlagen der Krebsergengung. *Klin. Wschr.* 22: 532–540.
11. Bryan, W.R. and Shimkin, M.D. 1943. Quantitative analysis of dose-response data obtained with three carcinogenic hydrocarbons in strain C3H male mice. *J. Nat. Cancer Inst.* 3: 505–531.
12. Berenblum, L. 1945. Systems of grading carcinogenic potency, *Cancer Res.* 5: 561–564.
13. Shubik, P. 1950. Studies on the promoting phase in the stages of carcinogenesis in mice, rats, rabbits and guinea pigs. *Cancer Res.* 10: 13–17.
14. Van Duuren, B.L., Sivak, A., Katz, C., Seidman, I., and S. Melchionne. 1975. The effect of aging and interval between primary and secondary treatment in two stage carcinogenesis in mouse skin. *Cancer Res.* 35: 502–505.
15. U.S. Congress. 1960. House Committee on Interstate and Foreign Commerce, Hearings, Color Additives. 86th Cong. 2d sess. 61.
16. Mantel, N. and Bryan, W.R. 1961. "Safety" testing of carcinogenic agents. *J. Nat. Cancer Inst.* 27: 455–470.
17. Sielken, R.L. June 27, 1986. An Individualized Response Model For Quantitative Cancer Risk Assessment. Report presented at the Joint Meeting of The Risk Assessment Subcommittee of the American Industrial Health Council and The Health and Safety Committee of the Chemical Manufacturers Association, Washington, DC.
18. Gaylor, D.W. and Kodell, R.L. 1980. Linear interpolation algorithm for low-dose risk assessment of toxic substances. *J. Environ. Path. and Toxicol.* 4: 305–312.
19. General criteria for assessing the evidence of carcinogenicity of chemical substances: Report of the Subcommittee on Environmental Carcinogenesis. National Cancer Advisory Board. 1977. *J. Nat. Cancer Inst.* 58 (No. 2): 461–462.
20. Report of the NTP Ad Hoc Panel on Chemical Carcinogenesis, Testing and Evaluation, Board of Scientific Counselors, National Toxicology Program, U.S. Dept. of Health and Human Services, PHS. Aug. 17, 1984.
21. *Federal Register* 38: 19226. July 19, 1973.
22. Food-Producing Animals, Criteria and Procedures for Evaluating Assays for Carcinogenic Residues. Feb. 22, 1977. *Federal Register* 42 (No. 35).
23. Mantel, N. Improved Mantel-Bryan Procedure for "Safety" Testing of Carcinogens. 1975. *Cancer Res.* 35: 865–872.
24. Chemical Compounds in Food-Producing Animals, Criteria and Procedures for Evaluting Assays for Carcinogenic Residues. March 20, 1979. *Federal Register* 44 (No. 55).
25. Blank, C.H. 1974. The Delaney Clause: Technical Naivete and Scientific

Advocacy in the Formulation of Public Health Policies. *Calif. Law Rev.* 62: 1084–1120.
26. *Federal Register* 37: 3060. Feb. 11, 1972; *Federal Register* 37: 15426. (Aug. 2, 1972).
27. *Federal Register* 38: 10116. April 24, 1973; *Federal Register* 38: 33072. November 30, 1973.
28. *Federal Register* 38: 10458. April 27, 1973; *Federal Register* 39: 9540. March 12, 1974.
29. *Federal Register* 49: 13873. April 9, 1984.
30. *Federal Register* 50: 1843. Jan. 14, 1985.
31. *Federal Register* 47: 14141. April 2, 1982.
32. Chemical Carcinogens; A Review of the Science and Its Associated Principles, Office of Science and Technology Policy. March 14, 1985. *Federal Register* Part II, 10371–10442.

2. De Minimis and the Threshold of Regulation

Alan M. Rulis

Food and Drug Administration

INTRODUCTION

This chapter addresses the concept of a "threshold of regulation" for food additives. The Food and Drug Administration (FDA) has appreciated the need for such a concept since the early years after the passage of the 1958 Food Additives Amendment to the Federal Food, Drug and Cosmetic Act (the FD&C Act, or the Act). The approach put forward in this chapter is based upon the premise that, through an examination of a sufficiently large sample of toxicological data from both classical toxicological feeding studies and from carcinogenicity bioassays, some global delimiters of risk and exposure can be determined to define levels of human exposure and/or levels of migration of substances to food that can be said to fall below some "threshold of regulation." When this is the case, the substance in question would not necessarily need to undergo the rigors of the premarket safety evaluation requirements of the Act. Instead, the particular use of the substance could be acceded to by the FDA after an abbreviated review of pertinent information, thus avoiding the need for the submission and agency approval of a food additive petition covering the use of the substance. (Implicit in such a process is the absence of any indication that the substance in question is in fact a carcinogen or other potent toxin. Known carcinogens would need to be handled in a considerably more formal manner, and would be subjected to more formal risk assessment and risk management decisionmaking.)

THE PROBLEM

Section 201(s) of the FD&C Act defines a food additive as:

> . . . any substance the intended use of which results or may reasonably be expected to result, directly or indirectly, in its becoming a component or otherwise affecting the characteristics of any food (including

> any substance intended for use in producing, manufacturing, packing, processing, preparing, treating, packaging, transporting, or holding food. . . .[1]

Because this broad definition includes food packaging and other food-contact materials and their components that might migrate unintentionally into food, it is easy to see why the need has arisen for a policy concerning a "threshold of regulation." With the development of ever more sensitive and discriminating analytical capabilities, the analyst today can confirm the presence in food, or in food-simulating solvents, of extremely small amounts of substances. In some cases, these amounts may be as small as a few parts per billion (ppb).

The FDA is often presented with situations in which a chemical substance used or present in a food-contact material may indeed theoretically migrate to food, but is predicted to do so at levels that are so low as to arguably not satisfy the above statutory definition. Often such substances are not known to be carcinogens. Perhaps little else is known about their toxic potential either. The question that often arises for these very low level migrants is, "Is it necessary for the agency to request and approve a full-blown petition allowing the presence of the substance in food?" Over the years, the agency has made many threshold-of-regulation decisions on a case-by-case basis, although some people have cited inconsistencies in these decisions and some have called for a more formal policy statement by the agency, indicating the principles and procedures it would follow in granting threshold-of-regulation approvals. If the agency wishes to continue to use its discretion not to require a petition in certain cases, is there some consistent basis on which this can be done, while still protecting the public health in the (unlikely) event that the substance in question turns out to be a carcinogen or other potent toxin? (At exceedingly low exposure levels—say, less than 10 ppb—it is likely that potential carcinogenesis would be almost the only toxic phenomenon capable of producing any concern.)

Stating the problem another way, we may ask, "How can the agency determine under a general policy that a given use of a food-contact material or a component thereof, not presently known to be a carcinogen, does indeed not require full petition review and regulation, and will, even in the worst case, present no unwarranted risk?" Furthermore, we might ask, "How can the agency build consistency and scientific credibility into the threshold-of-regulation decisionmaking process?" It is with one form or another of this problem, and with the above questions, that the agency, the regulated industry, and the courts have grappled for nearly three decades.[2–8] Arguably, requiring the regulation of low ppb levels of substances places significant and costly burdens on both the agency and the affected industries, while the human exposure to such low levels of substances may have negligible public health significance. Conversely, however, even small exposures to highly toxic substances or to potent carcinogens might sometimes be found to produce significant human risk. The agency must navigate carefully between these two potentially unpleasant consequences.

Obviously, no simple solutions to this problem will materialize overnight. However, recent progress in understanding the global trends and distributions of risk and exposure may help the agency to sufficiently clarify the scope of available options, so that a consistent policy can finally be achieved. This chapter addresses some areas of recent thinking about the problem at the FDA and offers some recommendations for making further progress.

A PROBABILISTIC APPROACH

The complexities of the biological interactions of any chemical with a living system are enormous. To construct a threshold-of-regulation policy that anticipates all potential adverse responses that a compound may induce, including potential carcinogenesis, is clearly impossible. Fortunately, however, the present toxicological literature contains an enormous amount of data on the toxicity of chemical substances. The premise of this chapter is that a proper summarization and analysis of these data may provide important needed insights into the threshold-of-regulation question.

For example, Figure 1 displays potency data on a subset of 343 oral carcinogens from animal studies recently compiled by Gold et al.[9] For the purpose of the present discussion, "potency" is defined as the slope of a straight line connecting the point representing TD_{50} dose (toxic dose to 50% of the test animals) of Gold et al. with the point representing zero risk and zero dose. (For a more complete description of the definition, refer to Flamm et al.[10] Figure 1 shows that when carcinogen potencies are analyzed by grouping them into ranges and plotting them as a probability distribution, they form, on a semilogarithmic scale, a curve that is remarkably Gaussian (normal) in shape. The existence of such a curve for carcinogen potencies demonstrates not only the well known fact that chemical carcinogens present greatly different potencies (probabilities of inducing cancer per unit level of intake), but also that a single chemical carcinogen selected at random can be predicted a priori to have a potency that falls within fairly well defined limits. It seems that such knowledge ought to be applicable to the threshold-of-regulation question. Therefore, let us now try to apply this knowledge to the problem at hand, namely the delineation of options for determin-

Figure 1. Histogram and nonlinear least squares best fit Gaussian to the potencies of Gold et al.[9]

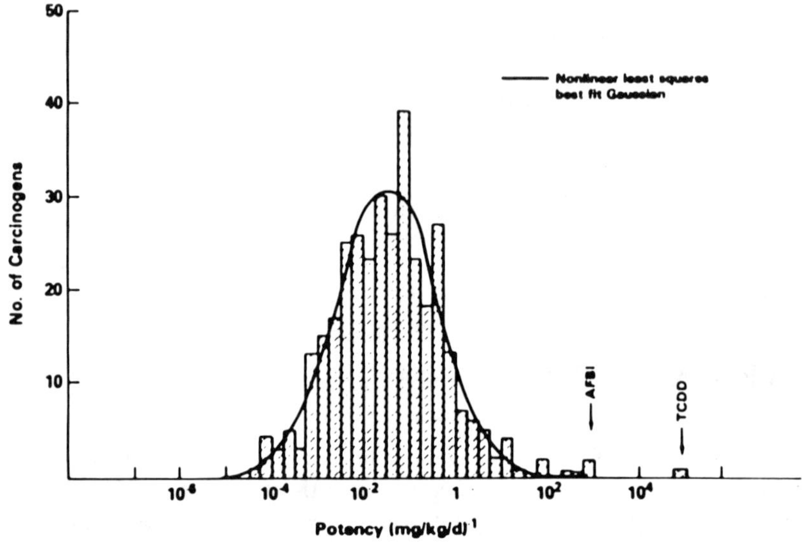

ing exposures to chemicals that might be said to fall below a threshold of regulation.

First, in order to be useful, the Gaussian distribution of Figure 1 must be transformed into an exposure distribution at a constant assumed risk of 1×10^{-6} per lifetime. (This process has been previously described.[10]) (A risk level of 1×10^{-6} per lifetime has been chosen because it is the upper bound level of risk identified by the agency recently as de minimis for the purpose of regulating the carcinogen methylene chloride as a coffee decaffeinater.)[11] The transformation applied to the potency distribution of Figure 1 results in curve number 5 of Figure 2. We call such a curve a "risk equivalent exposure distribution." It describes the relative probability that a carcinogen selected at random from the universe of known carcinogens will be one that presents a risk of 1×10^{-6} per lifetime at the exposure level indicated on the horizontal axis across the top of Figure 2. As can be seen, about half the area of curve 5 falls on either side of the 1 ppb line. Therefore, at a human dietary intake of 1 ppb, about half of all carcinogens are predicted, under the very simplified assumptions of this analysis, to present a risk of greater than 1×10^{-6} per lifetime; the other half, lower than 1×10^{-6}.

Figure 2. Relative probability distributions for various groupings of toxicity data. Curves are arbitrarily vertically scaled and are described in the key. Upper abscissas refer to various measures of intake. "PF" refers to the "packaging factor" defined in the text. Lower abscissa values are negative base-10 logarithms of effect levels in mg/kg/day for nonacute data, mg/kg for acute data, and TD_{50}s for the carcinogen data of Gold et al.[9] Below lower abscissa are minimum lethal doses for selected toxins for comparison purposes.

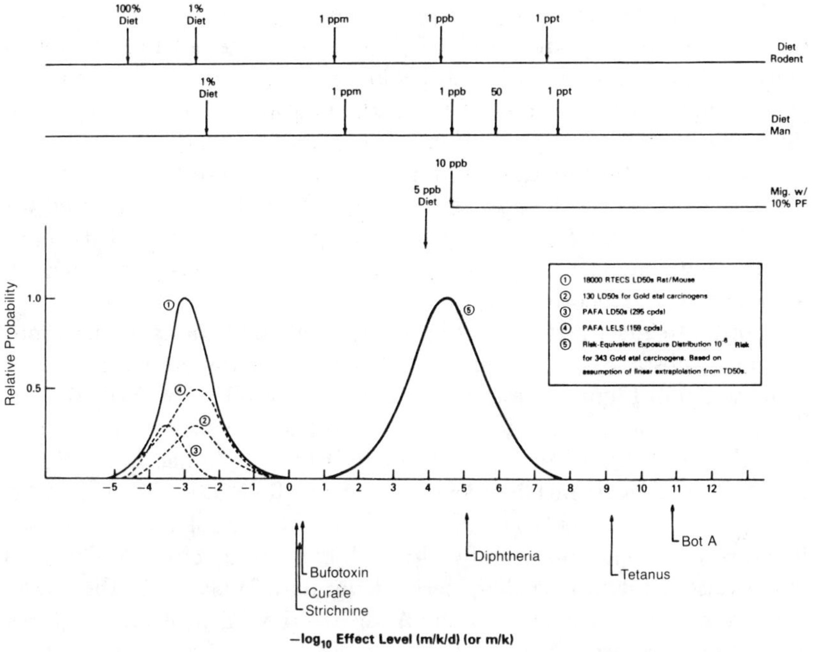

KEY: 1 18000 RTECS LD_{50}s Rat/Mouse; 2 130 LD_{50}s for Gold et al. carcinogens; 3 PAFA LD_{50}s (295 cpds); 4 PAFA LELS (159 cpds); 5 Risk-Equivalent Exposure Distribution 10^{-6} Risk for 343 Gold et al. carcinogens. Based on assumption of linear extrapolation from TD_{50}s.

A threshold-of-regulation level that precludes almost any measurable lifetime risk from carcinogenesis (except for risk from substances like aflatoxin B-1 or 2,3,7,8-tetrachlorodibenzo-p-dioxin [TCDD]), say—one part per trillion (ppt) in the human diet—would in all likelihood be so low that it could not be reliably measured analytically, and certainly could not be easily and consistently enforced. To insist on such a stringent level of risk protection would, in effect, preclude any reasonable and practical threshold-of-regulation policy.

The use of a 5-ppb dietary level as a threshold of regulation presents an interesting example, because this level has recently been suggested by some as an appropriate level to use for deciding on a threshold for regulation.[12] About 60% of the area under curve 5 of Figure 2 falls to the right of the 5-ppb line. Therefore, it can be predicted on a probabilistic basis that should a substance permitted under a threshold-of-regulation decision unknowingly be a carcinogen, it would, under the present analysis, have roughly a 60% chance of presenting greater than a 1×10^{-6} per lifetime level of risk at the 5-ppb level of exposure.

Some have suggested that an appropriate threshold-of-regulation level of exposure would be the 5-ppb human dietary exposure as above, but "modulated" by known classical (noncarcinogenic) toxicity information. For example, one could add to the 5-ppb requirement the additional constraint that the LD_{50} of the migrant be no lower than 10^5 times the threshold-of-regulation level (5-ppb). In other words, any candidate for a threshold-of-regulation decision whose LD_{50} falls to the right of about −1 on the abscissa of Figure 2 would be disqualified for consideration at the 5-ppb exposure level on the grounds that its (noncarcinogenic) toxic potency is too high. Certain aspects of such an approach have recently been put forward.[12]

In order to analyze this type of proposal, data on toxicity of numerous chemical substances were compiled. The curves numbered 1, 2, 3, and 4 on the left of Figure 2 describe these data. Curve 1 (arbitrarily scaled) depicts a nonlinear least squares best fit of a Gaussian curve to data compiled from 18,000 oral (rat or mouse) LD_{50}s contained in the Registry of Toxic Effects of Chemical Substances (RTECS). Curve 2 (arbitrarily scaled) represents a similar envelope of LD_{50}s for 130 compounds identical to the carcinogens of Gold et al. comprising curve 5. Curves 3 and 4 (also arbitrarily scaled) derive from data analyzed in the FDA's Priority-Based Assessment of Food Additives (PAFA) project.[13-15] Curve 3 represents the probability distribution of LD_{50}s for 295 regulated food additives analyzed in the PAFA project. Curve 4 represents an envelope of lowest effect levels from subchronic or chronic feeding studies on food additives studied in PAFA.

As can be seen, the "realm" of classical toxicity (except for pesticides, economic poisons, or some exquisitely toxic substances arrayed along the abscissa for comparison) is fairly well delineated by these curves, with no sizeable probability of such effects occurring in rodents at doses lower than about 1 mg/kg/day (to the right of about 0 on the abscissa). (Such knowledge is actually not new at all, having been observed and duly noted by Frawley in 1967, using a different data base.[2])

The noteworthy point for our purposes, however, is that the requirement that the acute toxic dose of a substance be no less than five orders of magnitude higher than the benchmark 5-ppb dietary intake level is no requirement at all, because virtually all chemicals can meet that criterion. To see this, just step five decades to the left of the 5-ppb arrow in Figure 2. Analysis of relationships between acute toxicity and carcinogen potency in recent work in progress at the FDA reveals that use of such comparisons is of dubious practical value when applied to threshold-of-regulation questions where the acceptable upper bound risk value is set at 1×10^{-6} or lower. Recently, Wilson et al.[16] have dealt with this relationship and have demonstrated its potential utility in the decisionmaking process. They too, however, note that such an approach is of limited value when the acceptable risk level is set at 1×10^{-6} per lifetime or lower.[16]

So far, we have shown how compilations of potency and toxicity data can be used to rule out certain threshold-of-regulation scenarios. They can also be used to support other choices. For example, suppose we define a threshold-of-regulation level to be 50 ppt in the human diet. Such a level, although quite low, would correspond to a migration level to food of 1.0 ppb, assuming a 5% "packaging factor." (A "packaging factor" may be defined as the weight fraction of the human diet likely to be packaged in a given material.) One ppb, while on the borderline of reliable analytical detection, is not outside the realm of measurability. This level divides curve 5 of Figure 2 in such a way as to exclude about 85% of carcinogens that produce greater than 1×10^{-6} per lifetime risk, if held at that level of exposure. If we conservatively assume that one out of five chemicals assented to by the agency under a threshold-of-regulation policy at this exposure level is unknowingly a carcinogen, then on a purely probabilistic basis we can argue that 95 out of every 100 such threshold-of-regulation decisions will result in no more risk than 1×10^{-6} per lifetime. Furthermore, that level of risk is a maximum, with the vast majority yielding far lower levels of potential risk.

SUMMARY AND RECOMMENDATIONS

A probabilistic approach has been described for analyzing large amounts of toxicological data on numerous chemical substances. The application of such an analysis to the problem of the threshold of regulation for food-contact materials was described.

While such an approach certainly does not address the immense biochemical complexity inherent in the interactions of individual chemical substances with living systems, it may prove to be a useful tool for FDA in evaluating the viability of options for, and in providing a scientific basis of support for, choices of threshold-of-regulation migration levels. Once the agency is able to support a given migration level or range of levels as acceptable for a threshold of regulation, it will be in a better position than ever before to make threshold-of-regulation judgments on a consistent basis.

Once the principles of such a policy have been established, it would remain to devise appropriate procedures for implementing the policy. It has been suggested that the agency could make such threshold-of-regulation judgments based on a sufficient summary of both chemical and toxicological information provided by an applicant. Certainly some sort of abbreviated agency review of summary information may be an important aspect of such procedures. In order to be able to efficiently review its accumulated decisions over the years, the agency could keep records of all threshold-of-regulation actions. Such accumulated decisions could form the basis for a data base of information on low level migration of chemical substances to food.

It appears that there presently exists an adequate scientific basis of data and information on which to construct a threshold-of-regulation policy relating to food-contact substances. By applying this information to the problem and implementing appropriate policies and procedures, the agency will be in a position to better delineate the boundaries of the food additive definition of the FD&C Act as it relates to food-contact substances.

REFERENCES

1. Federal Food, Drug and Cosmetic Act, as amended, (Title 21 U.S. Code). 1958. U.S. Government Printing Office, Washington, DC.
2. Frawley, J.P. 1967. Scientific Evidence and Common Sense as a Basis for Food-Packaging Regulations. *Food. Cosmet. Toxicol.* 5: 293–308.
3. Checchi, A.A. 1959. Developments Under the National Pure-Food Law Affecting the Packaging Industry. *Food Drug Cosmetic Law Journal* 14: 527–533.
4. Checchi, A.A. 1959. Food Additives Procedures and Policies. *Food Drug Cosmetic Law Journal*, 14: 591–596.
5. Rankin, W.B. 1959. Incidental Food Additives. *Food Drug Cosmetic Law Journal* 14: 768–773, 777.
6. Kirk, J.K. 1960. Food Additive Developments. *Food Drug Cosmetic Law Journal* 15: 755–760.
7. Harvey, J.L. 1962. Food Additives and Regulations. *Food Drug Cosmetic Law Journal* 17: 272–281.
8. DC Cir. 1979. Monsanto v. Kennedy, 613 F. 2d 947.
9. Gold, L.S., et al. 1984. A Carcinogenic Potency Data Base of the Standardized Results of Animal Bioassays. *Environ. Health Pers.* 58: 9–314.
10. Flamm, W.G., Lake, L.R., Lorentzen, R.J., Rulis, A.M., Schwartz, P.S., and Troxell, T.C. Carcinogenic Potencies and Establishment of a Threshold of Regulation for Food Contact Substances. Submitted for Publication in Risk Assessment.

11. Cosmetics: Proposed Ban on the Use of Methylene Chloride as an Ingredient of Aerosol Cosmetic Products. Dec. 18, 1985. *Federal Register* 50: 51551–51559.
12. Threshold of Regulation for Packaging Materials Proposed. Dec. 2, 1985. *Food Chem. News* 29–30.
13. Rulis, A.M., Hattan, D.G., and Morgenroth, V.M. 1984. FDA's Priority-Based Assessment of Food Additives: Preliminary Results. *Regul. Toxicol. Pharmacol.* 4: 37–56.
14. Rulis, A.M., and Hattan, D.G. 1985. FDA's Priority-Based Assessment of Food Additives: General Toxicity Parameters. *Regul. Toxicol. Pharmacol.* 5: 152–174.
15. Hattan, D.G. and Rulis, A.M. 1986. FDA's Priority-Based Assessment of Food Additives: Specific Toxicity Parameters. *Regul. Toxicol. Pharmacol.* 6: 181–191.
16. Zeise, L., Wilson, R., and Crouch, E. 1984. Use of Acute Toxicity to Estimate Carcinogenic Risk. *Risk Analysis* 4: 187–199.

3. The Difficulties and the Possibilities of Epidemiologic Investigations of Low Risks Attributable to Food Constituents

Frank Cordle
D. Jesse Wagstaff

Food and Drug Administration

INTRODUCTION

Although epidemiology may have first been used by Hippocrates, the definition or meaning has undergone some significant changes since then. First described as the study of epidemics and, later, as the study of the determinants of differences in disease distributions in human populations, epidemiology today refers essentially to the activities of the epidemiologist.

Historically, the development of epidemiology has been, to a large degree, a reflection of these changes in disease patterns, and it is interesting to follow this evolution in light of the objectives of this book. An attempt will be made to illustrate this evolution with a description of four investigations: two where the disease entities were recognized but no causal relationship in the sense that we would recognize it today was established, and two modern day environmental contamination episodes in which exposure is known but ultimate outcome is not.

One of the extraordinary things that has occurred over time has been the regularity with which different types of diseases have replaced others. Thus, largely due to preventive measures through public health protection, the diseases of infectious origin have in time been replaced by the modern epidemics of the chronic, degenerative diseases, such as elevated blood pressure, arteriosclerotic heart disease, stroke, various types of cancer, mental disorders, arthritis, diabetes, and more recently, while perhaps not epidemic, concern for potential adverse health effects resulting from exposure to a variety of environmental insults.

Two classic studies illustrate superbly the early development of epidemiologic thinking and methods. These are the studies of John Snow[1] during the cholera epidemics in England during the 1800s, and the work of Goldberger[2] on pellagra in the United States during the early 1900s.

Cholera is one of the major types of acute infectious disease which spreads very rapidly, has a high death rate under most circumstances, and produces an intense diarrhea and loss of body fluids, with death due to the loss of fluids and the resulting electrolyte imbalance. It is a disease which occurs frequently in the Far East, but is rare in the western world. In the early 1800s, cholera epidemics were sweeping through London, killing tens of thousands of people, creating fear and panic among the population, resulting in people fleeing from the city to the country in hopes of escaping this particular form of epidemic.

At the time Snow began his investigations, there were various hypotheses about the causes of cholera and what was needed to prevent or treat the disease. These hypotheses were derived from the ancient teachings of the Greeks in the days of Hippocrates, some of the developments of Roman times, and some of the Arab teachings. Although there were variations on the theme, most of these beliefs held that many diseases, cholera included, were largely caused by exposure to foul-smelling gases derived from the bowels of the earth, i.e., the so-called miasmas, or bad air. These beliefs resulted in preventive measures such as moving people to higher elevations in order to get away from the environmental influences at lower levels.

However, Snow observed that cholera traveled along the great tracks of human intercourse, never going faster than people traveled. For example, new cases in a country always appeared first at a seaport where the ships arrived and emerged from the crews of ships coming from a country where cholera was common. Secondly, Snow observed that numerous cases of cholera developed shortly after contact with a previous case, leading him to believe that cholera spread from person to person, rather than from exposure to bad air.

If the disease could be communicated from the sick to the healthy, it probably had to be communicated by some material that passes from one person to another. In addition, since the disease appeared to extend itself to a variety of people in many locations, there must have been some method for the movement of the causative material to reach fresh victims. Snow suggested that a possible mechanism for this movement was the mixture of feces from cholera victims with the drinking water.

In the epidemic of 1848, Snow had the opportunity to test his ideas by using many of the methods we use in present day epidemiology, i.e., identifying the locations of deaths through the use of spot maps, the study of the differences between those who had cholera and those who did not, i.e., a case-control approach, and computation of population-based death rates as opposed to numbers of cases. All of these methods contributed to the now famous study of the area of the Broad Street pump and led to the actions which Snow took to demonstrate that contaminated water was in fact the principal means for the transmission of cholera. Later, studies of the different levels of contamination among the various water companies which supplied water to different parts of London confirmed these findings.

In a second example of the logical application of epidemiologic methods and thinking, the reverse occurred. This example is concerned with a disease, pellagra, that was thought for many years to be caused by an infectious agent. This belief resulted largely from the tremendous advances being made in the identification of the causes of so many infectious diseases, leading many to believe that all diseases were of an infectious origin. However, Goldberger was not convinced that pellagra was an infectious disease, based on some information from Spain and Italy that strongly implicated some foodstuff in the course of pellagra.

In the first studies, Goldberger observed children in orphanages in which the disease occurred frequently. Careful studies indicated that the younger children under six years, the children over twelve years, and the staff did not get the disease. The group at highest risk for pellagra was the six- to twelve-year-old children. Additional studies in this group identified the lack of milk and meat in the diet.

In order to confirm the lack of essential dietary factors in the diet of those most at risk of developing pellagra, Goldberger[3] carried out studies in seven mill towns in South Carolina where pellagra was occurring frequently. In these surveys, he found two major differences between those who had pellagra and those who did not. More of the pellagra families were poorer than those without the disease, and although the diets were somewhat similar, the families with pellagra had less milk and meat in their diet. In order to determine if poverty was a contributing factor, Goldberger observed those in the poor group with and without pellagra and found that those who did not have pellagra were able to get milk and meat. Other studies confirmed the absence of essential dietary factors in the etiology of pellagra and permanently laid to rest the suggestion that this disease was infectious in nature.

These examples of epidemiologic methodology and thinking have provided some of the early foundations for what one thinks of as traditional epidemiology, i.e., the epidemiology of a variety of the important diseases which have emerged in the past two or three decades, e.g., heart disease, cancer, diabetes, and stroke, to mention a few. In the traditional manner, epidemiologists deal with epidemics by interpreting patterns of disease, testing hypotheses, and assessing the risks and benefits of various options. Their methods involve the use of two categories of epidemiology often referred to as analytical and descriptive epidemiology. A description of some of these methods of epidemiology and a look at what we see as the prospective epidemiology of the future follows.

EPIDEMIOLOGIC METHODS

Descriptive

In order to test for possible causative factors related to specific diseases in humans, two general approaches are available. Descriptive studies are mainly observational and can suggest causal relationships although they can seldom prove them. For example, the observation that a sharp increase in lung cancer occurred twenty years after a rapid rise in cigarette consumption suggested that a cause and effect relationship existed. Further studies were required to prove that the cause and effect were directly related.

Analytic Studies

The second type of epidemiologic studies available are analytic studies which can be used to test specific hypotheses. Basically, there are four kinds of analytic studies: (1) an intervention study where manipulation of factor A (such as fluoridation of water) is followed by an alteration in factor B (such as a reduction in dental caries); (2) a controlled clinical trial such as comparing one therapeutic regimen to another one; (3) a cohort or prospective study which involves following a population exposed to a suspect carcinogen (i.e., aflatoxin) and comparing them with a nonexposed population to determine if the frequency of disease (i.e., liver cancer) is greater in the exposed group; and (4) the case-control or retrospective study.

An excellent example of a case-control study is that conducted for bladder cancer and artificial sweetener use in which the proportion of bladder cancer cases who use saccharin is compared to the proportion of controls who use saccharin. More frequency of exposure in the case group than the control group suggests a possible association between the suspect carcinogen and the disease. Ideally, the controls should be matched exactly to the case group with the sole exception of the suspect etiologic factor. In reality this is never attained because of the existence of several known risk factors which may or may not be associated with the suspect etiologic factor, because all of the possible risk factors are not known, or because the mechanics of selecting controls do not permit perfect matching.

These changes which have occurred in the incidence and prevalence of the chronic diseases compared to the infectious diseases have necessitated a reexamination of the efforts of various government agencies, including the regulatory agencies, in the area of public health protection and consumer safety. In the case of the Food and Drug Administration (FDA), both new and expanded activities are designed to deal with such issues as the safety of food additives and packaging materials, the safety of the use of animal drugs in food producing animals and the impact of environmental contamination, be it natural or industrial, on the food supply. Each of these areas of consideration has led to increased demands in the areas of toxicology, chemistry, and epidemiology.

The importance of the contribution of epidemiology in the decision-making process in the regulatory public health arena seems obvious. Without adequate assessment of factors that appear to increase the risk of disease in humans from chemical substances in the food supply, food may be destroyed, the public may be unnecessarily inconvenienced or inadequately protected, and the costs may be enormous. In short, epidemiologic efforts are directed toward the prediction of such risk at the community, state or national level.

As described above, in the traditional manner, epidemiologists deal with epidemics by interpreting patterns of disease, testing hypotheses, and assessing the risks and benefits of various options. In perhaps a more pragmatic and contemporary manner, in particular in a regulatory agency such as FDA, epidemiology, first of all, combines with toxicology, residue analysis, food consumption, and risk assessment to provide counsel and guidance in the regulatory decisionmaking process. In previous reports, Cordle and Kolbye;[4] Cordle;[5] and Cordle, Locke, and Springer[6] have described some of the methods and necessary data and information involved in regulatory management. These include data on residue levels of substances such as PCBs in food, the amount of potentially contaminated food consumed and the length of time consumed, the numbers and characteristics of the individuals consuming the food, available animal toxicology data, and epidemiological data on the human consequences of exposure to the substances if such exist.

In a second, and equally important role, epidemiology provides the more traditional methods for followup, i.e., the prospective epidemiologic approach to determine the potential consequences of exposure.

The lack of good examples of dose-response relationships in man, even in the area of occupational exposure, is remarkable. Even where dramatic exposure to environmental contamination has occurred through ingestion of contaminated food, efforts to establish an acceptable dose-response relationship in man have been met with considerably less than unqualified success.

However, epidemiological concern should not be limited to these obvious dramatic outbreaks of disease due to high exposures to toxic chemicals, but should be equally concerned with the effects of lower, more prolonged and sometime insidious exposures, such as exposure to PCBs, PBBs, dioxin and dibenzofurans, to mention a few.

Doll and Peto[7] have also described some general points of consideration concerning the current status of the use of epidemiology to determine risk in a variety of settings. This description seems particularly appropriate for those in the regulatory process and their expectations from epidemiology. They reported that when positive epidemiologic results show an association between exposure or some other variable and cancer, based on a valid study, they tend to dominate any decision to be made on carcinogenicity. However, even risks that will ultimately kill, for instance, 1% or more of the exposed population may be overlooked or attributed to chance unless a very large-scale investigation is undertaken. In these circumstances, too, when the cancer rates among exposed people are only a moderate multiple of those among the unexposed, problems of interpretation may become acute, and it may be extremely difficult to disentangle the various contributions of biased information, confounding of two or more factors, and cause and effect.

In other words, unless epidemiological studies have been carried out in reasonably large, well-defined groups of people who have been heavily exposed to a particular substance for two or three decades without apparent effect, they can offer no guarantee that continued exposure to moderate levels will in the long run be without some increase in risk.

However, this is not to say that human evidence can ever be totally dispensed with. It is always relevant, but the weight that can be given to it varies greatly with the duration and intensity of the exposure experienced by individuals. Positive evidence is always important. Negative human evidence may mean very little, unless it relates to prolonged and heavy exposure. If, however, it does, and is consistent in a variety of studies (correlation studies over time, cohort studies of exposed individuals, and case-control studies of affected patients), and if the laboratory evidence is also limited in its scope (for instance, to a particular type of tumor in a few species), negative human evidence may justify the conclusion that for practical purposes the agent need not be treated as a human carcinogen.

Epidemiology has, at present, an undeserved poor reputation among administrators and regulatory decisionmakers who first have artificially limited themselves by wondering which chemicals to restrict, and then find that almost none of the few thousand chemicals they are worried about have been adequately studied by epidemiologists.

However, this is to condemn epidemiology for failing to achieve ends that it has neither the resources nor data to meet. Epidemiology starts, not with a series of rodents under carefully controlled laboratory conditions for time, exposure, and outcome over a relatively short lifetime, but rather with people in an uncontrolled environment and with little possibility of controlling lifetime exposure to a variety of chemical substances. Epidemiology is, admittedly, more likely to overlook many undetectably small effects of various chemicals than laboratory studies, but it is much less likely to overlook the large determinants of contemporary cancer rates and trends, especially if these are important environmental pollutants or dietary contaminants.

It is quite likely that the majority of humans are exposed to a large number of chemical substances in quite small amounts over an extended period of time, rather than to large doses such as those described in the Minamata or Yusho episodes. The importance of the possible cumulative effects of these small doses versus the importance of the effects of a single large dose, or relatively large doses over a short period of time, is the subject of considerable scientific debate.

Whatever the outcome of such debate, it is clear that one of the major problems associated with exposure is that reliable information is frequently lacking on current exposure, much less on how long and at what level such exposures have taken place in individuals in the past. While a considerable amount of information is available from a variety of sources on the consumption of various foodstuffs, few quantitative data are available for many of the residues of substances found in specific foods that may be of particular interest. An additional problem is that the available data permit only a general assessment to be made of the average diets consumed by any specific group of people. Implicit in this description of the inadequacies of food residue data are similar inadequacies in the kinds of data needed to assess occupational, environmental, or a variety of other types of exposure.

Surveillance of disease in most countries has usually been based on mortality statistics, the deficiencies of which are well known. Furthermore, while mortality statistics may be useful for diseases that are highly and/or rapidly lethal, they are less so for others of low mortality and/or long incubation periods or periods of latency as in the case of cancer.

In the United States, the Great Lakes present a unique situation for long-life contaminants which can bioaccumulate in living organisms. These lakes act as reservoirs for atmospheric and terrestrial pollutants such as the polychlorinated biphenyls, the polybrominated biphenyls, dioxin, dibenzofurans, metals, and other industrial chemicals. The relatively slow turnover and removal of contaminants plus biomagnification in the entire ecological system make the aquatic environment of the Great Lakes a significant source of potential human exposure and adverse health effects which are largely unknown.

The discussion to follow summarizes the other two examples mentioned previously of episodes of known human exposure but with unknown consequences. These describe some of the toxicologic and epidemiologic data and information concerning PCB and dioxin residues in food and some of the epidemiology studies which are being used to determine outcome of such exposures in the state of Michigan.

A considerable amount of attention has been directed toward exposure to PCBs which occurred in Japan as the result of the consumption of contaminated rice oil (the Yusho episode). A series of epidemiological studies has been carried out on this exposed population and the results of these studies are available in a variety of publications. Short-term epidemiological studies have revealed less than the toxicity that might be expected based on animal studies. In addition, studies of PCB exposure in Michigan, in particular in the sportfishermen, have also revealed less than might be expected from animal studies.[6,8] It should be pointed out that these studies may represent some excellent examples of the difficulties of epidemiological investigations of low risks attributable to food constituents. Do we in fact have a sufficient study population for the investigation of the potentially low risk outcome of such exposure?

In any estimates of human risk derived from the extrapolation of animal data, close attention should be given not only to the levels of exposure to the various Aroclors in a variety of animal studies but also to the way in which the exposure relates to human experience. For example, studies in monkeys have reported signs and symptoms similar to those of Yusho after 2 months of exposure to Aroclor 1248 at levels of 2.5 and 5.0 ppm in the diet or 125 and 250 μg/kg body weight/day, respectively. Reproduction problems were reported in these monkey studies at each of these levels after 6 months of exposure.[9]

In comparison, problems of reproduction have been observed in rodents only at considerably higher levels of exposure, e.g., in rats fed 7.2 and 37.0 mg/kg body weight/day of Aroclor 1254 or 100 and 500 ppm in the diet.[10] Rats fed 500 ppm or 35.4 mg/kg body weight/day of Aroclor 1260 also exhibited reproduction problems.[11] No problems were observed with Aroclor 1260 at levels of 5 ppm (0.39 mg/kg body weight/day), 20 ppm (1.5 mg/kg body weight/day), or 100 ppm (7.4 mg/kg body weight/day).

In one study, the carcinogenicity of PCBs in rats (Fischer strain) fed Aroclor 1254 at 25 ppm (1.9 mg/kg body weight/day), 50 ppm (3.8 mg/kg body weight/day), or 100 ppm (7.16 mg/kg body weight/day) was reported negative under the test conditions.[12] Although some malignancies were observed, there was no statistical difference between test animals and controls. In another study, female Sherman rats fed Aroclor 1260 at 100 ppm (7.4 mg/kg body weight/day) exhibited a statistically significant difference between test animals and controls for hepatocellular carcinomas.[11]

In contrast, there appears to be little evidence of human exposure to these levels in the United States especially for the consumption of fish.

Even in the Yusho experience in Japan where clinical signs and symptoms were observed, the average level of consumption of PCB residues in the rice oil is estimated at 0.75 mg/kg body weight/day.[13]

In the United States, estimates of the daily intake of PCBs for eaters in the 50th percentile are 8.46 μg/day or 0.72 μg/kg body weight/day (based on a 70-kg individual) assuming no PCB tolerance; 7.57 μg/day or 0.11 μg/kg body weight/day assuming a tolerance of 5 ppm, 5.59 μg/day or 0.08 μg/kg body weight/day assuming a tolerance of 2 ppm, and 3.30 μg/day or 0.05 μg/kg body weight/day assuming a tolerance of 1 ppm.[6] Estimates of the intake of PCBs for eaters in the 90th percentile are 0.32 μg/kg body weight/day assuming no tolerance and 0.29, 0.21, and 0.13 μg/kg body weight/day assuming a tolerance of 5, 2, and 1 ppm, respectively.

In Michigan sportfishermen, who are presumed to be among the higher consumers of fish with PCB residues, the average intake has been reported at 1.7 μg/kg body weight/day with a range of 0.09 to 3.94 μg/kg body weight/day.[8]

Thus problems of interpretation arise in comparing the levels of the various Aroclors which have produced effects in animals, ranging from 125 μg in monkeys to the milligram levels in rodents, with the exposure estimates in humans from fish consumption. For example, estimates of the lifetime human risk of cancer and reproduction problems for exposure in the 90th percentile of fish eaters, i.e., 0.29, 0.21, and 0.13 μg PCB/kg body weight/day, indicate risk from exposure well below the average Michigan exposure and certainly well below the levels of exposure in the Yusho (rice oil) incident in Japan.[6]

The outcome of human exposure to the tetrachlorodibenzodioxins (TCDD) presents a somewhat similar puzzle to that of the PCBs when comparing animal toxicity to man. Although 22 incidents of human exposure to TCDD in connection with the manufacture of chlorinated phenols have been reported worldwide since 1949,[14] there remains a scarcity of reliable information concerning the results of these exposures.

In a recent report of the mortality experience of a cohort of workers exposed to TCDD in Nitro, West Virginia, in 1949,[15] described some of the signs and symptoms observed in the exposed population shortly after the accident occurred. Employees who worked in the area of 2,4,5-trichlorophenol (TCP) production or were involved in the cleanup began to develop symptoms immediately following exposure to the material, which was discharged from the autoclave. Symptoms included eye and respiratory tract irritation, headache, dizziness and nausea, and a severe irritant reaction of the exposed skin. After these initial symptoms subsided, chloracne and other symptoms became evident. A total of 12 more severely affected workers were examined on three occasions during the period 1949–1953. Another 26 persons with chloracne that was apparently unrelated to the accident were also examined in 1953.

The clinical symptoms included acneform lesions; severe pains in muscles of upper and lower extremities, shoulders, and thorax on exertion; fatigue; nervousness and irritability; decrease in libido; dyspnea; vertigo; and intolerance to cold. All of the cases showed evidence of chloracne. For the six workers examined during 1949 and 1950, another examination was carried out in 1953, and at that time six additional workers involved in the accident were also examined. The findings in this later examination indicated a general regression of both the cutaneous and noncutaneous symptoms which had been present earlier. No specific levels of exposure could be determined.

In other reports of industrial exposure to TCDD from Great Britain, the Netherlands, West Germany, and Czechoslovakia, chloracne was the most common and prominent sign observed following exposure. In some reports, liver function tests indicated liver damage, whereas in other reports they did not. Two major problems encountered in all of these studies were the lack of a clear identification of those exposed, other than their subsequent development of chloracne, and the absence of any measures for the levels of exposure that might have taken place.

Pazderova-Vejlupkova et al.[16] reported results of a 10-year followup study of workers exposed to TCDD between 1965 and 1968 during the production of 2,4,5-trichlorophenoxyacetic acid (2,4,5-T). In this study group of 55 individuals (originally 80 of the 400 persons engaged in the production became ill), the first indications of illness were feelings of sickness, fatigue, weakness in the lower extremities, and the formation of chloracne. Subsequent examinations indicated that about 20% had mild hepatic lesions.

During the 10-year followup study of these exposed individuals, most of the patients did not experience all of the symptoms and signs of intoxication, and some patients showed the same symptoms and signs as others, but in different combinations. It is assumed that in this type of intoxication all the systems and organs mentioned in the study were simultaneously affected, although some were affected only slightly. This assumption is supported by several facts. Fluorescence of liver tissues in ultraviolet light, which is a sign of pathological porphyrin metabolism, was present in all cases of necropsy and biopsy, i.e., in persons for whom long-term monitoring of porphyrin excretion in urine was carried out and in whom δ-aminolevulinic acid values were constantly within normal limits. Probably a slight subclinical lesion was present in each of these patients. Further evidence was furnished in repeated neurological examinations. Polyneuropathy of the lower extremities was manifest in some patients only in the third or fourth year of illness. There is definite clinical and electromyographic evidence that the results of the first examinations conducted when the illness commenced showed that the patients were entirely normal.

Additional study results have been reported recently[15,17,18] describing the mortality of employees engaged in the manufacture of 2,4,5,-T. In two of these studies, cohorts of employees were assembled on the basis of their exposure to TCDD, which was indicated by the presence of chloracne; the third cohort consisted of individuals employed over the same time period. Unfortunately, each of these three cohorts comprised a limited number of individuals, e.g., the chloracne groups contained 121 and 49 individuals, and the employee group contained 204.

In each of these studies, there does not appear to be an apparent excess in total mortality rates or in deaths from malignant neoplasms. It must be pointed out, however, that each of these studies does have limitations both in size of the population studied and in other methodological areas, such as exposure levels.

In summary, we have presented some of the difficulties associated with epidemiologic investigations of low risks attributable to food constituents. Obviously, there are a substantial number of other examples that could be cited. As Doll and Peto[7] have stated, "even risks that will ultimately kill, for instance, 1% or more of the exposed population, may be overlooked or attributed to chance unless a very large-scale investigation is undertaken." In other words, unless epidemiological studies have been carried out in reasonably large, well-defined groups of people who have been heavily exposed to a particular substance for two or three decades without apparent effect, they can offer no guarantee that continued exposure to moderate levels will in the long run be without some increase in risk.

REFERENCES

1. Snow, J. 1936. On the mode of communication of cholera. In Snow on Cholera. The Commonwealth Fund, New York. (Originally published 1855, Churchill, London.)
2. Goldberger, J. 1964. The cause and prevention of pellegra (*Publ. Health Rep.* 29:2354, 1914). In Goldberger on Pellagra (M. Terris, Ed.). Louisiana State University Press, Baton Rouge.
3. Goldberger, J., Wheeler, G.A., and Sydenstricker, E.A. 1964. A study of the relation of family income and other economic factors to pellagra incidence in seven cotton-mill villages of South Carolina in 1916 (*Publ. Health Rep.* 35:2673, 1920). In Goldberger on Pellagra (M. Terris, Ed.). Louisiana State University Press, Baton Rouge.

4. Cordle, F. and Kolbye, A.C. 1979. Food safety and public health Interaction of science and law in the federal regulatory process. *Cancer* 43:2143–2150.
5. Cordle, F. 1981. The use of epidemiology in the regulation of dioxins in the food supply. *Reg. Toxicol. Pharmacol.* 1:379–387.
6. Cordle, F., Locke, R., and Springer, J. 1982. Risk assessment in a federal regulatory agency: An assessment of risk associated with the human consumption of some species of fish contaminated with polychlorinated biphenyls (PCBs). *Environ. Health Pers.* 45:171–182.
7. Doll, R. and Peto, R. 1981. The causes of cancer: Quantitative estimates of avoidable risks in the United States today. *J. Nat. Cancer Inst.* 66:1191–1308.
8. Humphrey, H.E.B., Price, H.A., and Budd, M.L. 1976. Evaluation of changes of the level of polychlorinated biphenyls (PCBs) in human tissue. Final report of FDA Contract 223-73-2209.
9. Allen, J.R. and Norback, D.H. 1976. Pathobiological responses of primates to polychlorinated biphenyl exposure. National Conference of Polychlorinated Biphenyls, Chicago. Pub. No. 560/6-75-004, EPA, Washington, DC.
10. Kimbrough, R.D. 1974. Toxicity of polychlorinated polycyclic compounds and related chemicals. *Crit. Rev. Toxicol.* 2(4):445–448.
11. Kimbrough, R.D., Squire, R.A., Linder, R.E., Strandbert, J.C., Mondali, R.J., and Bubse, V.W. 1975. Induction of liver tumors in Sherman strain female rats by polychlorinated biphenyl Aroclor 1260. *J. Nat. Cancer Inst.* 55:1453–1459.
12. National Cancer Institute. 1978. Bioassay of Aroclor 1254 for possible carcinogenicity. Pub. No. (NIH 78-838), DHEW, Washington DC.
13. U.S. Department of Health, Education, and Welfare (DHEW). 1976. Final report, DHEW subcommittee on the health effects of polychlorinated biphenyls and polybrominated biphenyls, Washington, DC.
14. Holmstedt, B. 1980. Prolegomena to Seveso. *Arch. Toxicol.* 44:211–230.
15. Zack, J.A. and Suskind, R.R. 1980. The mortality experience of workers exposed to tetrachlorodibenzodioxin in a trichlorophenol process accident. *J. Occup. Med.* 22:11–14.
16. Pazderova-Vejlupkova, J., Nemcova, M., Pickova, J., Jirasek, L., and Lukas, E. 1981. The development and prognosis of chronic intoxication by tetrachlorodibenzo-p-dioxin in men. *Arch. Environ. Health* 36:5–11.
17. Ott, M.G., Holder, B.B., and Olson, R.D. 1980. A mortality analysis of employees engaged in the manufacture of 2,4,5-trichlorophenoxy acetic acid. *J. Occup. Med.* 22:47–50.
18. Cook, R.R., Townsend, J.C., Ott, M.G., and Silverstein, L.B. 1980. Mortality experience of employees exposed to 2,3,7,8-tetrachlorodibenzo-p-dioxin. *J. Occup. Med.* 22:530–532.

BIBLIOGRAPHY

Allen, J.B., Carstens, L.A., and Barsotti, D.A. 1974. Residual effects of short-term, low-level exposure of non-human primates to polychlorinated biphenyls. *Arch. Environ. Contam. Toxicol.* 2:86–95.

Altman, N.H., New, A.E., McConnell, E.E., and Ferrell, T.L. 1979. A spontaneous outbreak of polychlorinated biphenyl (PCB) toxicity in rhesus monkeys (*Macaca mulatta*): Clinical observations. *Lab. Anim. Sci.* 29:661–665.

Brown, J.M. 1976. Linearity vs non-linearity of dose response for radiation carcinogenesis. *Health Phys.* 31:231–245.

Crump, K.S., Hoel, D.G., Langley, C.H., and Peto, R. 1976. Fundamental carcinogenic processes and their implications for low dose risk assessment. *Cancer Res.* 36:2973–2979.

Crump, K.S., Guess, H.A., and Deal, K.L. 1977. Confidence intervals and test of hypotheses concerning dose response relations inferred from animal carcinogenicity data. *Biometrics* 33:437–451.

Dennis, D.S. 1976. Polychlorinated biphenyls in the surface waters and bottom sediments of the major drainage basins of the United States. National Conference on Polychlorinated Biphenyls, Chicago. EPA Pub. No. 560/6-75-004, Washington, DC.

Devesa, S.S. and Silverman, D.T. 1978. Cancer incidences and mortality trends in the United States: 1935–1974. *J. Nat. Cancer Inst.* 60:545–571.

Guess, H., Crump, K., and Peto, R. 1977. Uncertainty estimates for low dose rate extrapolation of animal carcinogenicity data. *Cancer Res.* 37:3475–3483.

Hendricks, J.D., Putnam, T.P., Bills, D.D., and Sinnhuber, R.O. 1977. Inhibiting effect of a polychlorinated biphenyl (Arocolor 1254) on aflatoxin B_1 carcinogenesis in rainbow trout (*Salmo gairdneri*). *J. Nat. Cancer Inst.* 59:1545–1551.

International Agency on Research in Cancer. 1978. Monograph on the carcinogenic risk of chemicals to humans. Vol. 18, Polychlorinated Biphenyls and Polybrominated Biphenyls, Lyon, France.

Ito, N., Nagasaka, H., Ari, M., Makuira, S., Sugihara, S., and Hiro, K. 1973. Histopathologic studies on liver tumorogenesis induced in mice by technical polychlorinated biphenyl and its promoting effect on liver tumors induced by benzene hexachloride. *J. Nat. Cancer Inst.* 51:1637–1646.

Jelinek, C.F. and Corneliussen, P.E. 1976. Levels of PCBs in the food supply. National Conference on Polychlorinated Biphenyls, Chicago, Pub. No. 560/6-75-004, EPA, Washington, DC.

Keplinger, M.L., Francher, O.E., and Calandra, J.C. 1971. Toxicologic studies with polychlorinated biphenyls. *Toxicol. Appl. Pharmacol.* 19:204–205.

Kimbrough, R.D. and Linder, R.E. 1974. Induction of adenofibrosis and hepatomas of the liver in BALB c/J mice by chlorinated biphenyls (Aroclor 1254). *J. Nat. Cancer Inst.* 53:547–552.

Kuratsune, M., Yoshimura, T., Matsuzaka, J., and Yamaguchi, A. 1972. Epidemiologic study on Yusho: A poisoning caused by ingestion of rice oil contaminated with a commercial brand of polychlorinated biphenyls. *Environ. Health Pers.* 1:119–128.

Kuratsune, M., Masuda, Y., and Nagayama, J. 1976. Some of the recent findings concerning Yusho. National Conference on Polychlorinated Biphenyls, Chicago. Pub. No. 560/6-75-004, EPA, Washington, DC.

Kutz, F.W. and Yang, H.S. 1976. A note on polychlorinated biphenyls in air. National Conference on Polychlorinated Biphenyls, Chicago. EPA Pub. No. 560/6-75-004, Washington, DC.

Linder, R.E., Gaines, T.B., and Kimbrough, R.D. 1974. The effects of PCB on rat reproduction. *Food Cosmet. Toxicol.* 12:63–77.

Makuira, S., Aoe, H., Sugihara, S., Hirao, K., Ari, M., and Ito, N. 1974. Inhibiting effects of polychlorinated biphenyls on liver tumorgenesis in rats treated with 3′-methyl-4-dimethylaminoazo-benzene, n-2-fluorenylactamide, and diethylnitrosamine. *J. Nat. Cancer Inst.* 53:1255–1257.

McConnell, E.E., Hass, J.R., Altman, N., and Moore, J.A. 1979. A spontaneous outbreak of polychlornanted biphenyl (PCB) toxicity in rhesus monkeys (*Macaca mulatta*): Toxicopathology. *Lab. Anim. Sci.* 29:666–673.

McNulty, W.P. 1976. Primate study, National Conference on Polychlorinated Biphenyls, Chicago. EPA Pub. No. 560/6-75-004, Washington, DC.

Nagayama, J., Masuda, Y., and Kuratsune, M. 1975. Dibenzofurans in Kanechlors. *Japan. J. Hyg.* 30:126–129.

National Marine Fisheries Service (NOAA). 1976. Seafood Consumption Study 1973–1974. National Purchase Diary, U.S. Department of Commerce, Washington, DC.

National Research Council. 1978. Nitrates: An environmental assessment. Environmental Studies Board, Commission on Natural Resources, National Academy of Sciences, Washington, DC.

National Research Council. 1978. Chloroform environmental assessment. Environmental Studies Board, Commission on Natural Resources, National Academy of Sciences, Washington DC.

Office Technology Assessment. 1981. Assessment of technologies for determining cancer risks from the environment. U.S. Government Printing Office, Washington, DC.

Preston, B.D., Van Miller, J.P., Moore, R.W., and Allen, J.R. 1981. Promoting effects of polychlorinated biphenyls (Arocolor 1254) and polychlorinated dibenzofurans-free Arocolor 1254 on diethylnitrosamine-induced tumorgenesis in the rat. *J. Nat. Cancer Inst.* 66(3):509–515.

Schwartz, L. 1936. Dermatitis from synthetic resins and waxes. *Int. J. Pub. Health* 26:586–592.

Veith, G.D. 1975. Baseline concentrations of polychlorinated biphenyls and DDT in Lake Michigan fish, 1971. *Pestic. Monit. J.* 9:921–929.

Wickizer, T.W., Brilliant, L.B., Copeland, R., and Tilden, R. 1981. Polychlorinated biphenyl contamination of nursing mothers' milk in Michigan. *Am. J. Publ. Health* 71(2):132–137.

Zitko, V., Hutzinger, O., and Chor, P.M.K. 1972. Contamination of the Bay of Fundy & Gulf of Marine Area with PCBs, PCTs, and chlorinated DBF and DBD. *Environ. Health Pers.* 1:47–50.

4. Food Safety, Risk Assessment, and Democracy: Achieving Public Understanding and Acceptance of What the Experts Say and Do

Michael R. Taylor

King & Spalding

INTRODUCTION

Food safety regulation is one of democracy's great challenges. Like other areas of health and safety regulation, it affects every citizen and addresses our most fundamental personal aspirations—assuring long and healthy lives for ourselves and those we love.

It is thus no wonder that much of what government does in the name of food safety is controversial and, in some instances, elicits intense, emotional responses from the public, which so often seems to find the government's actions insufficiently protective, overly protective, or simply incomprehensible.

Those of us engaged professionally in food safety regulation often find the process no more satisfying. We feel frustrated at the inability—even, it seems sometimes, the unwillingness—of politicians, the media and individual members of the public to understand what we are doing and why. There are many reasons for this lack of understanding, a complete study of which could productively occupy the best minds from many scientific and academic disciplines for years to come.

Some of the most important reasons, however, have to do with risk assessment and the way it is currently used by public health officials to reach and explain their decisions. This brief chapter will sketch out the basic problem of understanding that exists between public health experts and the public, as affected by risk assessment, and suggest in general terms some things that could be done to address it. I hope to explain why it is vital to work persistently on the problem, but why we can never expect to solve it entirely.

THE GAP IN UNDERSTANDING AND PERSPECTIVES

Food safety regulation is such a challenge for democracy because there is such a large gap between the simplicity of the public's fundamental desires and understanding of the problem and the complexity of the way experts perceive and deal with food safety questions. This gap has several features.

First, the public's goal is simple, perhaps excessively so. It does not typically think in terms of "risk" or "degree of risk" when it comes to food. If a food substance is "unsafe," it should be removed. This seemingly simple goal has been codified through the political process in very rigorous statutes requiring virtually every component of the food supply to meet a high standard of safety and empowering the government to take actions necessary to remove components that are unsafe. The public cares little about the details of how this is accomplished, so long as it is accomplished.

While the public sees simplicity and ignores the details, the experts in the various regulatory agencies,* whose task it is to achieve the goal of safety, see much complexity and must worry a great deal about the details of how food safety can be assured. And the task is becoming more complex every day. Scientific advances have made it possible to ask increasingly numerous and difficult questions about the safety of substances in food, especially about their chronic hazards, and have required increasingly complex analyses to answer those questions—analyses that are far beyond the ken of lay citizens. Moreover, in contrast to most laymen, the experts have become accustomed to thinking about the food supply in terms of risk. A substance is "safe" if its risks are sufficiently low. The emergence of quantitative risk assessment as an accepted analytical tool for estimating potential human cancer risks has only increased the experts' tendency to think in terms of risk.

Another important feature of the gap between expert and lay perspectives on risk is that public health officials are concerned about and address risk in the aggregate, as a question of *public* health. They operate at the macro level. Individual members of the public, in contrast, view the problem at the micro level. They want to know what the evidence about a substance means to them individually. Will I get can-

*The Food and Drug Administration's Center for Food Safety and Applied Nutrition has the principal responsibility for assuring the safety of the nation's food supply, but important roles are also played by the U.S. Department of Agriculture (regarding meat and poultry products) and the Environmental Protection Agency (regarding pesticide residues).

cer from this food additive? Is it "safe" for me? To the average citizen, notions of aggregate or population risk are highly abstract and probably meaningless.

Finally, there is the problem of uncertainty. Scientists and other experts engaged in food safety regulation recognize and accept the biological and statistical uncertainty inherent in their analyses. No substance can be proven harmless to an absolute certainty, and all test systems are statistically imperfect surrogates for the human population. Careful scientific and public health officials do not expect or demand from a body of data more than it is capable of delivering.

The lay public has the opposite tendency. It assumes scientists are capable of delivering clear-cut, yes or no answers on questions of safety and demands nothing less. Uncertainty is difficult to accept because it conflicts with the lay image of science as the domain of fact, and it defeats the desire for absolute safety.

Given these differences in perspective and understanding—complexity vs simplicity, degree of risk vs safety, macro vs micro-level risk and acceptance of uncertainty vs demand for certainty—it is easy to see the problems posed by quantitative risk assessment. As currently practiced and presented to the public, it compounds all the preexisting problems of understanding. Quantitative risk estimates are typically presented to the public with little or no explanation of the upper bound nature of the estimates and the uncertainties that attend the entire assessment. Indeed, the typical presentation of the estimate(e.g., 8.2×10^{-6}) has the appearance of great precision and seems to be speaking directly to the individual's real, personal chances of contracting cancer. Complexity and uncertainty are obscured, and the layman's preconceived notions of simplicity and certainty are confirmed.

The reality of the matter is, of course, quite different. Quantitative risk assessment is merely an analytical tool public health officials use to put the potential hazards of substances in some rational perspective at the macro level. As such, it has become essential to the regulatory process, but no one claims that it provides an accurate picture of any individual's true risk.

Obviously, then, the experts see and rely on quantitative risk assessment as one thing. The public sees and relies on it as something entirely different.

Does this gap in understanding matter? The answer seems clearly to be yes, for at least two reasons. First, due to the public's conception of what quantitative risk estimates mean, the mere attachment of a risk number to a substance—almost regardless of what the number is—gives rise in some quarters to great anxiety and a demand for regulatory action. The number is (mis)taken as describing the scientific fact that some individuals *will* contract cancer and die from consumption of the substance. The agencies can then be forced to waste their scarce resources taking or considering actions that have little or no public health justification.

Second, when the government decides not to take action in every such case, it triggers protest and controversy. Cynicism sets in, and that essential public understanding of and support for public health programs erodes.

RECOMMENDATIONS

What can be done? My thesis is that we can never fully close the gap in perspective and understanding between experts and the lay public. The public will always (and understandably) view risk at the micro rather than macro level, and it will never come to grips with scientific complexity and uncertainty the way experts typically do. Citizens, politicians, and the media can gradually be educated concerning the nature of risk, but, because of its subject matter, food safety regulation will always be controversial; and in the end experts will always be misunderstood. There are some things that can be done, however, to minimize the problem of understanding, and, in a democracy, it is the duty of government to do all it can to make its activities comprehensible and to win the public's support. Most of what I suggest is obvious.

First, regulatory agencies should reduce their reliance on quantitative measures in making their regulatory decisions and in explaining them to the public. This does not mean quantitative assessments should not be made. As already noted, they are often vital. Agencies should, however, be clear in their own minds and make clear to the public that intelligent food safety decisions rely on the exercise of *scientifically well-informed public health policy judgment*. There are no quantitative formulas that yield correct regulatory results.

Second, risk assessment should routinely incorporate both quantitative and qualitative evaluations that take advantage of all the data. The strength of the evidence of carcinogenicity, the degree of certainty or uncertainty concerning the relevance of the animal results to humans, and any special features of the dose-response relationship, including the possibility of a threshold, should all be considered as part of the risk assessment and should be made known to decisionmakers. When the data permit, a range of risk estimates should be calculated, including the "worst plausible" and "most likely" upper bound estimates.

Third, when presented to the public, quantitative risk estimates should be accompanied by as much explanation as possible of their upper bound nature, the range of uncertainty, and the role the quantitative estimate played in reaching the ultimate decision. Much of this will not be passed along to the public by the news media, and much will not be understood, but the information should nevertheless be made available.

Fourth, regulatory agencies should be provided legal tools that expressly direct the exercise of judgment in making decisions about specific substances and in setting regulatory priorities. The Food and Drug Administration's efforts to deal with and avoid the sometimes irrational impact of the Delaney Clause has certainly done nothing for that agency's credibility with the public and has diverted FDA from activities that might have yielded greater public health gain. Similarly, without any express mandate to set priorities and allocate its resources to addressing the most significant public health risks, FDA is often forced to deal with the "crisis of the week," a phenomenon that is propelled at least in part by the results of quantitative risk assessment and the public's misperception of what they mean.

Fifth, the more we emphasize the inherently judgmental nature of food safety decisions, the more important it becomes to enhance the credibility and real capacity of the institutions we rely on to make those judgments. This means assuring that the agencies have the personnel and other resources they need to do the best possible job. It also means insulating the agencies from even the appearance of being influenced by partisan politics or other narrow interests. This does not mean that agencies should disregard the larger interests of the public they serve. Maintenance of a sound, broadly supported public health program requires that regulatory decisions be perceived as making common sense. Agencies should thus be aware of and consider the practical impact of what they do, so long as that is consistent with the public health mandate.

Finally, but perhaps of greatest importance, is the need to improve the scientific basis for food safety decisions. This means, among other things, improving the scientific quality of risk assessment. There is nothing more destructive of public confidence in a regulatory decision than for reputable, well-credentialed scientists to stand up in public and take vehement and diametrically opposed positions on what the agency has done. This is a regular occurrence today, and it is made possible in large part by the scientific immaturity of risk assessment. The gaps in our understanding of the relevance of high-dose animal results to true human risk at low levels of exposure make it possible for any trained scientist to poke holes in the agency's analysis. This problem will be resolved only as the gaps in understanding are filled, and this will come about only through a serious, sustained commitment to the research necessary to improve methods of risk assessment.

The methylene chloride case provides a recent example of the problems we now face. FDA's own document on methylene chloride candidly acknowledges that, if EPA had performed the risk assessment, it would have attributed to methylene chloride a potency 26 times greater than that derived by FDA due solely to the differences in choices and assumptions made by the agencies in conducting their assessments.* This is the difference between a one-in-one-million estimate of the upper bound risk, as derived by FDA, and a one-in-forty-thousand estimate of the upper bound risk, which presumably would have been derived by EPA. Such differences, which flow directly from our lack of knowledge, have obvious potential to yield inconsistent decisions. More importantly, however, they defeat any efforts one might make to enhance public understanding of and confidence in the regulatory process.

CONCLUSION

A common trait of modern democracies is the granting to the state and its experts of great powers—such as the power to regulate the safety of food. Those experts are then bombarded with technical criticism, legal challenges, and expressions of popular outrage. This comes with the territory.

The task of the expert is to do the best job he or she can, explain it as well as possible, and then learn to live with the imperfect, often raucous results. Science, food safety regulation, and democracy are inherently messy affairs. We must never stop working to improve them, but we have to accept that controversy is an unavoidable and healthy part of the process.

---

*50 *Federal Register* 51551, 51555. Dec. 18, 1985.

5. Risk Assessment for Effects Other Than Cancer

Joseph V. Rodricks
Vasilios Frankos
Duncan Turnbull
Robert G. Tardiff

ENVIRON Corporation

INTRODUCTION

Risk assessment is now a routine component of decisionmaking as it relates to carcinogens. Although substantial controversy exists regarding both the conduct of carcinogen risk assessment and its use in decisionmaking (where its uncertainties are frequently ignored), it remains an effective and systematic device for organizing and presenting what little information and knowledge we have about low-level exposure to carcinogens. There appears little doubt that its use will continue and we shall continue to seek ways to improve its conduct and use.

An exploration of risk assessment for noncarcinogens, which is the subject of this chapter, should perhaps begin with the observation that it is hardly a new topic. Ever since we began evaluating toxicity data for purposes of recommending limits on human exposure, we have been engaged in risk assessment. We are, of course, referring to risk assessment in its broadest sense—the attempt to predict the likelihood that the toxic properties of a substance will become manifest in humans under conditions of exposure far removed from the conditions under which toxic effects can be observed and measured, either in humans or under experimental conditions. In the case of carcinogens, the results of a risk assessment are typically expressed in quantitative terms as probabilities, although it should be emphasized that any such expression, without accompanying qualifications, substantially distorts our state of knowledge. Risk assessment for noncarcinogens is different in concept and form, but it is nevertheless a risk assessment.

In the remaining sections of this chapter, we shall discuss risk assessment for noncarcinogens as it is currently practiced, identify some of its limitations, and examine the question of whether there is any basis for adopting for noncarcinogens the form of risk assessment now used for carcinogens.

NONCARCINOGENS

The term "noncarcinogen" is ambiguous, and should perhaps be dropped. In operational terms, the term can be applied to a substance that has not induced excess neoplasms when subjected to what are currently considered appropriate tests for carcinogenicity. This is perhaps the best definition, but it could be misleading because we can never be sure such a "noncarcinogen" would not produce excess neoplasms under some other test conditions. Perhaps the rather awkward label "not shown to be carcinogenic in appropriate tests" is a better descriptor.

This type of descriptor is necessary to distinguish substances that are noncarcinogens because they have not been adequately tested from those that have. The former are better described as "not having been subjected to tests appropriate for the identification of carcinogens."

Although distinguishing the two types of noncarcinogens is important, the question of risk assessment for such substances can best be explored by organizing the investigation according to toxic endpoint. Thus, the exploration concerns risk assessment for endpoints other than carcinogenicity. Under this approach, the risk assessment questions and methodologies we explore are also applicable to the noncarcinogenic effects of proven carcinogens.

BASIC CONCEPTS

Thresholds

Risk assessment for noncarcinogenic forms of toxicity, as it has been and is now conducted, depends on acceptance of the threshold hypothesis for every such form, just as risk assessment for carcinogens depends, at least in the regulatory context, on rejection of the same hypothesis. One of several well-known statements of the threshold hypothesis is the following:

> One of the most important principles of occupational health . . . is that exposure to a toxic agent may be permitted up to some limit of tolerance above zero, within which [an individual] can cope successfully with the insult with no significant threat to health.[1]

The notion of a threshold for toxic agents is based both upon empirical observations and knowledge of the biological processes underlying the production of toxicity. Thus, for many agents it is completely clear that there is a range of exposure levels within which no identifiable adverse health effects are produced in humans or experimental animals. Moreover, it is clear from experimental studies that for many toxic agents or their active metabolites, a minimum concentration at target sites must be achieved before a measurable response is produced. Even then, the measured response may have no adverse consequences for health, usually because the host has the capacity to adapt to or recover from the response. Of course, the target site concentration of the toxic agent or its active metabolite may be increased to the point at which a response is produced of such magnitude that recovery is either very slow or impossible; under such conditions we have clearly exceeded the threshold level and toxicity is produced.

Some scientists reject this notion for carcinogens. It is hypothesized that any finite concentration of a carcinogen at the ultimate target site (DNA) will increase the probability of a neoplastic transformation. The increased probability may be extremely small and without practical consequences over a wide range of low concentrations, and there may exist many host defenses that contribute to nonlinearity in the quantitative relationship between the administered dose and the ultimate target site concentration, but the nonthreshold hypothesis cannot be rejected for these reasons.

The consequences for risk assessment of assuming a threshold or nonthreshold hypothesis are significant, but before they are discussed it will be useful to note some of the problems and limitations of both these hypotheses.

Limitations in Measurement of Thresholds

Except in a few, limited circumstances, there are no means available for direct measurement of thresholds. For any individual, human or animal, it is assumed that exposures produce no effects up to a given level of exposure, that biological changes or effects having no permanent adverse consequences appear at some dose (the effects threshold), and that adverse effects begin to appear at a still higher dose (the threshold for adverse effects). Above the latter threshold, frank toxicity occurs, the severity and incidence of which increases with increasing dose.

In most cases (and probably in all cases involving repeated chronic exposures) experimental animals are relied on to estimate these thresholds and the dose-response relationship above them. There are several reasons why true thresholds cannot be identified using this experimental device:

1. For practical reasons, only a few doses can be used in experimental studies. While these doses may fall above and below the true threshold doses, it is only by chance that any will precisely match the true threshold doses (and this chance is very small). The maximum dose at which no biologically and statistically significant effect is observed is called the "no-observed effect level" (NOEL), and the maximum dose at which no effect considered adverse to health is observed is labeled the "no-observed adverse effect level" (NOAEL), but the relationship between these and the true "effects threshold" and the true "threshold for adverse effects" is not known.

2. The NOEL and NOAEL are in part functions of the experimental sample size. Every animal bioassay has a limit of detection that is dependent upon the number of animals used, and the failure to find effects may be due to this detection limit and not to our having uncovered a threshold.

   It is very interesting to note that this argument is frequently used by proponents of the "no-threshold" hypothesis for carcinogens. There are many cancer bioassays in which NOELs are identified (at least for the production of neoplasms). Proponents of the no-threshold hypothesis point out that the existence of such a NOEL does not necessitate rejection of the no-threshold hypothesis, because the NOEL could simply reflect the detection limit of the bioassay (usually in the risk range of 5–10 percent). It is thus interesting that the problem of experimental detection limits is not generally thought to provide evidence for a no-threshold hypothesis for other types of toxic endpoints.

3. The measured NOEL and NOAEL also depend upon the thoroughness of the clinical and pathological examinations conducted on the experimental subjects. There is often uncertainty regarding the health significance of the various changes observed in the dose region just above the NOEL, leading to uncertainty in identification of the NOAEL.

For these three reasons the relationships between the NOEL, NOAEL and the true threshold doses are unknown, perhaps even unknowable. And, particularly because of the second and third reasons, there is also substantial doubt about whether the mere observation of a NOEL or NOAEL for any toxic effect can be taken as proof that a true threshold dose exists: the existence of a NOEL or NOAEL is consistent with the threshold hypothesis but is not sufficient to prove it.

Theoretical Limitations

Although true thresholds cannot be measured with reliability, it is patently obvious for numerous agents and effects that they exist. This is particularly true for acute effects, if not for those that are delayed. But is there a basis for accepting the threshold hypothesis for any effect of any chemical when the only information available concerns the production of some form of toxicity in experimental animals—i.e., when there is no information available on the mechanism of toxicity? We suggest that, based on the type of evidence usually considered necessary to establish a scientific hypothesis, the answer must be no. Without knowledge of the mechanism of toxicity of a specific agent, it seems impossible to demonstrate the reliability of the threshold hypothesis for the effects it produces. At the same time, it should be emphasized that there appears to be no way to establish the absence of a threshold for a carcinogen unless its mechanism of action is understood.

It thus seems that, in cases in which the mechanisms of toxic or carcinogenic action are not understood, it is not possible to establish or reject the threshold hypothesis or the no-threshold hypothesis, at least with the degree of certainty usually sought in scientific proof. There are numerous reasons to believe that thresholds must exist (existence of repair mechanisms, detoxification mechanisms, etc.), but generalizations to all agents and all effects is not possible.

IMPLICATIONS FOR RISK ASSESSMENT

The Traditional Approach and its Limitations

For all noncarcinogenic forms of toxicity, it has become the custom to assume the threshold hypothesis for purposes of risk assessment. Under this hypothesis, risk is zero at levels below the threshold dose, and increases with dose thereafter. If the true threshold dose could be identified for each effect of an agent, it would be possible to identify the full range of exposures over which no risk exists and the dose at which the risk begins to increase.

Several problems arise in attempts to identify the "risk-free" range of exposures:

1. As already noted, it is not possible to measure the true threshold dose, even for experimental animals.

2. There is evidence that the location of the true threshold dose is species-specific and is likely different for humans and test animals for most agents.

3. The true threshold dose is not constant within a given population but varies among members of a population. The distribution of thresholds in the human population is agent- and chemical-specific, and probably takes different forms for different subpopulations.

Given these uncertainties, it has become common to incorporate various "safety factors" to overcome them. Generally, the NOEL or NOAEL for a substance is divided by a "safety factor" to yield an exposure level, usually called the Acceptable Daily Intake (ADI), that is assumed to be protective—i.e., the ADI is assumed to be at or below the threshold for the most susceptible members of the population to be protected. The ADI is used as a yardstick to measure the safety of residues of direct and indirect food additives, pesticides, drinking water contaminants, and other environmental agents. Formal ADIs are not established for carcinogens, although acceptable intake levels are commonly established for carcinogens based on the policy that some low level of risk is acceptable or tolerable.

The magnitude of the safety factor used in specific cases depends on a number of factors: the quality of the database, the nature of the toxic effect, the duration of exposure for the experimental subjects relative to that expected in the population to be protected, the care with which the NOEL or NOAEL has been identified, and any available data concerning intra- or interspecies variabilities in response.

Safety factors were originally introduced in the early 1950s for substances such as food additives and food residues of pesticides to which the general population might be exposed daily for their entire lifetimes. The safety factor approach was originated by Lehman and Fitzhugh[2] of the FDA who indicated that variability in sensitivity to chemicals (expressed as differences in dose causing similar responses) across several species was usually in the range of two or threefold and did not appear to exceed tenfold. They also indicated that the variability among extensively outbred individuals and individuals of all ages and degrees of susceptibility (e.g., persons in the general population) appeared also to be less than one order of magnitude. They consequently founded the 100-fold safety factor as a general method of dealing with the uncertainties of extrapolation. This incorporated a factor of 10 when extrapolating from animals to humans and an additional factor of 10 to account for differential sensitivities within the human population. When this 100-fold safety factor is applied to the highest experimental animal NOEL, it is considered to approximate a NOEL for humans in the general population, and becomes the ADI.

In 1962, the FAO/WHO Expert Committee on Food Additives[3] outlined limitations associated with the use of the 100-fold safety factor for establishing ADIs. Factors such as animal species, strain, and sex differences, variations in susceptibility among persons exposed in the general population, and insufficient experimental animal data were identified as important considerations when identifying the ADI. For example, food residues (or any other environmental pollutants) may be ingested by persons of all ages, by the healthy and the ill, and under various conditions and patterns of exposure. Consequently, the FAO/WHO no longer consistently uses the 100-fold safety factor and sometimes modifies the factor due to a lack of available data on a substance under consideration. Since 1962, the FAO/WHO Expert Committee on Food Additives[3] has used the terms "conditional" and "unconditional" ADIs. Conditional ADIs incorporate a safety factor larger than 100 because of uncertainties or limitations in the available animal data or in regard to the identity and purity of the substance under question.

The Center for Food Safety and Applied Nutrition of the FDA has expanded the original 100-fold safety factor approach in the derivation of chronic ADIs in the absence of chronic toxicity data. Where only subchronic animal data are available, but are available in two species, the FDA has recommended a factor of 1000, incorporating an additional factor of 10 to account for the added uncertainty in estimating a chronic ADI from subchronic data.[4] In cases where subchronic data are available for only one species, a 2000-fold safety factor has been recommended; the additional twofold factor is intended to account for possible interspecies differences.

Another aspect of the experimental data that may influence the selection of safety factors is the inherent nature of the effects observed. Two types of observation are critical: (1) the severity of the effect and (2) the reproducibility of the effect among various animal species. Generally, severe effects (i.e., those that may significantly interfere with function or are life-threatening) have been treated with more caution than have less severe effects. For example, FDA generally uses a larger safety factor for chemicals that cause serious birth defects than for chemicals that do not cause such effects.[5] Also, substances that produce marked qualitative or quantitative differences in response among several animal species have been considered more worrisome (with respect to extrapolation to humans) than those producing similar responses in several species. The latter concern is based on the notion that in the case in which marked differences in response are observed among several animal species, human responses cannot be as confidently predicted as they can in instances in which responses are similar in several animal species.* Alternatively, when the same type of response is seen in several species, we can be more sure that the response is likely to occur in humans.

Although many factors influence the magnitude of the safety factor, it must be recognized that these factors are nevertheless generic; that is, there are rarely data available in specific cases that allow selection of accurate factors. Thus, as we proceed from experimentally derived NOELs or NOAELs, which are themselves of unknown relation to the true experimental thresholds, to estimate ADIs, we engage in a number of extrapolations beyond the data. While each type of extrapolation would seem to be required, there is little empirical basis for assigning specific numerical factors to account for the expected differences in threshold, or to account for the uncertainty in the relation between the NOEL and the true threshold. Moreover, all of the extrapolations depend upon acceptance of the threshold hypothesis, which cannot be rigorously proved in every specific case.

So, like risk assessment for carcinogens, risk assessment for noncarcinogenic effects depends upon a large number of assumptions. As a result, ADIs probably have considerable uncertainty—we do not know how they relate to the true population threshold (assuming there is one in all cases). It is thus the case that we do not know what fraction of the exposed population may be exposed above the threshold level when actual exposure is at the ADI.

Are the Limitations in the NOEL/Safety Factor Approach a Serious Problem?

Several points should immediately be made:

1. We know of no evidence that ADIs, when based on adequate data,** are not health protective. Of course, it is extremely difficult to test their protective value epidemiologically, and there is probably no fully adequate test for any substance, at least in respect of chronic effects.

*Note that this variability might also be interpreted to mean that we should be less concerned about human toxicity. The choice to interpret it as of potentially greater concern can be seen, at least in part, as a policy choice.

**We distinguish the problem of inadequate data (e.g., the thalidomide case) from the problem of using NOELs and generic safety factors.

2. A limited survey of available data on inter- and intraspecies variability suggests that the factors ordinarily used for these extrapolations (10 × for each) are probably fully protective and are likely overprotective.[6] Of course, the data base available for this type of study is extremely limited at the present time.

3. In many cases, an additional safety factor exists because actual human exposure is only a fraction of the ADI.

It thus appears there are reasons to believe that ADIs established according to the procedures described earlier, are protective of human health (i.e., they present, at most, very low risks), although they are only very weakly supported reasons if judged according to the usual scientific criteria for proof.

The various uncertainties associated with establishing ADIs have been known for many years. It is thus curious that there is so little criticism of them in the scientific literature and so little concern over their value as health protective devices. There are several possible reasons for this relative silence:

1. *Custom and Authority*—The procedures have been in use, in their modern form, since the 1940s. They were introduced by regulatory and public health authorities and have been used by every national and international organization charged with recommending acceptable intakes of dietary and environmental agents.

2. *Intuitive Appeal*—It is very difficult to believe that a significant risk could be associated with an exposure level several orders of magnitude below a NOEL observed in experimental tests designed to stress experimental subjects to ensure the production of toxicity.

3. *Simplicity*—The NOEL-safety factor approach is relatively easy to understand, apply, and explain to individuals without technical training.

4. *Apparent Absence of Risk*—Risk is not an explicit component of the NOEL-safety factor approach. Although no one can be sure that some individuals in a population will not be at risk (i.e., that their thresholds are lower than the ADI), the NOEL-safety factor approach includes no method for estimating that risk. Risk never enters the discussion, so little unease is created.

5. *Practicality*—The operational need for a means to establish safe levels for chemicals is readily satisfied by use of the NOEL-safety

factor approach. Further, very few chemicals have had to be restricted because they could not be used in conformance with the prescribed ADI.

Should the NOEL-Safety Factor Approach be Changed?

Notwithstanding all the shortcomings in the NOEL-safety factor approach, we see little basis at present for departing from it. There are several suggestions in the literature for adopting the form of risk assessment now used for carcinogens for noncarcinogenic effects.[7,8] These authors suggest that a procedure that explicitly recognizes risk and uncertainty is more faithful to our true state of scientific understanding than is one that hides uncertainties in the form of "arbitrary safety factors." The concerns of these authors have merit, and we shall return to them in the closing paragraphs. First, we shall discuss why we believe major alterations in the NOEL-safety factor approach are not presently called for.

One possible alternative to the present approach is to reject the assumption of a threshold for all noncarcinogenic effects and to adopt a dose-response model for low dose extrapolation. Rejection of the threshold model would be based primarily on the notion that we cannot prove the existence of a threshold, at least for populations if not for individuals. Behaving as if no threshold exists under such circumstances is held to be more in keeping with basic scientific principles—i.e., science can deal with and measure effects, but it cannot ever prove a negative statement. The threshold statement is such a negative, and so cannot be proved with the degree of rigor with which positive statements can be proved.

This view, which certainly has merit, must be balanced by the observation that there is no known theoretical basis by which all chemicals, except perhaps some carcinogens and mutagens, could be said to pose a health risk at all finite exposure levels. Without such a basis, it seems unnecessarily rigid to reject the threshold model simply because we cannot accurately identify the locations of thresholds in populations. Unless a theoretical basis for rejection of the threshold hypothesis can be found, there appears to be no reason to adopt no-threshold models. This is not to conclude that there is no effect of any substance for which the threshold hypothesis is incorrect; rather, it is only to note the absence of a theoretical basis at the present time for rejection of the hypothesis for all substances. (What seems likely, but which remains unverified, is that, just as some carcinogens probably act through threshold mechanisms, some other toxicants act through nonthreshold mechanisms.)

Several mathematical models for low dose extrapolation (e.g., probit, logit) assume that each individual in a population has a *tolerance* for a chemical — a threshold of dose that must be attained before a response is elicited.[9] These models differ from one another in assuming that the distribution of tolerances for a toxic chemical in a large population is expressed by different mathematical distributions (e.g. , log-normal in the case of the probit model). These distribution assumptions are equivalent to rejecting the possibility of identifying the threshold dose for a large population. It should be noted, however, that while the probit and logit models do not assume zero risk (risk being the fraction of a population whose tolerances are exceeded) at any finite dose, they approach zero risk much more quickly than other models, and achieve "practical threshold" doses at finite exposure levels.

These models would seem to have a greater attractiveness than the other no-threshold models because of their assumption of at least individual thresholds. These models, unlike the NOEL-safety factor model, can also incorporate dose-response data in a systematic way. At least theoretically, these tolerance-distribution models would seem to enjoy advantages over other models for low-dose extrapolation of noncarcinogenic effects. It will be necessary, however, to investigate their applicability before recommending their routine use, and this has not yet been done to a significant extent.

Because of the long history of use of the NOEL-safety factor approach, because it appears consistent with current hypotheses of toxic action, and because there are no alternatives of demonstrated superiority, it would appear desirable to continue its use. There are, however, several possible improvements that can be envisioned, some available now, some only after additional research efforts are undertaken. We close by noting some of them:

1. Although there is no apparent way to understand the relation between NOELs, NOAELs and threshold doses, there are means to reduce the dependence of the NOEL upon experimental sample size. Use of dose-response data to estimate a statistical lower confidence limit on an ED_{01} or ED_{10} eliminates the dependence upon sample size of these measures.[8,10] These estimates are made in ways that "reward" experimentation with larger sample sizes and "penalize" those with small sample sizes. The ED_{01} and ED_{10} are measures of response rates at or just below the detection power of most experiments, and so they, and confidence limits on them, can be estimated without significant extrapolation. Crump (1984)[10] has labeled these "benchmark doses." In order to estimate the ADI, Crump recommends dividing the estimated lower confidence limit of the benchmark dose by an arbitrary safety factor. Gaylor, on the other hand, is more explicit. He recommends strict use of the ED_{01} as the benchmark dose and calculates the

safety factor as the estimated upper confidence limit on the risk at the benchmark dose divided by the desired risk in the population (such as 0.000001). The safety factor approach of Gaylor is analogous to the linear extrapolation technique for carcinogens proposed by Gaylor and Kodell,[11] and used by FDA. To quote Gaylor, "both techniques assume that as the dose is decreased, the risk decreases proportionately more rapidly." As noted we believe the use of linear, low dose extrapolation models for noncarcinogenic effects is not now warranted, but do support consideration of the use of benchmark doses.

Although use of such a benchmark dose has attractive features, its routine use will require that toxicologists collect and report dose-response data in more quantitative forms than they currently do. This is particularly problematic for continuous data. We think it desirable that toxicology data be collected and reported in a form that permits modeling and estimation of the ED_{01} or ED_{10} values. Under the NOEL-safety factor system, it is not apparent that dose-response data are used at all. It is claimed that dose-response data are used in selecting safety factors, but we have not found anything in the literature except very general statements about this matter. Moving toward the use of benchmark doses could make a practical reality of the old dictum that dose-response is critical to safety determination.

2. The various selections of safety factors are based on very little empirical information. For this reason, the selections should be seen largely as policy choices, and should be described as such. We urge the adoption of procedures for the evaluation of noncarcinogenic effects similar to those used for carcinogens, wherein attempts are made to distinguish those components of risk assessment that are based on scientific consensus from those that are based on science policy.[12,13]

3. At present, little systematic consideration is given to the relative severity, or seriousness, of the toxic effect observed—although a subjective distinction between adverse and nonadverse effects, which would also differ in severity, does occur in the selection of NOELs or NOAELs. As a result, similar ADIs for different substances may be set on the basis of the occurrence of a variety of substantially different toxic effects which may be indicative of substantially different degrees of functional impairment. For example, two chemicals may be assigned identical ADIs on the basis of NOAELs of 0.1 mg/kg/day in chronic animal studies. However, one chemical may cause only slight hypertrophic changes in the liver at 1 mg/kg/day, while the other may cause necrosis of the kidney tubules and heart muscle, and degenerative changes in the liver at the same dose level. Clearly, errors in determination of the NOAEL for the latter chemical could lead to

> unacceptably severe effects which would be of greater concern than the former despite the fact that traditional procedures yield identical ADIs for both. Further study might allow an objective and systematic mechanism for differentiating between ADIs set on the basis of toxic effects representing different degrees of severity and functional impairment.

Whether the empirical basis for safety factor selection can be improved is unclear at this time. Information on intraspecies variation in response in the human population has been collected for a few substances, mostly drugs, and there is also a substantial data base concerning variation in the pharmacokinetic behavior of exogenous compounds among individuals. It would seem appropriate to examine all such data exhaustively to determine whether any generalizations can now be made, either for certain classes of chemicals or for certain classes of effects. We know of no such exhaustive review in the context of chemical safety evaluation and suggest that one be undertaken. Such an evaluation might also greatly assist development of a research agenda to increase understanding of human variability in toxic response.

Interspecies extrapolation would seem to be in a similar state of uncertainty. Although a great deal has been written on this subject, there appear to have been few attempts at a creative synthesis of available data.

Until these issues of inter- and intraspecies variability are scientifically untangled and clarified, it would seem appropriate that care be taken to avoid confusing science and policy. One possible approach that might help achieve this objective is to redefine slightly the role of the risk assessor. Thus, the risk assessor's goal might be to select appropriate dose-response information, identify NOELs, decide how the shape of the observed dose-response curve should influence low-dose risk, and estimate the magnitudes of the differences between the NOEL, the NOAEL, and the expected human exposure level. A separate science/policy analysis is then conducted to determine whether the magnitude of the difference is sufficiently large to protect public health and also to satisfy other requirements of applicable laws unrelated to health protection. Of course the risk assessor would bring forward any chemical-specific data available on intra- or interspecies variability, but he would avoid assigning generic safety factors in the absence of such specific data. Under this approach, selection of safety factors is in part a risk management decision. The factors are then true "safety" factors, distinguishable from those introduced based on empirical information on inter- or intraspecies variability.

REFERENCES

1. Hatch, T.F. 1971. Thresholds: Do They Exist? *Arch. Environ. Health* 22:687–689.
2. Lehman, A.J. and Fitzhugh, O.G. 1954. 100-Fold margine of safety. *Assoc. Food Drug Off. U.S.A. Bull.* 18:33–35.
3. FAO/WHO Expert Committee on Food Additives. 1962. Evaluation of the toxicity of a number of antimicrobials and antioxidants. World Health Organization Technical Report Series. Rep. 229. Geneva.
4. Kokoski, C.J. 1976. Written testimony of Charles J. Kokoski, Docket No. 76N-0070. DHEW. Food and Drug Administration, Washington, DC.
5. Frankos, V.H. 1985. FDA perspectives on the use of teratology data for human risk assessment. *Fund. Appl. Toxicol.* 5:615–625.
6. Dourson, M.L. and Stara, J.F. 1983. Regulatory history and experimental support of uncertainty (safety) factors. *J. Reg. Toxicol. Pharmacol.* 3:224–238.
7. National Academy of Sciences (NAS). 1980. Drinking water and health. Vol. 3. Washington, DC: National Academy Press. pp. 31–35.
8. Gaylor, D.W. 1983. The use of safety factors for controlling risk. *J. Toxicol. Environ. Health* 11:329–336.
9. Brown, C.C. 1984. High- to low-dose extrapolation in animals. In J.V. Rodricks and R.G. Tardiff, Eds. Assessment and Management of Chemical Risks. ACS symposium series 239. Washington, DC: American Chemical Society.
10. Crump, K.S. 1984. A new method for determining allowable daily intakes. *Fund. Appl. Toxicol.* 4:854–871.
11. Gaylor, D.W. and Kodell, R.L. 1980. Linear Interpolation Algorithm for Low Dose Risk Assessment of Toxic Substances. *J. Environ. Pathol. and Toxicol.* 4(Nos. 5,6):305–312.
12. National Academy of Sciences (NAS). 1983. Risk assessment in the federal government: Managing the process. Washington, DC: National Academy Press.
13. Office of Science and Technology Policy (OSTP). 1985. Chemical carcinogens; a review of the science and its associated principles, February 1985. *Federal Register* 50:10371–10442.

6. Toxicology Committee Conclusions and Recommendations

Toxicology contributes the essential scientific foundation for the ultimate public health policy judgment concerning what constitutes "safety" in a particular case. Toxicology performs this role best when it gives the decisionmaker the most complete picture possible of the potential human health effects of the substance in question. This means taking advantage to the extent our knowledge permits of all available data, including epidemiologic data, providing both quantitative and qualitative analyses of the data, and describing for the decisionmaker all of the assumptions and uncertainties that attend the assessment of risk.

The contribution of toxicology is constrained today by gaps in our understanding of how carcinogenic effects seen at high doses in animal studies relate to the probability of adverse effects in humans at much lower doses. Current methods of quantitative risk assessment are useful because they can identify reliable upper limits on likely human risk and thus support judgments in some cases that the human risk is insignificant. Current methods are quite imprecise, however, due to the assumptions necessarily relied upon to compensate for gaps in knowledge. This imprecision engenders controversy over agency actions, inconsistency in decisionmaking between agencies, and gradual erosion of public confidence in the regulatory process. There is thus a need for a well-planned effort, including research, scientific consensus building, and other activities, designed to enhance the precision of risk assessment by replacing the current assumptions with real knowledge.

This need was clearly recognized in the 1984 recommendations of the Toxicology Committee, which centered around improving risk assessment and enhancing its role in food safety decisionmaking. The Committee is pleased that significant progress has been made in these areas since 1984. Research and study projects to improve risk assessment are being undertaken by numerous government agencies and private organizations, and there has been an increased recognition of risk assessment's role in decisionmaking.

For example, in 1985 the Office of Science and Technology Policy in the White House published a major document stating the current scientific consensus in the area of carcinogenic hazard evaluation and risk assessment. This document, coupled with the National Academy of

Science's seminal 1983 "Redbook" (*Risk Assessment in the Federal Government: Managing the Process*), establishes a valuable blueprint for the scientifically sound conduct of risk assessment and its proper use in decisionmaking. The Environmental Protection Agency has also developed risk assessment guidelines.

At the regulatory level, the Food and Drug Administration has proposed a de minimis approach to implementing the Delaney Clause. This approach, which builds upon FDA's "Constituents Policy" for regulating trace carcinogenic components of additives, permits the agency to evaluate the human safety of food and color additives found to induce cancer in animals and approve those found to pose truly trivial risks.

These activities make it all the more important that risk assessment methods continue to improve. The following recommendations of the Toxicology Committee are based on the Committee's 1984 recommendations, but go beyond them. They are intended to suggest some more specific steps that can be taken toward improving the science of risk assessment and assuring that toxicology plays its full, appropriate role in food safety decisionmaking.

RECOMMENDATIONS

1. Current estimates of human carcinogenic risk are conservative upper limits because of the needs of regulatory agencies to assure public safety despite the inherent uncertainty in risk assessment. Such estimates are valuable for this purpose, but they can be misunderstood and may be regarded by the risk manager and the public as estimates of actual risk. Accordingly, we recommend that risk assessors attempt to provide, and that regulatory agencies make publicly available, both upper limits and most probable estimates of risks.

2. Improving the scientific basis for risk assessment is essential to the future success of our food safety regulatory efforts and will require a sustained program of well-focused research. We recommend the establishment of a blue-ribbon panel to provide leadership and coordination within the area of risk assessment research. Appointees to the panel should be recognized authorities in one or more areas of risk assessment and occupy positions of influence in the scientific community.

3. Advances in toxicology and analytical chemistry have revealed a vastly expanded universe of substances in or in contact with food whose safety and regulatory status must be addressed. In many

cases, however, human exposure to these substances is so slight that it is questionable whether scarce scientific and regulatory resources should be used to subject all of them to the same full-blown testing and review procedures typically applied to direct food and color additives. The question then becomes whether the public health can be protected adequately on the basis of judgments that certain categories of substances are associated with no more than de minimis risks and fall below the legal "threshold of regulation." Specifically, in the case of food packaging materials, the totality of existing toxicological data on the carcinogenic potencies as well as noncarcinogenic toxicities of potential migrants should be compiled and summarized so that upper bounds can be placed on migration levels and risks that are consistent with and define a "threshold of regulation."

4. Recognizing that interpretation of cancer bioassays must be made on a case-by-case basis, we still encourage the formulation and active discussion of general rules for interpreting the human significance of tumor data in laboratory animals. In many cases, additional research is needed to formulate general guidelines. The blue-ribbon panel mentioned in Recommendation 3 could identify research needs and facilitate formulation of guidelines. In the current situation, where many decisions will be made in the absence of guidelines, a weight of the evidence approach can often resolve problems of interpretation.

5. The role of epidemiology in food safety evaluation should be better defined. Epidemiologic data should be included on a weight of evidence basis in risk assessment, e.g., to validate upper limits of risk based on estimates from animal data. National human nutrition surveys should concentrate on exposure to food constituents and long-term health outcomes in the same individuals.

6. Consensus about risk assessment issues is needed between regulatory agencies and in the risk assessment area in general. Methods to bring parties together to strive for consensus need development. The blue-ribbon panel is one possible vehicle for consensus development. Alternatively, industry or public interest groups, academic groups, or government agencies could initiate consensus building activities.

7. The "OSTP Cancer Guidelines" should be updated as research makes change possible. The updating must be an open, consensus building activity that draws upon all interested scientists and offers

the revisions to public comment. The blue-ribbon panel, scientific organizations, industry or public interest groups, or government agencies acting alone or in concert could initiate consideration of these revisions.

8. Safety decisions necessarily involve the exercise of scientific judgment. Regulatory agencies should give appropriate emphasis to the qualitative as well as quantitative character of risk assessments. The OSTP Cancer Guidelines provide an appropriate balance in the characterization of risk assessments and should be followed.

9. The complexity and specialized nature of risk assessment requires that regulatory agencies be provided sufficient resources to carry out their own research and development activities and maintain the state-of-the-art scientific capacity required to interpret results produced by other scientists.

10. The 1984 recommendation (Recommendation 2) that suggested that "it should be possible to improve high-to-low dose extrapolation . . . by using techniques that incorporate more compound specific biological information . . ." has proved prophetic. Several recent examples of such improvements using pharmacokinetic and mechanistic data have been published, notably for formaldehyde and methylene chloride. By incorporating specific information on the nonlinear relationships between administered and effective target organ doses, and by using physiologically-based modeling, more accurate and reliable animal to human "scaleup" and more accurate estimates of animal risk appear possible. We support more work in this area on a wider variety of substances to enlarge the database and to better establish the validity of such procedures in risk assessment.

11. There appears no reason to change the current NOEL safety approach to noncarcinogenic risk assessment because of its long and satisfactory history of use, its apparent consistency with current hypotheses of toxic action, and because there are no alternatives of demonstrated superiority.

    Possible improvements in risk assessment for noncarcinogens can be envisioned from further research in the use of ED_{10} or ED_{01} benchmark doses (i.e., establishment of a NOEL in studies where a NOEL was not identified), inclusion of relative severity or seriousness of toxic effects, better understanding of inter and intraspecies variability, and incorporation of pharmacokinetic data.

MICROBIOLOGY

7. Principles and Applications of the HACCP Approach for the Food Processing Industry

John H. Silliker

Silliker Laboratories

STATEMENT OF PROBLEM

The problem relates to achieving the best means of assuring the microbiological safety and quality of foods. The responsibility for this is shared jointly by the regulator and the processor. In a broader sense, consideration of food safety and quality must start at the farm and receive attention at all stages in the food chain to the ultimate consumer. However, the present discussion is limited to processing.

Food laws governing processing are of ancient origin, e.g., the dietary laws handed down by Moses were intended to assure clean and safe foods. At a later time laws were enacted to prevent deceptive practices. With this, responsibility for enforcement passed from religious to civil authorities. It should be emphasized that, despite the ancient origin of food laws, our knowledge of the association of microorganisms with food spoilage dates from the early nineteenth century, and their association with foodborne disease came at a still later time.

Historically, the inspectional approach has been used to determine compliance with applicable food law. This has proven to have serious shortcomings. Not the least of these lie in the laws which the inspector is charged with enforcing and with which the processor must comply. The laws, whether of local, national, or international origin, contain vague terms relative to stated requirements—terms that fail to specify what constitutes compliance with the requirement. This is seen even in food laws of recent origin, such as the umbrella GMPs and Codex Codes of Hygienic Practices. Lack of specificity and indication of the relative importance of requirements within these laws leaves interpretation of compliance to the discretion of the inspector. Further, the inspector may fail to distinguish between important and relatively unimportant requirements. Factors that are critical to safety may be overlooked or underestimated. The wording of the laws frequently fails to distinguish

Exhibit I. Umbrella GMP, Sanitation of Equipment and Utensils. (Section 110.37(c), Good Manufacturing Practice in Manufacturing, Processing, Packing, or Holding Human Food. FDA, 1969.)

(c) Sanitation of equipment and utensils. All utensils and product-contact surfaces of equipment shall be cleaned as frequently as necessary to prevent contamination of food and food products. Nonproduct-contact surfaces of equipment used in the operation of food plants should be cleaned as frequently as necessary to minimize accumulation of dust, dirt, food particles, and other debris. Single-service articles (such as utensils intended for one-time use, paper cups, paper towels, etc.) should be stored in appropriate containers and handled, dispensed, used, and disposed of in a manner that prevents contamination of food or food-contact surfaces. Where necessary to prevent the introduction of undesirable microbiological organisms into food products, all utensils and product-contact surfaces of equipment used in the plant shall be cleaned and sanitized prior to such use and following any interruption during which such utensils and contact surfaces may have become contaminated. Where equipment and utensils are used in a continuous production operation, the contact surfaces of such equipment and utensils shall be cleaned and sanitized on a predetermined schedule using adequate methods for cleaning and sanitizing. Sanitizing agents shall be effective and safe under conditions of use. Any facility, procedure, machine, or device may be acceptable for cleaning and sanitizing equipment and utensils if it is established that such facility, procedure, machine, or device will routinely render equipment and utensils clean and provide adequate sanitizing treatment.

(d) Storage and handling of cleaned portable equipment and utensils. Cleaned and sanitized portable equipment and utensils with product-contact surfaces should be stored in such a location and manner that product-contact surfaces are protected from splash, dust, and other contamination.

between factors critical to food safety and those that are of an aesthetic nature.

Illustrative of the above: starting in the late 1960s, the U.S. Food and Drug Administration issued a number of regulations covering good manufacturing practices (GMPs). These regulations were an attempt to standardize FDA's enforcement of section 402(a) of the Food, Drug and Cosmetic Act and to provide food processors with better guidance about the requirements of this section of the Act. The "umbrella" regulations were made effective in April, 1969. Exhibit I reproduces section 110.37(c), sanitary operations and sanitation of equipment and utensils. Note the phrases "cleaned as frequently as necessary," "appropriate containers," "where necessary to prevent introduction of undesirable

Exhibit II. Low-Acid Canned Food GMP, Mercury Thermometer. (Section 113.40 (b) (1), Thermally Processed Low Acid Foods Packaged Hermetically Sealed Containers GMP. FDA, 1973a.)

(b) Equipment and procedures for pressure processing in water in still retorts—(1) Indicating mercury-in-glass thermometer. Each retort shall be equipped with at least one mercury-in-glass thermometer that has a temperature range of not more than 100°F in the processing range on a scale at least 7 inches in length or a temperature range of not more than 150°F on a scale at least 9 inches in length. The scale divisions shall be no more than 2°F. Thermometers shall be tested for accuracy against a known accurate standard thermometer upon installation and at least once a year thereafter or more frequently as may be necessary to ensure their accuracy. Bulbs of indicating thermometers shall be located in such a position that they are beneath the surface of the water throughout the process. On horizontal retorts this entry should be made in the side at the center, and the thermometer bulbs shall be inserted directly into the retort shell. In both vertical and horizontal retorts, the thermometer bulbs shall extend directly into the water a minimum of at least 2 inches without a separable well or sleeve. Thermometers shall be installed where they can be accurately and easily read. A thermometer that has a divided mercury column or that deviates more than 1°F from the standard shall be repaired or replaced. The mercury thermometer—not the recorder chart—shall be the reference instrument for indicating the processing temperature.

microorganisms," "adequate methods of cleaning and sanitizing," etc. Thus, despite the modern origin of the umbrella GMP, vague and general terms are used to guide both the food processor and the responsible regulatory authority. This led the U.S. courts to hold that the umbrella GMPs do not have the force of law.

Subsequently, the U.S. Food and Drug Administration adopted detailed GMP regulations for specific food commodities. These included low-acid foods packed in hermetically sealed containers, acidified foods, cacao products and confectionery, smoked and smoked-flavor fish, frozen breaded fish, and processed and bottled drinking water. These GMPs have the force of law. In contrast to the vague terms contained in Exhibit I, Exhibit II shows requirements contained in Section 113.40 of the thermally processed low acid foods packaged in hermetically sealed containers GMP, which relates to equipment and procedures for pressure processing in steam-in-still retorts—indicating mercury-in-glass thermometers. It will be noted that here the requirements are highly specific. The processor knows exactly what he must do, and the regulator can readily determine adherence to the requirement.

Such specificity with respect to requirements in food laws is rare. Indeed, the U.S. regulation (GMP) relating to low-acid foods in hermetically sealed containers evolved because existing regulations were ineffective in the prevention of microbiological hazards in these products. It focused on the critical points in the production of thermally processed canned foods, the control of which is essential to assuring a safe product. This is the basis of the Hazard Analysis Critical Control Point system (HACCP), which appears to be the most rational approach to the control of microbiological hazards in foods.

There are other problems with the traditional inspectional approach. The inspector makes a sporadic visit to observe a food processing operation to determine whether the processor is operating in compliance with applicable law. His observations relate, almost exclusively, to what is occurring on the day of his visit, more particularly, to what is occurring at a particular point in time when he observes a segment of the total operation. To be sure, meat and poultry operations are subject to continuous inspection, thus obviating this criticism. But despite continuous inspection, a major portion of foodborne illness is traced to meat and poultry.

A chapter in the Proceedings of the Second National Conference on Food Protection, "Economic Loss Resulting from Microbial Contamination of Foods," clearly shows that the economic losses from food spoilage and foodborne disease are enormous. In addition, these undesirable effects of microorganisms in terms of loss of sources of human (and animal) nutrition as well as attendant human suffering cannot be overestimated. Clearly the traditional approach to the control of microbiological hazards has produced results falling far short of the ideal.

EVALUATION OF THE PROBLEM

The HACCP approach to the control of microbiological hazards in foods was first presented at the 1971 Conference on Food Protection. Though the current application of HACCP differs considerably from that presented at the 1971 conference, the concept of a new approach to the control of microbiological hazards was first presented at the conference and "recommended for widespread use."

Following the 1971 conference, some companies in the food industry used the principles of HACCP in connection with their quality assurance programs, but these were in the minority. In the early seventies, microbiological problems with low-acid canned foods, particularly but not limited to mushroom canning, led to the low-acid canned food regulations promulgated by the U.S. FDA. These regulations embodied the principals of HACCP: (1) hazard analysis led to the identification of critical control points, which were points at which control over identified hazards could be exercised, and (2) monitoring methods were established whereby it could be determined whether the critical control points were indeed under control. Such monitoring involved the use of tests that yielded results with sufficient speed to permit correction of failures to effect control. Accordingly such tests involved, primarily, visual observations and physicochemical measurements. Microbiological testing was of little value in monitoring critical control points, as the time required to obtain results of such tests did not permit rapid reaction to situations where a critical control point was not under control.

The application of HACCP to low-acid canned foods has been successful for a number of reasons: (1) industry and government, working cooperatively, developed monitoring procedures for critical control points, (2) FDA required operators of retorts, aseptic processing packaging systems, and container closure inspectors to be under the supervision of a person who had satisfactorily completed the prescribed course approved by the FDA Commissioner, (3) FDA inspectors were trained in the elements of the HACCP system, and (4) the use of HACCP was mandated by law.

The HACCP system, as applied to low-acid canned foods, is a great improvement over the traditional inspectional approach to the control of microbiological hazards. Its focus is indeed on the microbiological hazards associated with the production and ultimate use of food. Aesthetic factors, unrelated to microbiological hazards, play no role in the system. Identification of critical control points is integral to the system. These are points at which control can be exercised.

Monitoring of critical control points permits on-the-line evaluation of the effectiveness of control. To make such on-the-line control possible, the procedures used for monitoring must yield results with sufficient rapidity that adjustments can be made before serious microbiological hazards develop. It is for this reason that monitoring, with a few exceptions, involves visual and physicochemical measurements. Microbiological testing is simply too time consuming to be an effective means of monitoring critical control points. A notable exception is raw material inspection. When the microbiological status of a raw material is a critical control point, microbiological testing is the only acceptable monitoring procedure, e.g., the testing of cocoa and dried milk for *Salmonella* before their use as raw materials in chocolate production, the testing of sugar or starch for thermophilic spore-forming bacteria before use as ingredients of canned foods. With food destined for high-risk populations, e.g., relief foods, the finished product constitutes a critical control point that must be monitored by microbiological analysis. In such cases, the finished product remains under the physical control of the processor until the results of microbiological monitoring indicate that the product is safe to ship. Here, in a sense, the holding of the product pending the result of microbiological tests becomes a part of the process. Similarly, dried egg albumen is held in a "hot room" until microbiological tests indicate it is free of *Salmonella*. Such microbiological testing as a means of monitoring critical control points is unusual in the HACCP system.

The HACCP system offers benefits to the regulator, the processor, and the consumer. Both the regulator and the processor concentrate on factors directly related to the control of microbiological hazards. They ignore extraneous considerations which can interfere with concentration on the objective of controlling microbiological hazards. The regulator can, through review of monitoring results at critical control points, review the effectiveness of control methods at a given plant with historical perspective, rather than base his conclusions upon what is seen on the day of inspection. If monitoring results indicate that critical control points have been under control, inspection efforts can be more economically expended elsewhere. Through monitoring critical control points, the processor is in a position to control his operations on a continuous basis, rather than wait for hazards to develop. The consumer benefits by receiving a product produced under conditions wherein microbiological hazards have been controlled.

As previously indicated, the HACCP approach was first introduced at the 1971 Conference on Food Protection. In 1973, the use of the HACCP approach was mandated by law in connection with the production of low-acid canned foods. Despite the obvious benefits of the HACCP approach, its use by industry in the control of hazards other than in low-acid canned foods has been spotty, and the HACCP approach to FDA regulatory inspections is primarily confined to low-acid canned foods.

In the Proceedings of the Second National Conference on Food Protection, the Microbiology Committee Recommendations included the following: "The Hazard Analysis Critical Control Point (HACCP) program should be stressed and extensively applied to all aspects of the food manufacture, distribution, service, and retail sectors." Yet, no discussion regarding the HACCP approach is to be found in the body of the report.

In 1985, two highly important reports were published by the National Research Council. The first of these, "An Evaluation of the Role of Microbiological Criteria for Foods and Food Ingredients," embraced HACCP as the most rational approach to the control of microbiological hazards in foods. It recommended that the use of HACCP throughout the U.S. food industry be required by regulation. This matter is discussed in more detail in Chapter 8. The second report, "Meat and Poultry Inspection: The Scientific Basis of the Nation's Program," discussed the application of HACCP to meat and poultry processing. It indicated that the FSIS Administrator had informed the

Committee that the concept of HACCP is now part of the strategy of FSIS. The committee encouraged FSIS to "move as vigorously as possible in the application of the HACCP concept to each and every step in plant operations, in all types of enterprises involved in the production, processing, of meat and poultry products."

Thus, two committees of the National Research Council have, since the 1984 conference, strongly urged adoption of the HACCP approach by the American food industry. Significantly, the agencies sponsoring these studies are those responsible for food regulation, namely, the Food and Drug Administration, the Department of Agriculture, the Department of Defense, and the National Marine Fisheries Service.

If HACCP is to be used throughout the food industry, certain barriers must be overcome. These are discussed in detail in Chapter 10 of the National Research Council report on microbiological criteria. Among these barriers are the following:

1. The hazard analysis of a food process requires technical sophistication as does identification of critical control points and the establishment of monitoring systems. This will require the mobilization of technical manpower within the various segments of the food industry. Though large companies may have the technical knowledge to establish HACCP programs, smaller companies do not. They will thus need aid from within their industry.

2. Those responsible for regulation must be trained in the concepts of the HACCP system. FDA inspectors, for example, were trained in the HACCP approach to low-acid canned foods, but training with respect to HACCP inspections of other types of food operations has not occurred since 1975.

3. There must be training of food processing technicians in the elements of HACCP.

4. For HACCP use to be broadly utilized, it is likely that it will have to be required by regulation. The first of the two National Research Council reports recommends this, indicating that regulations should not specify the details of application, that such details should be the prerogative of industry.

In addition to the four items discussed above, adversarial attitudes and lack of cooperation between regulatory agencies and the food industry have been a barrier to the expansion of HACCP. A particularly sensitive issue has been regulatory access to records. If HACCP is to work, the regulator must have access to records of monitoring results at critical control points. This is the basis for judging whether microbiological hazards have been controlled. On the other hand, records relating to manufacturing practices, having nothing to do with monitoring results at critical control points, may be viewed as proprietary information with no relevance to microbiological safety and quality. This issue is discussed in some detail in Chapter 10 of the National Research Council report on microbiological criteria. If the use of HACCP is mandated by law, this issue will be addressed, as occurred with respect to low-acid canned foods.

SOLUTIONS TO THE PROBLEM

At this time, both of the National Conferences on Food Protection, as well as the reports of two National Research Council committees, have recommended that the American food industry adopt the HACCP approach. As indicated in the report of the Subcommittee on Microbiological Criteria, such expansion of HACCP will probably require that it be mandated by law. Therefore, it seems evident that the next move requires action by the regulatory agencies responsible for food regulation. If HACCP is required by regulation, there is little doubt that the industries involved will take the necessary actions to implement its use. In the absence of legal requirements that HACCP programs be adopted by industry, it is doubtful that the present situation will change. It is clear that this is a "two-way street." The regulator must accept the HACCP approach as a cost effective route to determine that a food operation is under control. The processor must identify critical control points in his operations and establish monitoring systems to determine whether critical control points are under control. Implicit in this scheme, the processor must establish action systems to respond when monitoring indicates an out-of-control situation. This approach has been effective with respect to low-acid canned foods. There is no reason that it should not be effective in other segments of the food industry. The need is evident.

CONCLUSION

It is recommended that the federal agencies responsible for the regulation of foods embrace the recommendations of the two National Research Council Committees with respect to HACCP. Specifically it is recommended that the use of HACCP be mandated by law and that the primary basis for the judgement of a food processor's operations should be the adequacy of his HACCP program and the monitoring results at critical control points.

8. Practical Procedures for Using the Hazard Analysis Critical Control Point (HACCP) Approach in Food Service Establishments by Industry and Regulatory Agencies

John J. Guzewich

New York State Department of Health

STATEMENT OF THE PROBLEM

The food service industry is a major industry in the United States with sales equaling nearly 5% of the U.S. Gross National Product. According to the National Restaurant Association, in 1986 food service will be a $185.8 billion industry involving an estimated 561,000 food service establishments employing 8 million people. There will be an estimated 45 billion food service meals served this year. Industry figures show that the average person eats 3.7 meals per week at a food service establishment. Consumers will spend 40% of their food budget on meals eaten outside the home.

According to the Centers for Disease Control (CDC),[1] there were 656 outbreaks of foodborne disease reported in 1982 involving 19,380 cases of illness. In 60% of the 218 outbreaks where a place of mishandling was reported, the food involved was mishandled in a food service establishment. Data reported between 1973 and 1978 show that food service establishments account for 70% of reported outbreaks where the place of mishandling was known. It is important to realize that foodborne disease reporting is a tip of the iceberg situation. One scientist has estimated that there are five million cases of foodborne disease at food service establishments each year. The economic cost of these outbreaks has been estimated from $1 billion to $10 billion. If one computes a relative risk of developing a food service related illness from these estimates, one in each 9,000 (5 million illnesses/45 billion meals) meals results in a foodborne illness.

Data collected over 18 years by the CDC consistently point up foodborne disease problems across the United States. The major contributing factors, vehicles and agents in foodborne disease outbreaks have been known for many years. In the past 18 years, the five most commonly reported contributing factors of foodborne disease outbreaks, ranked in predominance of reporting, were: improper cooling, lapse of a day or more between preparation and serving, improper hot storage, infected food handler, and inadequate reheating. Food service establishment regulatory programs have attempted to reduce the incidence of foodborne disease; however, the reported number of outbreaks, the types of vehicles and agents identified, and relative ranking of various contributing factors have changed very little.

Food service regulatory agencies are facing dwindling manpower and other resources in the years to come at a time when Americans are expected to increase the number of meals they eat that are prepared outside the home. This combination of events has left many local regulatory sanitarians frustrated and looking for new ways to cope. They will need to develop approaches that focus their limited resources on those foods and establishments with the greatest risk of causing foodborne illnesses. Food service companies are hiring their own quality control personnel with the expertise necessary to develop their own food protection procedures and inspection programs.

Although this chapter will use the term "food service," the observations, conclusions, and recommendations can also apply to retail and convenience food stores that prepare the same foods that are served in restaurants.

THE HAZARD ANALYSIS CRITICAL CONTROL POINT APPROACH

The Hazard Analysis Critical Control Point (HACCP) approach was developed in the 1970s for use in the food processing industry. The HACCP approach assesses the flow of food through the establishment and, following needed changes, provides a mechanism to frequently monitor these operations. A full HACCP evaluation* begins with a review of the menu to identify any potentially hazardous food items subject to bacterial contamination, bacterial growth, or microbial survival during preparation. The entire preparation process is then observed. Next, the process is diagrammed, keeping in mind contamination, survival, and growth of microorganisms capable of causing

*A full HACCP evaluation can be a lengthy process involving several hours to complete; however, full HACCP evaluations are only done for training purposes or to assess unfamiliar food preparation practices. Trained staff can complete HACCP objectives in time frames compatible with their routine workloads.

foodborne illness—this is hazard analysis. Critical control points are identified. All processes have control points, but only critical control points are identified: those that preclude contamination, survival, and growth of microorganisms; for example, rapid cooling of a large quantity of beef stew. Any deficiencies are corrected based on prioritization, and practical management monitoring steps are established. A critical control point is an operation (practice, procedure, process, or location) or a step of an operation at or by which preventive or control measures can be exercised that will eliminate, prevent, or minimize a hazard that has (or several hazards that have) occurred prior to this point.

Outcome specifications in a process need to be established at critical control points and they must be monitored. Monitoring means checking that a processing or handling procedure at each critical control point meets established criteria. It involves systematic observation, measurement, and/or recording of the significant factors for prevention or control of hazards. The monitoring procedures that are chosen must enable action to be taken to rectify an out of control situation or to bring the product back into acceptable limits either before the start-up or during the operation. For example, "portion beef stew into 2-inch-deep pans and immediately place in cooler." To be effective the procedures should set parameters for what is acceptable performance of a process and describe what to do when the tolerance is exceeded; that is, the action to take to prevent a problem from developing. For example, if a manager observes a potentially hazardous food cooling in a deep pot, such as the beef stew, that food should be immediately heated to 165F and then rapidly cooled. Subsequent regulatory inspections focus on the establishment's implementation of monitoring critical control points by observing if required process steps are carried out. The measure of compliance is outcome-based.

HACCP is implemented by industry and monitored by regulatory agencies. HACCP principles are applied during all regulatory work. An HACCP program differentiates between items of public health concern and those of aesthetic concern. It provides management with tools and monitoring points; it can be used to efficiently and effectively protect the public's health. HACCP should not be viewed as complicating food service operations or regulations. In its practical sense, HACCP simplifies public health protection by identifying the critical control points and providing meaningful ways to monitor them. This outcome will make management's job and the regulator's job simpler.

The HACCP approach has already succeeded in the low-acid canned foods industry because: (1) industry and government worked cooperatively identifying critical control points and monitoring points, (2) the FDA required training of industry personnel, (3) the FDA trained its own staff, (4) the use of HACCP was mandated by federal regulation, and (5) the nature of food processing tends to be linear and repetitive allowing for a greater dependence on mechanical monitoring of critical control points. State and local regulatory agencies and the food service industry will have to address all of the factors for HACCP to be successful in the food service industry.

HACCP fits the bill in the food service industry as well. New York state began implementing an HACCP food service program in 1985. Their experience has shown that sanitarians quickly learn the concept, and they just as quickly learn how to identify critical control points and monitoring procedures in a minimal amount of time. Food service managers are reluctant at first, but they soon become enthusiastic as they learn new things about their operation and they see regulators paying most of their attention to the food outcomes rather than to "things" (structural violations). HACCP has been recommended by food processing, government, and university scientists for 15 years; it has been successful in other food industries; and it can meet the needs of both government and industry. The next sections of this chapter address why HACCP has not been used in food service, and how it can be implemented now.

EVALUATION OF THE PROBLEM

Three of the recommendations of the Second National Conference for Food Protection addressed the use of HACCP in food service:

1. The HACCP program should be stressed and extensively applied to all aspects of the food manufacturing, distribution, service and retail sectors.
2. Foodborne disease outbreak investigations should include hazard analysis critical control point evaluations to identify factors that contribute to outbreaks, rather than inspections for sanitary code violations.
3. Food protection agencies should adopt the Hazard Analysis Critical Control Point approach in their regulatory programs. Inspection personnel should be trained to evaluate realistic hazards, identify critical control points, verify operators' ability to check critical control points and train industry personnel.

The National Academy of Sciences, National Research Council (NRC) recently recommended that government agencies responsible for the control of microbiological hazards in foods should promulgate appropriate regulations that would require industry to utilize the HACCP system in their food protection programs.

The U.S. Food and Drug Administration (FDA) completed a survey on the use of HACCP by state and local food service regulatory agencies in the spring of 1986.[2] The FDA found that 23 state and 8 local agencies out of 2,700 total agencies are currently interested in HACCP. According to the National Restaurant Association, in six large food service companies they contacted in a limited survey, 9,000 individual outlets are receiving an HACCP-based self-inspection. These numbers indicate that only a small start has occurred. There are many reasons for this slow beginning.

1. There is a need for training. People who are going to perform HACCP evaluations will require classroom training in food microbiology, epidemiology, the principles of HACCP, and field exercises in conducting HACCP evaluations. The problem with training is the cost in terms of personnel time and related expenses to prepare and attend training courses. Health agencies with limited resources and small food service companies find it difficult to incur such expenses.

2. There is a misimpression that HACCP requires lengthy field inspections and writeups, both of which are too costly. While initial training exercises and experimentation might be lengthy, this is a short-term effect. New York's experience has been that trained staff can conduct an HACCP evaluation in 2 to 3 hours, which is the length of time involved in inspecting large establishments using the traditional approach. Subsequent inspections require less time as inspectors focus on implementation of monitoring procedures.

3. There will have to be a change from the traditional regulatory approach of inspecting all establishments at the same frequency using an inspectional method that pays a disproportionate amount of attention to sanitation and aesthetic items. This approach is institutionalized by using standardized inspectional techniques. Many local sanitary codes require that a certain number of inspections be performed within the term of the permit or license. Most permit or license terms are of the same duration regardless of the degree of public health risk the establishment menu poses. Local program effectiveness is often judged by the number rather than quality of inspections that are performed.

4. Perennial government staffing shortages—whether due to multiple program responsibilities for food inspection personnel or simple shortfalls in staffing levels—make it difficult to find the time to undergo training and experiment with new concepts.

5. Many people in the food service industry do not understand that HACCP can be an excellent management tool for them. They see HACCP as a new and costly requirement that might involve government-mandated recordkeeping. While some food service chains

have set up recordkeeping systems of their own, government-mandated recordkeeping would set back the program by yielding "dry lab" records and violations for failure to keep records rather than failure to implement monitoring procedures. HACCP monitoring procedures provide food service managers with a few easily understood steps that can be delegated to workers and easily watched thereafter. For example, all liquid, potentially hazardous foods should be cooled in shallow pans. This HACCP practice and any HACCP practice can be written into standard menus. Such procedures are easily observed. HACCP helps management develop and implement preventative practices using known methodologies. These improvements will result in better quality food and reduced spoilage and loss, thereby yielding increased profits.

6. Some health agencies state that they currently conduct critical item inspections or quality inspections and, therefore, HACCP is not needed for them. The difference in these approaches is one of perspective. Critical item and quality inspections tend to focus on finding key violations in regulations, i.e., they compare the establishment to a standard. Regulations essentially consider every requirement to be a critical control point as every violation can result in a penalty. In terms of disease prevention, however, only a few points are critical. HACCP, on the other hand, looks at the flow of food. The frame of reference is oriented to the establishment's operations and what is most important in those operations. The most important items in the flow of food invariably are embodied in the regulations and, therefore, HACCP ties back into the regulation by starting with the food. HACCP applies the standards in an establishment-specific, process-specific manner. HACCP seeks the outcome of the standard rather than the standard itself. The standard has not changed, therefore, only the means to achieving compliance with it.

7. Some health agencies are concerned that efforts to implement HACCP will necessarily cause neglect of consumer expectations. This viewpoint is based on the premise that consumers are more concerned with aesthetic aspects of sanitation than about more sophisticated public health concerns like slow cooling of foods. The HACCP approach should not result in the abandonment of consumer concerns over cleanliness. Dirty establishments should be cited for violations and prosecuted when they fail to comply regardless of the inspectional approach used. The HACCP philosophy calls for dealing decisively with genuine sanitation problems by not spending disproportionate time on minor violations in establishments in substantial compliance with the regulations.

8. Many individuals express concern that changes in menus and employee turnover in food service establishments will make HACCP difficult to implement. These same problems make all food service

requirements difficult to implement. On the other hand, HACCP monitoring procedures will boil down to the few most important things to do in an establishment. This simplification of what needs to be done should improve the chances that the requirements will be followed.

SOLUTIONS TO THE PROBLEM

HACCP in food service will prove cost effective because routine inspections that emphasize monitoring procedures require less time and are more public health effective. With HACCP implementation, government/industry relationships improve and operators have a clearer idea of what is expected of them. Sanitarians who use HACCP have greater job satisfaction. Morbidity due to certain contributing factors, agents, and vehicles will decline thereby reducing overall foodborne morbidity. HACCP implementation requires a plan as outlined below:

1. Close cooperation must be maintained between the food service industry and regulatory agencies. The National Restaurant Association has recently developed its SAFE (Sanitary Assessment of Food Environment) program, which is a demonstration of that association's interest and ability to become a leader in the application of HACCP principles in the food service industry.

2. Support must be generated for HACCP among environmental health directors, health commissioners, and legislative bodies that oversee food service regulatory programs. This can be accomplished by explaining the approach, how it can reduce morbidity, and how it can lead to a more cost-effective regulatory program.

3. Training programs should be developed for regulatory inspectors, food service managers, and the general public. These should include theoretical components on food microbiology, epidemiology, and HACCP principles as well as practical exercises. Initial efforts in this area require an up-front short-term commitment of resources in training, equipment purchases, etc., during the startup phase. However, after HACCP is understood and new insights have been gained, the time required and resources needed to conduct HACCP will be significantly reduced while inspections will be more efficient and effective.

4. Methodologies must be developed to conduct HACCP evaluations in a short period of time by grouping similarities in food preparation practices. For example, a technique called recipe HACCP breaks food preparation into six categories: (a) thick meats, (b) thin meats, (c) sauces, soups, gravies, (d) vegetables and fruits, (e) casseroles and salads, and (f) bakery and pastry. Each of these food groups has characteristic critical control points and monitoring points regardless of the indi-

vidual recipe. With this basic knowledge, a trained person can evaluate the flow of each appropriate food in a given establishment and tailor the concept to the unique aspects of that establishment. The approach should not be allowed to be oversimplified into a ritualistic checksheet system, however.

5. High-risk establishments should be identified by local health agencies according to the risks of a foodborne disease outbreak posed by their operations. Factors such as menu, food handling practices, advanced meal preparation, patron susceptibility, and number of meals served should be taken into account.

6. Critical control points should be identified for all activities. The object of an HACCP program is to identify practical monitoring of critical control points that are designed to produce preventive outcomes. This is particularly important in high-risk establishments.

7. Necessary equipment should be obtained. At the very least, the equipment needed to conduct HACCP evaluations should include a thermometer capable of measuring foods of various depths and a pH meter. If additional funding is available, an automatic recording thermometer and a water activity meter will permit even more thorough and useful HACCP evaluations. Appropriate laboratory support should include the capacity to test for total aerobic plate count, common pathogens, pH, and a_w.

8. If not already in existence, foodborne disease surveillance programs should be developed. These programs are needed to identify immediate public health problems as well as measure changes in foodborne disease trends that might identify new critical control points. Surveillance programs can also be used to measure the effectiveness of regulatory programs including the HACCP approach. Improved surveillance can be expected to identify more outbreaks, which should not be confused as a failure of HACCP but as a better indication of the true incidence of foodborne disease hazards and critical control points.

9. Traditional inspection programs will have to change. These changes should include permit terms and inspection frequencies based on risk assessment. Inspections must emphasize the use of critical control points. Critical control points and monitoring procedures should be posted in kitchens as a tool in training employees and as a reference for employees, managers, and regulators. Subsequent enforcement is facilitated because the operator and regulator will already have agreed on specifically how code compliance is to be achieved. Therefore, it is difficult for the operator to plead ignorance of the rules.

10. Regulations must be adopted requiring the use of HACCP by the food service industry. As was seen in the low-acid canned food industry, success of the program was due in part to a regulation requiring use of HACCP. Without such regulations, implementation will be spotty and

inconsistent. These regulations must be outcome or performance standards so that violations relate to failure to perform (i.e., cool food) rather than failure to comply with a ritual such as recordkeeping.

11. The effectiveness of HACCP programs should be measured through the critical control points. Goals can be set to implement critical control point monitoring in a certain percentage of establishments over time. Thereafter, the number of establishments actually monitoring critical control points is a measure of effectiveness. Those establishments not monitoring critical control points should be subject to immediate legal action for operating with clear public health hazards. Long-term program effectiveness should be reflected in changing trends in reported foodborne disease outbreaks. Actual HACCP inspection findings in individual establishments should be assessed to determine if the HACCP analysis, plan, or implementation are failing to provide acceptable process outcomes.

RESEARCH NEEDS

The use of HACCP in food service establishments will create the need for research in three areas:

1. HACCP evaluations often identify serious cooling problems. Rapid cooling of potentially hazardous foods requires the use of certain food containers, procedures and refrigeration equipment and techniques. Since these methods will require changes in current procedures and the purchase of equipment, scientifically developed criteria must be established to provide additional flexibility on the safe cooling of various foods. Additional research on alternate cooling methods will also be needed so that critical control points can be better tailored to various types of food service operations.

2. Additional research will also be required into the practical use of chemical properties of foods, i.e., pH and water activity, as critical control points requiring monitoring in the food service industry. For example, acidification of cold salads with vinegar or reduction of water activity of cream pastry fillings with sugars can cause such modified foods to no longer be potentially hazardous. While these alternatives are understood, few regulators are ready or willing to accept these modifications in lieu of time/temperature requirements at the present time.

3. Cold-growing pathogens such as *Listeria monocytogenes* and *Yersinia enterocolitica* have the potential to cause foodborne disease in foods that have been cooled in accordance with current regulations. Research is needed into appropriate control strategies for these organisms in light of their preference for growth in conditions provided by compliance with current regulations.

RECOMMENDATIONS

1. The implementation of the HACCP approach must be a top priority in the food service industry and among food service regulators. Industry and regulatory representatives must work together in the implementation of the program.

2. Regulators must receive training developed and provided cooperatively by regulators and industry in food microbiology, the epidemiology of foodborne disease, and HACCP. Food service managers must learn the broader concepts needed to understand and monitor critical control points.

3. Government must adopt regulations that require the use of monitoring procedures by the food service industry.

4. Regulators must take enforcement action when monitoring procedures of critical control points are not followed.

5. Additional research must be carried out by regulators and industry to provide greater flexibility in the cooling criteria for potentially hazardous foods, the equipment and procedures needed to meet these cooling criteria, and the practical use of pH and water activity as critical control points.

6. The FDA, state agencies, universities, and trade and professional groups need to work together to give local agencies and industry the guidance they need to get started in an HACCP program. Training courses and reference materials have to be made available. Consultative resources need to be identified and made available to those who need them.

7. Research is needed into control methods for cold-growing pathogens such as *Listeria monocytogenes* and *Yersinia enterocolitica*. Industry and regulatory groups must financially support this research.

REFERENCES

1. Centers for Disease Control (CDC). 1982. Foodborne Disease Surveillance. Annual Summary. Issued September 1985. U.S. Dept. of Health and Human Services, Public Health Service, CDC, Atlanta, GA.
2. William Wollschlager. 1986. Division of Retail Food Protection, U.S. Food and Drug Administration (FDA). Personal communication.

9. Microbiological Criteria for Foods and Food Ingredients: A Review of the Report of the National Research Council

Carl Vanderzant

Texas A&M University

Over the last 20 years, interest in and controversy over the application of microbiological criteria to classify foods as either microbiologically acceptable or microbiologically unacceptable have grown steadily, particularly relative to microbiological quality standards. A 1964 report of the National Research Council,[1] *An Evaluation of Public Health Hazards from Microbiological Contamination of Foods*, concluded that microbiological criteria for foods should (1) accomplish what they purport to do, i.e., reduce public health hazards; (2) be technically feasible, i.e., attainable under conditions of good commercial practice; and (3) be administratively feasible. It was also concluded, at that time, that it would be premature to formulate on a broad scale microbiological limits or standards for foods that are meaningful from the standpoint of public health. This conclusion was reached because of a number of weaknesses in (1) our knowledge of the hazards from microbiological contamination of foods, (2) our understanding of potential hazards from newer foods and technologies, (3) our knowledge of what is technically feasible commercially, and (4) methodology to assure compliance with such criteria. Although some of the same problems exist today and most likely in the future, there can be no doubt that much progress has been made in each of these areas. Concurrently with that progress, an increasing number of microbiological criteria have been proposed and/or established with accompanying disagreement and controversy over their value. The most comprehensive study on microbiological criteria for foods was carried out under the auspices of the National Research Council of the National Academy of Sciences.

In September 1980, four federal agencies (the National Marine Fisheries Service (NMFS), the U.S. Department of Agriculture (USDA), the

Food and Drug Administration (FDA), and the U.S. Army Natick Research and Development Center) requested that the National Research Council convene a panel of experts to formulate general principles for the application of microbiological criteria for foods and food ingredients and to provide recommendations for a unified, coordinated approach to the subject by policy-setting agencies. This report entitled *An Evaluation of the Role of Microbiological Criteria for Foods and Food Ingredients*[2] was prepared by the Food and Nutrition Board Subcommittee on Microbiological Criteria in response to that request. It is available from the National Academy Press, Washington, DC. The report (436 printed pages) is made up of eleven basic chapters each with extensive but pertinent literature citations, an executive summary, seven appendices and approximately 170 specific recommendations. In this report the NRC Subcommittee attempted to bring the uses of microbiological criteria into perspective with respect to their place in a total program for microbiological control of foods in the United States. The structure of the book (report) is as follows:

Executive Summary
Recommendations
Chapter 1. Introduction
Chapter 2. Definitions, Purposes, and Needs for Microbiological Criteria.
Chapter 3. Selection of Foods for Criteria Related to Safety.
Chapter 4. Selection of Pathogens as Components of Microbiological Criteria.
Chapter 5. Selection of Indicator Organisms and Agents as Components of Microbiological Criteria.
Chapter 6. Consideration of Sampling Associated with a Criterion.
Chapter 7. Consideration of Decision (Action) to be Taken When a Criterion (Limit) is Exceeded.
Chapter 8. Current Status of Microbiological Criteria and Legislative Bases.
Chapter 9. Application of Microbiological Criteria to Foods and Food Ingredients.
Chapter 10. Expansion of the HACCP System in Food Protection Programs.
Chapter 11. Plans of Action for Implementation of the HACCP System and of Microbiological Criteria for Foods and Food Ingredients.
Appendixes
A. Summary Responses to Specific Contract Items
B. General Principles for the Establishment and Application of Microbiological Criteria for Foods

Chapters 1 through 10 of the 1985 NRC report will be discussed only briefly. Emphasis will be placed on the recommendations of Chapter 11, "Plans of Action for Implementation of the HACCP System and of Microbiological Criteria for Foods and Food Ingredients," because this chapter focuses on the essence of the charge to the NRC Subcommittee namely "to formulate general principles for the consideration of microbiological criteria for foods and food ingredients and to provide recommendations for a unified, coordinated approach."

Chapter 1 deals primarily with the sources of microorganisms in foods, the activities of microorganisms in foods, and approaches of microbiological control in foods, through education, inspection, and microbiological testing. The NRC Subcommittee concluded that the Hazard Analysis Critical Control Point (HACCP) system, first presented at the 1971 National Conference on Food Protection provides a more specific and critical approach to the control of microbiological hazards in foods than that provided by traditional inspection and quality control approaches.

Chapter 2 presents the Subcommittee's view on definitions of microbial criteria (standards, guidelines, specifications) and purposes and needs for microbiological criteria in the United States. A standard is part of a law or ordinance and is a mandatory criterion. A guideline is a criterion used to assess microbiological conditions during the processing, distribution, and marketing of foods and functions as an alert mechanism to signal whether microbiological conditions prevailing at critical control points or in the final product are within normal range. They signal when there are microbiological problems that require attention. A specification is used in purchase agreements between buyers and vendors. The NRC Subcommittee recommends that microbiological standards should be considered only when (1) there is evidence to indicate that a problem exists between a food and outbreaks of foodborne disease and that the standard will alleviate the problem, (2) exceeding the limits is evidence that the food contains decomposed ingredients, and/or is evidence of preparation, packaging, or storage of the food under grossly poor conditions, or (3) there is no jurisdiction over processing and distribution practices, such as with certain imported foods, and the standard would eliminate a health hazard and/or reject products produced under questionable conditions. The components of a criterion are specified, namely (1) the identity of the food, (2) the contaminant of concern (the microorganism, toxin or other agent), (3) the analytical method to be used, (4) the sampling plan, and (5) the limits considered appropriate. It is emphasized that a microbiological criterion should be established and implemented only when there is a need and if it can be shown to be effective and practical. The NRC Subcommittee also concluded that microbiological quality standards such as recently proposed for seafoods should be reviewed and evaluated according to the plan proposed in Chapter 11 of the NRC report.[2]

In Chapter 3, the NRC Subcommittee recommends that the following factors provide the primary bases for deciding whether a microbiological criterion related to safety for a given food would or would not serve a useful purpose. Of special importance is epidemiological evidence that the food in question is a significant vehicle of disease. Also important are various hazard considerations which include (1) susceptibility of the food to contamination by pathogens, (2) the opportunity for the survival of pathogens, (3) the likelihood of microbial growth during manufacture, storage, distribution, and preparation for serving, (4) whether or not the food is to be heated before consumption, and (5) the susceptibility of probable consumers to infectious agents or toxins.

Chapters 4 and 5 treat in great detail the agent (contaminant) of concern, (i.e., the microorganism, toxin, or other agent) one of the components of a microbiological criterion. Pathogens are treated in Chapter 4; indicator organisms and other agents in Chapter 5. Each of these organisms or agents is discussed relative to its importance, status of the method available, and suitability to be included in a microbiological criterion.

Chapter 6 deals with the sampling plan, another essential component of a microbiological criterion. The concept of relating stringency of the sampling plan to the likely degree of hazard, as recommended in the NRC *Salmonella* Committee report[3] and by the International Commission on Microbiological Specifications for Foods[4,5] was endorsed.

Chapter 7 describes five categories of deficiencies that may be revealed by the application of microbiological criteria and the corrective actions that should be taken. Categories included are (1) evidence of existence of a direct health hazard, (2) evidence that a direct health hazard would develop, (3) indications that a product was not produced under conditions assuring safety, (4) indications that a raw material may adversely affect shelflife, and (5) evidence that a critical control point is not under control.

Microbiological criteria applied at the international, federal, and state or local levels are discussed in Chapter 8. The programs of the

Joint FAO/WHO Codex Alimentarius Commission, the European Economic Community (EEC), and the ICMSF are described. The Canadian food standards also are discussed. The role and activities of the Food and Drug Administration (FDA), the U.S. Department of Agriculture (USDA), the U.S. Army Natick Research and Development Center of the Department of Defense, and the National Marine Fisheries Service (NMFS) of the Department of Commerce relative to microbiological criteria for foods are outlined.

Chapter 9, which comprises 113 pages, provides recommendations for the need or lack thereof for microbiological criteria for each of 22 food products or groups of foods. They include: dairy products; raw meats; processed meats; raw poultry; processed poultry products; eggs and egg products; fish, molluscs and crustaceans; fruits and vegetables; fruit beverages; low acid canned foods; acid canned foods; water activity controlled canned foods; cereals and cereal products; fats and oils; sugar, cocoa, chocolate and confectioneries; spices; yeasts; formulated foods; nuts; miscellaneous additives; bottled water, processing water and ice; and pet foods. For each of these foods the following basic issues are discussed: (1) the sensitivity of the food relative to safety and quality, (2) the need for a microbiological standard or guideline, (3) assessment of information necessary for establishment of a criterion if one seems to be indicated, and (4) where the criterion should be applied. Because of space limitation, the NRC Subcommittee's recommendations on microbiological criteria for raw meats are presented as an example of this effort. They are as follows:

1. To produce raw red meat with optimum shelflife, critical control points such as slaughtering-dressing practices, sanitary condition of equipment and utensils, and control of refrigeration temperature should be monitored carefully. Microbiological guidelines are applicable to monitor some of these critical control points.
2. To achieve optimum shelflife of raw meats, the HACCP concept should be extended to include retail outlets, food service establishments, and homes, particularly as it pertains to handling practices and maintenance of adequate refrigeration.
3. Microbiological standards for raw meats are not recommended since they will prevent neither spoilage nor foodborne illness.
4. Microbiological criteria containing limits for pathogens in raw meats are not recommended.
5. To reduce the health hazards from raw meats the following should prevail:

   a. recognition that small numbers of pathogens may be present on raw meats
   b. strict adherence to good food preparation practices
   c. application of new or alternate production and processing practices that reduce the incidence of pathogens
   d. more effective education on foodhandling practices of food plant personnel, food service operators, and homemakers
   e. increased emphasis on inspection of food service establishments using the HACCP approach
6. Present inspections of foodprocessing facilities and operations by regulatory agencies such as the USDA should be made more cost effective by application of the HACCP concept.

The NRC Subcommittee embraced the Hazard Analysis Critical Control Point (HACCP) concept as an effective and rational approach to the assurance of safety and to the prevention or delay of spoilage in foods (Chapter 10). In the application of HACCP the use of microbiological criteria is at times the most effective means of monitoring critical control points. In other instances, monitoring of critical control points can best be accomplished through the use of physical and chemical tests, visual observations, and sensory evaluations. Thus, microbiological criteria may play an important role or no role at all depending upon the nature of the food or process.

Chapter 11 deals with plans of action for implementation of the HACCP system and of microbiological criteria for foods and food ingredients as detailed in the following sections.

IMPLEMENTATION OF HACCP

Because the application of the HACCP system provides for the most specific and critical approach to the control of microbiological hazards presented by foods, use of this system should be required of industry. Accordingly, the NRC Subcommittee believes that government agencies responsible for control of microbiological hazards in foods should promulgate appropriate regulations that would require industry to utilize the HACCP system in their food protection programs.

The regulations should identify the basic elements of the HACCP system and provide for ready availability of industry monitoring records that relate to critical control points and other appropriate information for review by regulatory inspection personnel.

The regulations should not specify details of the application of the

HACCP system. The development of such details should be the prerogative of industry.

The regulations should require adequate training of regulatory inspection personnel in the elements of the HACCP system so that their inspection activities focus on the review of monitoring records as the primary basis for assessing the adequacy of a food processor's control program.

The HACCP system should likewise be applied at points in the food-processing chain other than at the processing level, i.e., in production, storage, transport, retail sales, and at food service establishments.

Regulatory authorities should have the option of assessing the appropriateness of selected critical control points, the adequacy of the monitoring procedures, and the actions taken when results of monitoring indicate the need for action.

IMPLEMENTATION OF MICROBIOLOGICAL CRITERIA FOR FOODS AND FOOD INGREDIENTS

Introduction

The NRC Subcommittee believes that for the most part microbiological criteria can best be used as one of the components of the HACCP system for food protection, e.g., as one of the means of monitoring critical control points in food processing and distribution. However, microbiological criteria can be used independently of HACCP as a sole determinant of the acceptability of a food or a process, e.g., through attribute sampling and testing of finished products domestically produced and/or at point of import for presence of *Salmonella* or of staphylococcal enterotoxins. In any event, whether or not microbiological criteria are applied either through the HACCP system or independently, there are principles governing their development and application that must be adhered to if the criteria are to be meaningful in the interest of food protection. In the foregoing chapters of the NRC report, the Subcommittee set forth these general principles. However, there is need for a national plan of action by which these general principles would be applied uniformly in the process of developing and implementing microbiological criteria. Toward this end the following is directed.

The Food Industry

The NRC Subcommittee emphasizes that application of the HACCP system by industry is the most effective means of assuring the microbiological safety and quality of foods. However, it would be presumptuous to propose to industry the mechanism of a plan for development of microbiological criteria either as an integral part of the HACCP system or otherwise. The needs for microbiological criteria are far too varied, even within a single organization, for any other than corporate management to undertake that task. It is anticipated, however, that the principles and considerations set forth in the NRC report will provide guidance to those commercial organizations that need to use microbiological criteria in their food safety and quality control programs. Since it is industry's responsibility to provide safe food of acceptable quality, it should be industry's prerogative to design and implement the means by which such responsibility can be met to the satisfaction of regulatory agencies, or more broadly, the consuming public.

Government Agencies

The primary purpose for use of microbiological criteria by regulatory agencies is to supplement other means they have of assuring that foods produced under their jurisdiction are safe and of acceptable quality and to make certain that industry fulfills its responsibility. Since the processing and distribution of most foods fall under the regulatory jurisdiction of many state and municipal agencies, as well as being subject to federal regulations, there should exist a high degree of uniformity in the manner by which microbiological criteria (particularly those that pertain to finished product) are developed and implemented. Lack of such uniformity has resulted in the establishment of many criteria that are nonuniform, misapplied or unjustified, and highly controversial.

The magnitude of the impact of the implementation by regulatory agencies of microbiological criteria for foods that would apply nationally would demand widespread participation by concerned groups in the development of the criteria. Such participation would tend to assure that the best interests of all were served and that general acceptance of the criteria would be forthcoming. It is logical that concerned federal agencies assume leadership in fostering a multilateral approach to development of microbiological criteria for foods. In fact, it is evident from the survey by Wehr[6] that a preponderance of responses (50 of 64) received from states stated a preference for federal agencies to assume such leadership.

In view of the considerations given above, the NRC Subcommittee offers a national program for identification of foods for which microbiological criteria are needed, for development of criteria for those foods, and for appropriate implementation of such criteria in regulatory programs.

THE PLAN

1. The Center of Food Safety and Applied Nutrition of the Food and Drug Administration, an appropriate subsidiary body of the Department of Agriculture; the National Marine Fisheries Service; and the U.S. Army Natick Research and Development Center; being the primary federal agencies having the responsibility of assuring that foods available to their constituencies are safe and of acceptable quality, jointly should establish an ad hoc commission on Microbiological Criteria for Foods.

2. The objective of the commission should be to develop microbiological criteria for foods that will be responsive to need.

3. The Food and Drug Administration, having the broadest responsibility for food protection among the federal agencies, should take the initiative to arrange with the cooperation of the other three agencies identified above for appointment of members to the commission.

4. Membership of the commission should consist of appropriate personnel selected from (a) each of the four above-mentioned federal agencies, (b) state and municipal food regulatory agencies, (c) the food industry, and (d) academia. The number of members on the commission should be kept to a minimum in the interest of expediency in reaching decisions and economy of operations.

5. Initially, the commission should address (a) the selection of foods and food ingredients on the basis of need for microbiological criteria that could effectively supplement and be incorporated in food protection programs of federal, state, and municipal food regulatory agencies (The selection process should include, but not be limited to, a review of foods for which criteria currently exist and an evaluation of the suitability of these criteria.), and (b) the development of criteria for each food for which it was determined that microbiological criteria would serve a useful purpose.

6. The commission should appoint expert working groups to carry out the tasks of selecting foods and developing criteria. The expert working groups should present their recommendations to the commission.

7. The commission should present its recommendations to the federal agency having primary jurisdiction for the safety and quality of the food or food ingredient involved. The federal agency should take the

necessary steps to promulgate regulations embodying the recommended criteria and to promote uniform adoption of the criteria by state and municipal regulatory agencies if it deems it appropriate to do so. Insofar as possible, the Department of Defense should utilize microbiological criteria recommended by the commission in developing military specifications for foods and food ingredients.

8. The development, review, and modification of microbiological criteria for foods is a continuing task. Therefore, as soon as the initial work of the commission has been set in motion, it should take the steps necessary to organize itself as a continuing body elected by vote or other suitable means of the participating groups indicated above, except that the representative of each federal agency should be appointed by an appropriate official of the respective agency.

9. Finally, the commission should address the continuing need for research to (a) better understand the effects of current and emerging processing and distribution practices that may affect the microbial safety and quality of foods, and (b) improve and develop methods for the detection and enumeration of microorganisms and groups of microorganisms and their toxic agents that are important to the safety and quality of foods.

The NRC Subcommittee believes that the plan of action for implementation of the HACCP system provides for the most specific and critical approach to the control of microbiological hazards presented by foods. The plan of action for implementation of microbiological criteria that embodies the general principles and considerations presented in the NRC report will lead to a coordinated national program for the establishment and application of microbiological criteria for foods that will enhance food protection programs of federal, state, and municipal agencies as well as those of the military and industry. The NRC Subcommittee recommends that action be taken to implement these plans at the earliest opportunity.

ACKNOWLEDGMENT

Most of the material in this chapter is taken from “An Evaluation of the Role of Microbiological Criteria for Foods and Food Ingredients,” Copyright 1985 by the National Academy of Sciences, with permission of the publisher, National Academy Press, 2101 Constitution Avenue, NW, Washington, DC 20418.

Members of the NAS/NRC Subcommittee on Microbiological Criteria: D.H. Ashton, F.L. Bryan, D.L. Collins-Thompson, E.M. Foster, J.J. Jezeski, R.V. Lechowich, R.J. Marino, J.C. Olson, Jr., J.H. Silliker, D.F. Splittsstoesser (Vice-Chairman), C. Vanderzant (Chairman), and Margaret R. Stewart (Staff Officer).

REFERENCES

1. National Research Council (NRC). 1964. An Evaluation of Public Health Hazards from Microbiological Contamination of Foods. Food Protection Committee. National Academy of Sciences National Research Council, Washington, DC.
2. National Research Council (NRC). 1985. An Evaluation of the Role of Microbiological Criteria for Foods and Food Ingredients. Food Protection Committee, Subcommittee on Microbiological Criteria. National Academy of Sciences National Research Council, Washington, DC.
3. National Research Council (NRC). 1969. An Evaluation of the *Salmonella* Problem. Committee on *Salmonella*. National Academy of Sciences, Washington, DC.
4. International Commission on Microbiological Specifications for Foods (ICMSF). 1974. Microorganisms in Foods. 2: Sampling for microbiological analysis: Principles and specific applications. University of Toronto Press, Toronto.
5. International Commission on Microbiological Specifications for Foods (ICMSF). 1986. Microorganisms in Foods. 2: Sampling for microbiological analysis: Principles and specific applications. Second edition. University of Toronto Press, Toronto.
6. Wehr, H.M. 1978. Attitudes and policies of state governments. *Food Technol.* 32(1):63–67.

10. Selection of Microbiological Criteria Based on Hazard Analysis of Food

Donald A. Corlett, Jr.
Del Monte Research Center
Nabisco Brands, Inc.

INTRODUCTION

Selection of microbiological criteria for purposes of food safety requires a systematic analysis of the microbiological hazards associated with a specific food system. Failure to recognize the sources and causes of foodborne hazards may result in failure of critical control points in the production system. Two recent examples of foodborne illness illustrate this point. In 1985, in the Chicago area, 16,000 persons contracted salmonellosis from pasteurized milk. Last year there were 314 cases of listeriosis in California, and 40 of the deaths were linked to a brand of Mexican-style cheese.

These episodes raised questions concerning application of microbiological criteria:

- Why did a *Salmonella* outbreak occur in a modern dairy, where historically, microbiological criteria have been used?
- Would microbiological criteria have prevented the listeria outbreak in cheese? *Listeria monocytogenes* has only recently been recognized as a serious foodborne pathogen. Currently, there are other microorganisms that are emerging as causes of foodborne disease.
- Is there an effective approach for identifying the sources of hazardous microorganisms in foods and integrating their control into the elements of the food processing system?

The key issue raised by each of these questions is identification of hazardous microorganisms in order to implement preventative measures to insure their elimination. This is where the National Academy of Sciences National Research Council report[1] contributes a new approach that integrates risk assessment, control, and microbiological criteria. A major recommendation is use of the Hazard Analysis and Critical Control Point (HACCP) System which provides the bridge between process control and microbiological criteria.

The following comments provide views on selecting microbiological criteria based on HACCP, and consist of four topic areas:

- Microbiological criteria and definitions
- Causes and control factors for hazardous microorganisms in foods
- Risk assessment and the HACCP worksheet
- Establishing criteria and limits

MICROBIOLOGICAL CRITERIA AND DEFINITIONS

Microbiological criteria have been the subject of numerous debates over the years, primarily because of the inherent difficulty in applying microbiological limits to foods. Measurement of living microorganisms is difficult and includes detection of their toxins and chemical byproducts. The recent book, *An Evaluation of the Role of Microbiological Criteria for Foods and Food Ingredients*,[1] prepared by the Subcommittee on Microbiological Criteria, National Academy of Sciences, National Research Council, provides definitions of various types of microbiological criteria such as standards, guidelines, and specifications.

Key to this discussion are the definitions of the purpose of microbiological criteria, a criterion, and the components of a microbiological criterion:

- *Purposes of microbiological criteria*—To assess:
  1. The safety of a food
  2. Adherence to good manufacturing practices
  3. The utility (suitability) of a food or ingredient for a particular purpose
  4. The keeping quality (shelf-life) of certain perishable foods

- *Criterion*—A yardstick (standard) on which a judgment or decision can be made.

- *Components of a Microbiological Criterion*
  1. A statement describing the identity of the food or food ingredient
  2. A statement of the contaminant of concern, i.e., the

microorganisms or group of microorganisms and/or its toxin or other agent

3. The analytical method to be used for the detection, enumeration, or quantification of the contaminant of concern
4. The sampling plan
5. The microbiological limits considered appropriate to the food and commensurate with the sampling plan used

These three definitions provide the framework for development of microbiological criteria based on food process control factors and risk assessment of the food system.

CAUSES AND CONTROL FACTORS FOR HAZARDOUS MICROORGANISMS IN FOODS

Factors which cause foodborne disease outbreaks have been documented by Bryan[2] in the United States (Table 1) and Roberts[3] in England (Table 2). These abuses fall into six major areas:

- Contaminated raw food or ingredients
- Temperature abuse
- Time-temperature abuse
- Insanitation
- Cross-contamination
- Improper food handling or employee hygiene

Control of these factors is essential for safe preparation and processing of food. They are often interrelated and *a loss of control of one factor* may cancel out the control of the others. For example, if cross-contamination introduces *Salmonella* into a heat-pasteurized product, all preceding controls are negated.

Microbiological criteria may be utilized either directly or indirectly to monitor control factors in food processing, as follows:

- To *directly* determine the presence of hazardous microorganisms in a raw material or ingredient, in-process food, or finished product
- To *indirectly* monitor the effectiveness of a control element such as cooking, cleaning and sanitation of a piece of equipment, or employee hygiene, etc.

Thus, the *control* of the causes of microbiological hazards are the basis for establishing criteria.

Table 1. Factors Contributing to Outbreaks of Foodborne Disease,[a] United States, 1960–1976[2]

Factors	Percentages[b]
Improper cooling	46
Lapse of a day or more between preparing and serving	21
Colonized persons touching cooked foods	20
Inadequate thermal processing, canning, or cooking	16
Improper hot storage	16
Inadequate reheating	12
Contaminated raw food	11
Cross-contamination	7
Improper cleaning of equipment	7
Obtaining foods from unsafe sources	5
Use of leftovers	4

[a]Other factors were faulty fermentations, toxic species of plants or mushrooms mistaken for edible varieties, poor dry storage practices, storing high-acid foods in toxic containers, incidental additives, and intentional additives (for example, MSG).

[b]Note that percentages exceed 100% because foodborne diseases have multiple causation: foods must become contaminated, contaminants may survive processes, and frequently bacterial pathogens multiply to reach large numbers or to produce toxins.

Table 2. Factors Contributing to Outbreaks of Foodborne Disease,[a] England and Wales, 1969–1976[3]

Factors	Percentages of 1,000 Cases
Preparation of foods at least half a day before it was eaten	61
Improper cooling	72
Storage at room temperature	40
Improper refrigeration practices	32
Inadequate reheating	29
Contaminated incoming foods	19
Inadequate cooking	15
Improper thawing (and storage thereafter)	6
Cross-contamination	6
Improper hot holding	6
Colonized food worker (touching cooked foods)	5
Use of leftovers	5

[a]Other factors cited were: raw foods served and eaten, extra large quantities of food prepared, and contaminated canned foods.

ESTABLISHING MICROBIOLOGICAL CRITERIA BY RISK ASSESSMENT AND HACCP WORKSHEET

Hazard Classification Based on Risk[4]

A food and its raw materials and ingredients may be classified into hazard categories by means of a two-step procedure. The first step is *risk assessment*, and the second step is *assignment of hazard categories*. This may be carried out as follows:

Risk Assessment

The product is evaluated according to the following hazard characteristics, using + for yes and 0 for no:

HAZARD A — The product contains a "sensitive" ingredient or ingredients:

All ingredients have some degree of risk from a microbiological standpoint varying from high to essentially zero.

In general products of animal origin, meat and meat products, dairy products, egg products, and seafood — may be considered most sensitive.

Cereal products are next followed by vegetable products and lastly by fruit and fruit products.

HAZARD B — The manufacturing process does not contain a controlled processing step that effectively destroys harmful bacteria:

The processing step referred to here can be heat or chemical processing or the development of a food system that, by its chemical or physical composition, destroys harmful microorganisms.

HAZARD C — There is substantial potential for microbiological abuse in distribution or consumer handling that could render the product harmful when consumed:

The principal judgment criterion for abuse potential is whether the food product is a good medium for microbial growth in the state in which it is distributed or as normally prepared by the consumer. Consideration must be given to low levels of microbial contamination which have escaped control screening or processing designed to prevent contamination by harmful bacteria.

If all three hazard characteristics were present in a product, it would have a Hazard Class + + +. If a hazard is absent, an 0 would designate this as follows:

0 + + – No sensitive ingredient
+ 0 + – Product pasteurized
000 – No hazard involved

Assignment of Hazard Category

The three general food hazard characteristics can be combined into a set of eight configurations (of A, B, and C). *Rank* the food according to its potential consumer health hazard as follows:

CATEGORY I—This is a special category that must be recognized and applies to nonsterile products designed and intended for consumption by infants, the aged, or the infirm. This segment of consumers is particularly susceptible to the effects of harmful microorganisms. Products in this category will require food safety analysis and controls subject to the specific requirements for use of the food (e.g., baby food).

CATEGORY II—Food products subject to all *three* general hazard characteristics.

A. Hazard Class + + +

CATEGORY III—Food products subject to *two* general hazard characteristics.

A. Hazard Class + 0 +
B. Hazard Class + + 0
C. Hazard Class 0 + +

CATEGORY IV—Food products subject to *one* of the general hazard characteristics.

A. Hazard Class + 00
B. Hazard Class 0 + 0
C. Hazard Class 00 +

CATEGORY V—Food products subject to none of the general hazard characteristics.

A. Hazard Class 000

After a food has been placed in a hazard category, it is useful to take a closer look at all ingredients intended for use in a product as well as the proportion of each ingredient in a mixture and its handling. These may be assessed for risk and categorized just like products.

Table 3 illustrates how risk assessment and different hazard category assignment may be used for a variety of different foods. This useful

Table 3. Risk Assessment and Hazard Category Assignment for a Variety of Foods

Food	Contains Sensitive Ingredient(s)	Kill-Step	Potential for Consumer Abuse	Hazard Category
Infant formula	+	+	+	I
Raw frozen chicken	+	+	+	II
Frozen pizza	+	+	+	II
Frozen dinner with raw component	+	+	+	II
Frozen precooked dinner	+	o	+	III
Canned beans	+	o	o	IV
Canned peaches	o	o	o	V
Table salt	o	o	o	V

procedure is effective for quickly ranking foods according to their hazard characteristics and is the basis for establishing controls for safe preparation and processing, which may include critical control points for microbiological criteria for hazardous microorganisms.

Examples of Risk Assessment and Hazard Category Assignment

The HACCP Worksheet

A practical approach for the selection of microbiological criteria in conjunction with establishment of critical control points consists of utilizing a HACCP worksheet with the following headings:

1. Description of the food product and its intended use

2. Flow diagram with the following parts:
   - Raw material* or ingredient* receipt and handling
   - In-process preparation, processing, and fabrication steps
   - Finished product packaging and handling steps
   - Storage and distribution
   - Point of sale handling

3. Hazard categories of raw materials and ingredients, in-process foods, and finished product

4. Critical control points (including microbiological criteria)

*Raw material and ingredient Q.C. specifications should be attached to the HACCP worksheet because they contain the microbiological criteria.

This approach is illustrated by using frozen pizza as an example. The HACCP worksheet is a tool used in conjunction with development of other documents for quality assurance purposes (i.e., formula, specifications, preparation and processing procedures, and the quality control program).

HACCP Worksheet—Frozen Pizza

1. Description: Pizza that will be fabricated and frozen for retail sale. Product must be safe to eat without reconstitution, but directions on the package will state that product is intended to be heated prior to consumption by the consumer.

2. Flow Diagram and Critical Control Points (X)

Crust (1)
↓
Sauce depositor (3) ← Sauce preparation (2)
↓
Topping depositor (5) ← Topping preparation (4)
↓
Inspection (6)
↓
Packaging
↓
Freezing (7)
↓
Casing (8)
↓
Storage (9)

3. Hazard Categories

   A. Product: Hazard Category II(+ + + : contains sensitive ingredients, no kill step, and there is potential for consumer abuse.)

      Microbiological: *Salmonella*, *Staphylococci* (enterotoxin), enteropathogenic *Escherichia coli* (EEC).

B. Ingredients

Ingredients	Hazard Category	Microorganisms of Concern
Crust	III	*Staphylococci*
Sauce	IV	Low pH—none
Cheese	II	*Salmonella*
		Staphylococci (enterotoxin)
		E. coli (EEC)
Sausage	II	*Salmonella*
		Staphylococci (enterotoxin)
Spices	III	*Salmonella*
Vegetables	IX	*Salmonella*

4. Critical Control Points

Number	Type of Critical Control Point
1	Microbiological limits on dough
	Correct storage temperature and time
2,4	Microbiological limits on ingredients
	Sanitation
	Correct storage temperature and time
3,5	Sanitation
6	Equipment sanitation and employee hygiene
7	Correct internal temperature and time to freeze
8	Microbiological limits on finished product (for audit)
9	Correct internal freezing temperature

Completion of the HACCP worksheet establishes the critical control points for manufacturing a safe product. Critical control points 1, 2, 4, and 8 require microbiological testing for the presence of hazardous microorganisms. In the case of frozen pizza, these microorganisms are *staphylococci* (enterotoxin forming strains), *Salmonella*, and *E. coli* (enteropathogenic strains).

ESTABLISHING CRITERIA AND LIMITS

The discussion has related to the rationale for identifying the controls necessary to prevent microbiological hazards and selecting microorganisms of concern for a particular food. This approach has satisfied the requirements for the *first two* components of a microbiological criterion:

- A statement describing the identity of the food or food ingredient
- A statement of the contaminant(s) of concern, i.e., the microorganism, or group of microorganisms, and/or its toxin or other agent

This was illustrated in risk assessment, hazard category assignment, and the HACCP worksheet example for frozen pizza. Of equal importance, the critical factors needed to control hazardous microorganisms were identified—some of which happen to be microbiological criteria.

There are three more components of the microbiological criteria (in the case of pizza, for *Salmonella*, *S. aureus*, and enteropathogenic *E. coli*) which need to be developed including the analytical method of choice, the sampling plan, and the limits. These three components are covered in the excellent reference by the International Commission on Microbiological Specifications for Foods (ICMSF).[5]

CONCLUSION

The selection of microbiological criteria based on hazard analysis of food is an effective means to identify the potential presence of hazardous microorganisms in a food system. Criteria developed in this manner will be closely tied to the means of control and monitoring.

The HACCP concept fosters use of controls and criteria early in the food preparation sequence. That is, safety may be achieved by careful attention to ingredients and processing steps. Reliance on HACCP should ideally eliminate finished product microbiological testing except as a periodic auditing activity.

There are a number of issues that must be resolved, associated with use of HACCP and selection of microbiological criteria:

1. Survival parameters for emerging pathogens need definition. A case in point is the pasteurization requirement for destruction of *Listeria monocytogenes* in milk.
2. Training is needed to implement HACCP and particularly to focus on the relationship between control of hazardous microorganisms in food systems and use of microbiological criteria for monitoring.
3. Practical sampling procedures, more rapid and accurate test methods, and more definitive limit-setting procedures need

to be developed. This issue is not easily resolved and perhaps the Ad Hoc Commission, recommended in the NRC report, will benefit this area.

REFERENCES

1. National Research Council (NRC). 1985. An Evaluation of the Role of Microbiological Criteria for Foods and Food Ingredients. Food Protection Committee, Subcommittee on Microbiological Criteria, National Academy of Sciences, National Research Council. National Academy Press, Washington, DC, Ch. 2, pp. 55–64.
2. Bryan, F.L. 1978. *J. Food Prot.* 41:816–827.
3. Roberts, D. 1982. *Communicable Disease Reports* 82/2:3–4.
4. National Research Council (NRC). 1969. National Academy of Sciences National Research Council, Publication No. 1985.
5. International Commission on Microbiological Specifications for Foods (ICMSF). 1986. Microorganisms in Foods. 2: Sampling for microbiological analysis: Principles and specific applications. Second edition. University of Toronto Press, Toronto.

11. Rapid Methods for the Detection and Identification of Microorganisms in Foods

Nelson A. Cox
J. Stan Bailey

U.S. Department of Agriculture

D.Y.C. Fung

Kansas State University

Paul A. Hartman

Iowa State University

STATEMENT OF THE PROBLEM

Results obtained by using standard, conventional tube and plate tests to rapidly and accurately detect and identify pathogenic bacteria in foods present many inherent problems. Probably the most important is that the times required to obtain results are usually much too long; in many instances, foods are consumed before the laboratory analyses are complete.

PACKAGED MINIATURIZED IDENTIFICATION KITS

The age of rapid, miniaturized techniques for identification of bacteria received much impetus in the middle to late 1940s, when Bachman and Weaver[1] added concentrated inocula to small tubes of different media. From that time on, the concept steadily gained popularity. Increased use of plastic made possible the development of disposable multichambered trays, plates, and dishes of various shapes and sizes which were utilized by the manufacturers of packaged identification systems or kits. For approximately the past 15 years, these systems which require colony-forming units from solid plating media were our answer to rapid identification. While this does speed up the overall identification process, it is somewhat inaccurate to classify this as a "rapid" procedure. In the analysis of a food, 2 to 3 days is required to reach this point. In addition, the systems do not include appropriate tests for some important foodborne pathogens such as *Campylobacter jejuni*, *Bacillus cereus*, and *Listeria monocytogenes*. Therefore, for a variety of reasons, these packaged mini-kits are not the ultimate solution to the rapid recognition of pathogenic bacteria in foods.

In order for a procedure to be truly rapid, it must be able to detect the pathogen in a liquid medium or, even better, detect the organism directly from the tissues or food material in which it is present. This obviously eliminates the need to obtain a pure culture or colony-forming unit and, in doing so, may save 2 to 3 days of precious time.

IMMUNOASSAYS

With the introduction of "tagged" antibodies by Coons et al.[2] several immunoassays employing labeled antibodies have evolved. These include using various fluorescent dyes (FA), radioisotopes (RIA), and enzymes (EIA). These "tagged" antibody procedures have a wide spectrum of application, and therefore have become one of the best established procedures in microbiology.[3] The first application of this technology to detect the presence of salmonellae was reported by Thomason et al.[4] They showed that the O, Vi, and H antigens of *Salmonella typhi* could be stained individually or collectively by using specific antibodies labeled with fluorescein. In 1962, the first published report appeared in a Russian journal, in which the FA technique was used to detect salmonellae in milk.[5] Since that time, many published reports have described the application of FA to a variety of food and food products to detect the presence of salmonellae.

The FA technique involves the labeling of serum antibodies by chemical combination with fluorescent dyes, such as fluorescein isothiocyanate. These labeled antibodies, when applied to smears containing *Salmonella*, will attach to the cell walls of the organism and become visible when illuminated by light of an appropriate wavelength. *Salmonella* stained by FA appear under a fluorescence microscope as yellow-green rods against a dark background. In addition to the fact that FA does not require the use of pure cultures, another advantage is that FA can detect low numbers of the test organisms in smears containing high levels of extraneous organisms. However, considerable training and skill are needed in reading the test. There are several other limitations of the FA technique. It is a serologic test and therefore is limited to the specificity of the antibodies used. Cross-reactions are common among the *Enterobacteriaceae*; using high-titered antibodies will minimize but not eliminate these cross-reactions. Because of the known cross-reactivity with the *Salmonella* polyvalent reagent, the FA method is recommended as a presumptive screening test for the presence of *Salmonella*. All FA-positive results should be confirmed by conventional cultural and serological procedures.

Immunoassays using radiolabeled reagents were developed by Yalow and Berson.[6] Since that time, radioimmunoassays (RIA) to detect and to measure the concentration of a wide variety of organic substances including bacterial antigens have been developed. RIAs have certain inherent advantages and disadvantages.[7] Radioactivity can be detected rapidly, directly, and with a relatively high degree of sensitivity. The ionizing radiation which arises from nuclear decay should be of minor concern to the safety of laboratory technicians performing assays because of the low quantities of radiolabels used. Nevertheless, many people are reluctant to use a radiometric technique. The small number of radioisotopes which can be used, the need for complex and expensive equipment for measuring radioactivity, the chemical instability of certain radioactive levels, and the disposal of radioactive materials are other disadvantages. At present no radiolabeled immunoassays are being used in the food industry.

The disadvantages of radioisotopes in immunoassays have led to the development of alternative procedures with other labels including enzymes, coenzymes, fluorescent dyes, fluorogenic substrates, chemiluminescent precursors, bacteriophages, and metal ions.[8,9] Of these alternatives, immunoassays using enzyme-labeled tracers have been the most suitable, and this procedure has been referred to as enzyme immunoassay (EIA). The use of enzymes as tracers in immunoassays was first introduced by van Weeman and Schuurs[10] and Engvall and Perlmann.[11]

An EIA to detect *Salmonella* in foods was developed by Krysinski and Heimsch,[12] but the antibodies they used were not pure. Minnich et al.[13] developed an improved direct EIA procedure, but the purified Spicer-Edwards polyvalent anti-H IgG antibody was not ideal. Robison et al.[14] developed an EIA using a monoclonal antibody for *Salmonella*. The monoclonal MOPC 467 from an IgA myeloma was able to detect *Salmonella* from mixed cultures with no cross-reactivity to other enterics. Further testing showed that 3% of *Salmonella* serotypes were not detected by the MOPC 467 monoclonal. A second monoclonal producer, a hybridoma referred to as 6H4, was developed. This antibody showed reactivity against all the strains of *Salmonella* which the MOPC 467 did not detect. These two antibodies were combined on a metal bead carrier, and an EIA procedure using this system has received AOAC approval (Organon Technica, Charleston, South Carolina). The primary disadvantages of this system are that it requires 10^6 organisms per mL and the procedure which received AOAC approval requires 50 to 56 hours. Refinements in the procedure are needed so the test can be run in a minimum of 48 hours if it is to receive widespread acceptance. Important advantages of EIA are (1) radioisotopes are not needed, and (2) monoclonal antibodies can be made against most purified antigens. Therefore, this technology should be applicable to the identification of a wide variety of bacteria.

DNA HYBRIDIZATION

Recently, genetic engineering has made it possible to select specific pieces of DNA, clone them, and produce large quantities of desired DNA sequences. These DNA pieces can be employed through hybridization as specific tests for selected organisms or for DNA sequences that code for bacterial toxins or other virulence factors.[15]

Typically, a hybridization assay consists of the following steps:[16]

1. The bacteria or other organisms are applied to a solid support such as a nitrocellulose membrane filter. This can be done by simply spotting a bacterial culture onto the filter by touching the filter to colonies on agar plates or by actual filtration through the membrane.
2. The bacteria are lysed to release their DNA, and the bacterial DNA is denatured into separate strands.
3. The separated DNA strands are then fixed to the solid support so that they will not wash away later.
4. The filters are soaked in a hybridization cocktail that contains probe DNA. The probe DNA is a single-stranded sequence of DNA that has been labeled with a radioisotope or with a nonisotopic "reporter molecule" or "signal."
5. When a probe is mixed with test DNA on the nitrocellulose filter, the probe sequence finds its complementary sequence in the test DNA and forms hybrid molecules. These hybrids contain one labeled strand that can then be detected by autoradiography (if radiolabeled or by some other signal).

This technique has been developed for detecting *Salmonella*,[17] enterotoxigenic *E. coli*,[18] and *Yersinia enterocolitica*[18] in foods.

Gene Trak (Integrated Genetics, Framingham, Massachusetts), a hybridization assay for the detection of *Salmonella* in foods, is commercially available. The Gene Trak system is based on the work of Fitts[16,17] and utilizes a radiolabeled gene probe. Preliminary studies indicate a sensitivity of 5×10^6 *Salmonella* cells per mL of broth medium, no false negative samples, and 3% false positives when compared to standard BAM cultural procedures.

The primary advantage of DNA hybridization is that if the appropriate DNA sequences are selected for, the test should be highly selective with very low levels of false negatives and false positives. The disadvantages of DNA hybridization are (1) The only commercially available system uses a radiolabeled probe which many laboratories find objectional, (2) the sensitivity of the system (5×10^6 *Salmonella* cells/mL) is such that the cultural procedures required to ensure success of the test mean that the test will take at least 48 hours to complete from the time the assay begins, and (3) the reagents are expensive.

BARRIERS TO PROGRESS

1. Most food laboratories are reluctant to add a new technology, even if it has AOAC approval.
2. The BAM procedures to which all new rapid methodology must be compared in order to receive AOAC approval are not foolproof. The possibility that a new procedure might be more sensitive and produce positive results when the BAM was negative must be considered. Alternative procedures for certifying the accuracy of new assays should be considered.
3. Initial expenses (equipment, training) involved in some technologies minimize their use in small laboratories with limited budgets. In other instances, costs of reagents or kits are too high to be economical.
4. Many existing procedures are too complex or involve too many processing steps. This is true for most of the current EIA and DNA/DNA hybridization tests. Efforts must be made to develop homogeneous assays to replace the present heterogeneous, multistep assays.
5. Laboratories testing for all food pathogens may be reluctant to incorporate new technologies that are only capable of detecting a single pathogen from foods such as the *Salmonella* kits that are currently available.
6. There is little coordination between procedures used for the examination of foods, water, biologicals, etc. Laboratories that perform analyses on a variety of materials are often presented with a confusing array of different "official" and "unofficial" techniques and methods.

RESEARCH NEEDS

1. Research should be aimed at detection of low numbers of the various pathogens (10^3 or less), which may make possible the dream of a one-workday analysis from sample to confirmed result.
2. Usefulness of the existing technologies could be increased by inclu-

sion of more of the microorganisms that concern the food microbiologist and not only *Salmonella*.

3. Monoclonal antibodies and DNA probes should be increasingly used to detect toxins, viruses, and virulence factors of significance.
4. Better signals are needed to replace radioisotopes on antibodies and on DNA probes. Certain biotin/avidin and fluorescent technologies are available, but all have major shortcomings. The search should continue for better "reporter molecules" that include signal amplification, possibly through cascade systems, without amplifying background "noise."
5. Unique, new, more sensitive and specific assays are needed. Much room exists for further advancement and improvement in both sensitivity and specificity.
6. Cost containment is a substantial factor in any analytical laboratory. Emphasis should be placed on developing methods and kits that are inexpensive and economical.

RECOMMENDATIONS

1. Research efforts should be aimed at development of procedures to detect food pathogens within 8 hours or less and with acceptable levels of accuracy.
2. Research efforts should be directed to detect virulence determinants of food pathogens so that virulent and avirulent strains can be differentiated.
3. New technologies should be simplified or automated so they are inexpensive and not labor-intensive and do not disrupt the work flow in the food microbiology laboratory.
4. New technologies should be integrated into the computer reporting system of the laboratory so that results are rapidly and accurately stored, and the data are readily accessible.
5. Assays should be developed for all important microbial toxins. The assays should be simple, rapid (1 hour or less), sensitive, and quantitative. An ideal kit would analyze for a variety of toxins simultaneously, using food homogenates or extracts.
6. Strong efforts should continue in the development of more sensitive and specific test methods based on entirely new concepts and technologies. Basic research invariably precedes practical application.

REFERENCES

1. Bachman, R. and Weaver, R. W. 1947. A quick microtechnique for the detection of the reduction of nitrates to nitrites by bacteria. *J. Bacteriol* 54:28.

2. Coons, A. H., Creech, H. J., Jones, R. N., and Berliner, E. 1942. The demonstration of pneumococcal antigen in tissues by the use of fluorescent antibody. *J. Immunol.* 45:159.
3. Flowers, R. S. 1985. Comparison of rapid *Salmonella* screening methods and the conventional culture method. *Food Technol.* 39:103.
4. Thomason, B. M., Cherry, W. B., and Moody, M. D. 1957. Staining bacterial smears with fluorescent antibody. III. Antigenic analysis of *Salmonella typhosa* by means of fluorescent antibody and agglutination reaction. *J. Bacteriol.* 74:525.
5. Arkhangelskil, I. I. and Kartashova, V. M. 1962. Accelerated methods of detecting *Salmonella* in milk. *Veterinariya* 9:74.
6. Yalow, R. S. and Berson, S. A. 1959. Assay of plasma insulin in human subjects by immunological methods. *Nature* 184:1648.
7. Schall, R. F. and Tennoso, H. J. 1981. Alternatives to radioimmunoassay: labels and methods. *Clin. Chem.* 27:1157.
8. Thomason, B. M. 1981. Current status of immunofluorescent methodology for salmonellae. *J. Food Prot.* 44:381–384.
9. Schall, R. F. and Tennoso, H. J. 1980. Recent developments in heterologous enzyme immunoassays. pp. 127–139. In R. M. Nakamura, W. R. Dito, and E. S. Tucker II (Eds.) Immunoassays: clinical laboratory techniques for the 1980's A. R. Liss, New York.
10. van Weeman, B. K. and Schuurs, A. H. W. M. 1971. Immunoassays using antigen-enzyme conjugates. *FEBS Lett.* 15:232.
11. Engvall, E. and Perlmann, P. 1971. Enzyme-linked immunosorbent assay (ELISA). Quantitative assay of immunoglobulin G. *Immunochemistry* 8:871.
12. Krysinski, E. P. and Heimsch, R. C. 1977. Use of enzyme-labeled antibodies to detect *Salmonella* in foods. *Appl. Environ. Microbiol.* 33:947.
13. Minnich, S. A., Hartman, P. A., and Heimsch, R. C. 1982. Enzyme immunoassay for detection of salmonellae in foods. *Appl. Environ. Microbiol.* 43:877.
14. Robison, B. J., Pretzman, C. I., and Mattingly, J. A. 1983. An enzyme immunoassay using a myeloma protein for detection of salmonellae. *Appl. Environ. Microbiol.* 45:1816.
15. Klausner, A. and Wilson, T. 1983. Gene detection technology opens doors for many industries. *Biotechnol. News* 1:471.
16. Fitts, R. 1985. Development of DNA-DNA hybridization test for the presence of *Salmonella* in foods. *Food Technol.* 39:95.
17. Fitts, R., Diamond, M., Hamilton, C., and Neri, M. 1983. DNA-DNA hybridization assay for detection of *Salmonella* spp. in foods. *Appl. Environ. Microbiol.* 46:1146.
18. Hill, W. E., Paynew, W. L., and Aulisio, C. C. G. 1983. Detection and enumeration of virulent Yersinia *enterocolitica* in food by DNA colony hybridization. *Appl. Environ. Microbiol.* 46:636.

12. Automated Monitoring of Food Operations and Microbial Detection

Anthony N. Sharpe

Health and Welfare Canada

STATEMENT OF THE PROBLEM

Through deaths, productivity loss, medical care. crop disease, and spoilage, microorganisms cost North America more dearly than any other foodborne contamination. For example, Hauschild and Bryan[1] estimated that there were 740,000 and 150,000 cases annually in the United States and Canada, respectively, for salmonellosis alone between 1969 and 1978. Todd[2] estimated costs per case from $456 and up in salmonellosis outbreaks, to $40,508 in a staphylotoxicosis outbreak. We instinctively expect that automation of certain aspects of food microbiology would allow more efficient control of foodborne microbes. In view of the costly effects of these outbreaks, we should have seen large strides in this direction.

Unlike most other sciences, food microbiology depends almost entirely on manual analytical procedures. This severely limits the amount of microbiological testing possible (of raw materials, problem areas on process lines, food handling personnel, and regulatory surveillance, etc.). Monitoring of process parameters pertinent to microbiological safety (water activity, pH, temperature, etc.) is also very much a manual affair; I know of no systems in which microbiological data are automatically transferred to plant operation control.

Even where analyses are carried out according to accepted schemes such as 3-class attribute[3] or variables plans,[4] interpretation is often subjective or intuitive. There are no expert knowledge bases available yet on which product control—combining data from such plans with continuously telemetered process data—can be trusted to computers. In homes and catering places, where over 90% of food poisoning incidents begin, there are virtually no devices to warn of improper preparation procedures.

EVALUATION OF THE PROBLEM

Efforts at automating food microbiology seem to have been directed almost exclusively toward microbiological analysis and, until very recently, simply at providing more rapid or less labor-intensive substitutes for parts of plate count methods. Mechanical aids such as Spiral Platers are sometimes useful. In Europe, instruments for automatically screening milk for leucocytes and other abnormalities (e.g., Bactoscan) have achieved some popularity. A number of highly promoted "automated" techniques based on $^{14}CO_2$ release, ATP content, catalase content, etc., have been or are available. They have achieved only a very limited popularity and are, in any case, only electromechanical aids to a portion of the analysis. Only the impedance method can currently claim to be automated in the broad sense of the word. For sterility testing of ultra high-temperature (UHT) juices and shelflife testing of milk, it has been shown faster than traditional manual methods.[5–8] Generally, it has not been officially recognized as an acceptable substitute for plate counting.

In the United States, Stomachers are freely substituted for blenders. Spiral Platers are common where applications warrant. Electronic colony counters are having a hard time—none meet the accuracy and reproducibility criteria used in *Standard Methods for the Examination of Dairy Products*. Fluorescent antibody (FA) is more or less dormant and radioimmunoassay (RIA) is on the way out in favor of enzyme-labeled immunosorbent assay (ELISA) for a number of toxins. Impedance (see above and below) is under scrutiny for milk analysis. The Petri-Foss and Bactoscan instruments seem unused, though automated leucocyte counts are done under the Dairy Herd Improvement Association (DHIA) dairy program. The direct epifluorescence-filter technique (DEFT) has not received appreciable study. There have been many studies on biochemical test kits; they are fairly popular, though each kit has some problems.

In Japan, where the regulatory structure concerning food quality and official methods is fairly similar to the Association of Official Analytical Chemists (AOAC)/FDA system of the United States, the Stomacher and Spiral Plater are commonly used. Impedance and bioluminescence have only been used by a few labs in experimental trials. Biochemical test kits are only rarely used for food microbiology on account of their cost.

In Australia, automated techniques have not generally been adopted, mainly because of the high cost and the unavailability of technical support. The Stomacher has been included in the Australian standard methods for microbiological examination of food and is routinely used in both industry and regulatory labs. Biochemical test kits are reasonably well-accepted and used in many labs. The Petri-Foss, ATP, and impedance techniques have been evaluated by individual labs, but none has been adopted for routine use. A few Spiral Platers are in regular use. Interestingly, an automatic diluting and plating machine called the Auto Loop (made in New Zealand) is widely used in the dairy industry. Electronic colony counters are not popular.

Germany (at least in the dairy industry) is somewhat further ahead. About 10 million tests are made annually on bulk milk—approximately 4% of the milk is tested by electronic microcolony counting, 10% by the loop-plate method (manual or Petri-Foss), and about 62% by pyruvate determination using the Auto-Analyzer. Efforts are being made to introduce the Bactoscan method. Even though an automated counter is available for the DEFT, the technique is thought tedious compared with the Bactoscan.

A FEW MUSINGS

The engineering bottleneck in automated microbiology comes from the conservative and (in 1986—perplexing) desire to count microbes in order to determine food quality. This is perplexing because counts tell us little of what we need to know, i.e., how quickly food will spoil, how it will eventually affect the consumer—nothing about multiplication rates of particular species, inter-floral or flora-food components interactions, the ability of each new cell to produce or remove undesirable effects in the food, and the cumulative effect of these variables on the consumer.

At the same time, we could easily develop analytical methods that would produce good information, and these methods are fundamentally more suited to automation than traditional microbiological methods.

Sharpe[9] generalized these methods under the acronym SMART (sequential measurement until amplitude reaches threshold). They are metabolism-based. They depend on actually following the development of those microbial parameters considered important in a given product until enough information is acquired to make useful predictions. Though impedance is not strictly a parameter relevant to spoilage or hazard, the technique and its instrumentation come very close to those needed for carrying out SMART analyses (provided the data are used predictively, and not as plate count substitutes). So impedance illustrates how automated/computerized microbiological analyses could be with us if we were so minded.

This chapter dwells mainly on analysis, simply because the emphasis has mainly been there. However, manufacturing practice has much more effect than end product analysis on food quality. Regulatory microbiology may always be needed—perhaps it should be intensified. But, in food manufacture, the HACCP approach to quality assurance is the preferred approach, and it should be more strongly emphasized. Automated monitoring of CCP and automated control of plant operations need special attention in the future.

PROGNOSIS FOR AUTOMATION

Electronic colony counters have not become popular because they work satisfactorily only on a very limited range of subjects—the well-defined colonies, no spreaders, little food debris, clear agar, etc. While it only takes them a second to count, getting plates to the counter, inspecting them for suitability, and removing them, all detract from the overall time saving. Most microbiologists feel the net saving isn't worth it.

The Spiral Plater is very limited by its small sample volume. Beyond it, only the hydrophobic grid-membrane filter (HGMF) seems to have real potential for "automated" versions of traditional microbiological analysis. In principle it can handle all the analyses currently done by plating, most probably number (MPN), or presence/absence. A very recent manuscript by Sharpe et al.[10] describes a new approach to detecting colonies, which permits an electronic counter to perform creditably in all sorts of rotten situations without need for calibration. It only works for HGMF and should eventually stimulate this method.

It appears unlikely that any other "automated plate count" devices will come on the market. There are severe engineering and cost difficulties in making sufficiently versatile machines that slop about with reagents, dilutions, agars, etc., the way humans do so easily. Even if any good engineering solutions appeared, manufacturers would look very closely at the diminishing respectability of plate counting.

There seem to be no alternatives to plate counting, provided we remain committed to assessing food quality by CFU/g or g/CFU. Once we are prepared to get away from the numbers concept, the engineering and technological feasibilities of automated techniques open up considerably. A number of metabolism-based techniques (ATP, dye reduction, the SMART techniques) are, in principle, much more readily automated than are plate counts. And yet, for most of these approaches, the picture is fairly dismal. The problem is they do not provide results which are easily translated to CFU/g or g/CFU. The impedance method alone gains respectability; a combination of much money on research to provide a technical base and astute management focussing on a narrow range of satisfactory applications from which a positive image is developing.

There is another problem: No "alternative" methods—automated or not—have been remarkably less labor-intensive or more rapid than traditional methods. All of the techniques such as DEFT, ATP, Spiral Plater, impedance, HGMF, DNA hybridization, ELlSA, etc., require just as much front-end work (if not more) than a traditional count. Getting the microbes out of the specimen and into a suitable state for the technique to handle is always time-consuming, and if it still has to be done manually, much of the potential of a new technique is lost. The drawback is less marked for impedance.

A big attraction (or strength) of traditional microbiological methods is the way they can cope with almost any situation as it crops up. Provided a lab has a moderate selection of refrigerated plates and other essential reagents on hand, the microbiologist can turn almost instantly from a *Salmonella* analysis, to an *E. coli*, to a *S. aureus*, etc., simply by grabbing a slightly different set of plates or tubes. This capability does not come easily, i.e., cheaply, to machines. Most microbiology labs are staffed by one-half to two microbiologists or technicians at most. They have to carry out a range of analyses, and they have to get on with samples as they arrive in the lab. It is hard for such people to organize themselves into a factoryline approach to work—for example, by doing a long run of one type of analysis where OM studies or automation could be valuable.

SOLUTIONS

Analytical Microbiology

There may be no satisfying solutions. We are unlikely to see fully automated plate counts. Automated metabolism-based techniques could be easily developed, but since they would not provide results which are easily translated to CFU/g or g/CFU they would not be acceptable. Yet, bearing in mind our problems of relating CFU to quality (shelflife, food poisoning incidents, etc.) and the probability that metabolism-based measurements would at least be as good, while also being more easily automated, we should perhaps try to ease the way for acceptance of metabolism-based methods.

In the United States (though perhaps not so much in Canada), industrial microbiologists are reluctant to embrace novel techniques until they have regulatory approval (e.g., AOAC official action status) or until there is overwhelming (long-standing) evidence they produce equivalent or better data. Since most official methods lean toward the count, the acceptability of methods which do not yield counts is unfavorably biased, regardless of how much time and work they might save. A stronger effort to encourage official collaborative trials of nonplate count or automated methods might help.

Others would argue that automated methods must compete with traditional ones only on immediate merit; they should not need special assistance. For example, impedance people lament that (although there are published attractions to using impedance for milk quality and shelflife testing) they have given up trying to get it officially established because at least five regulatory bodies are involved in milk quality, and none of them will officially sanction to the method until the others have. Whether or not this is correct, it seems that if its attractions really were so apparent, there would not be such a difficult path to official acceptance. Undoubtedly, a little earnest lobbying and coordinated effort by high-profile microbiologists could improve the stature and acceptance of any alternative technique. However, it would be wrong to give any method special privilege simply because we feel the technology level in food microbiology should rival that of toxicological chemistry.

Control of Operations

Although not much seems to exist here (save perhaps some computerized retorting and other thermal history systems), this problem will likely solve itself in the next few years. Plant operations, generally, are coming increasingly under computer control. The impression is that microbiology does not yet figure prominently as a design consideration for computerized factories. This should change as the HACCP concept becomes more firmly established as *the* approach to industrial microbiology. The recent National Academy of Sciences publication[11] and, in particular, the forthcoming ICMSF publication which will deal with the subject at great length, should do much to encourage both HACCP-mindedness and development of automated monitoring systems. There is a large potential market, and instrument or systems manufacturers will become increasingly aware of it.

So-called "expert systems"—the forerunners of truly intelligent systems—are proliferating in all areas of life and will ultimately revolutionize the developed world. I know of no system so far that is dedicated to food manufacturing plant design, operation, maintenance, storage, or transport problems, but these cannot be far away. Perhaps one of the most significant steps food microbiologists could take during the next few years would be to ensure that their views and ways of thinking are soundly incorporated into such expert systems. In this way, good microbiology will be designed directly into manufacturing plants and processes, rather than being (as it is all too frequently at present) a secondary consideration to yield an immediate cost/profit with the microbiologist as a figure to turn to only when things go wrong.

REFERENCES

1. Hauschild, A.H.W. and Bryan, F.L. 1980. Estimate of Cases of Food- and Waterborne Illness in Canada and the United States. *J. Food Protect.* 43(6):435–440.
2. Todd, E.C.D. 1985. Economic loss from foodborne disease outbreaks associated with foodservice establishments. *J. Food Protect.* 48:169–180.
3. International Commission on Microbiological Specifications for Foods (ICMSF). 1974. Microorganisms in Foods 2. Sampling for Microbiological Analysis: Principles and Specific Applications. University of Toronto Press, Toronto, Canada.
4. Kilsby, D.C., Aspinall, L.J., and Baird-Parker, A.C. 1979. A system for setting numerical microbiological specifications for foods. *J. Appl. Bacteriol.* 46:591–599.
5. Bishop, J.R., White, C.H., and Firstenberg-Eden, R. 1984. Rapid impedimetric method for determining the potential shelf-life of pasteurized whole milk. *J. Food Protect.* 471–475.
6. Firstenberg-Eden, R. 1984. A collaborative study of the impedance method for examining raw milk samples. *J. Food Protect.* 47: 707–712.
7. Martins, S.B., Hodapp, S., Dufour, S.W., and Kraeger, S.J. 1982. Evaluation of a rapid impedimetric method for determining the keeping quality of milk. *J. Food Protect.* 45:1221–1226.
8. Zindulis, J. 1984. A medium for the impedimetric detection of yeast in foods. *Food Microbiol.* 1:159ff.
9. Sharpe, A.N. 1980. Food Microbiology: A Framework for the Future. Thomas, IL.
10. Sharpe, A.N., Diotte, M.P., Peterkin, P.I., and Dudas, I. 1986. Towards the truly automated colony counter. *Food Microbiol.* 3:247–270.
11. National Academy of Science. 1985. An Evaluation of the Role of Microbiological Criteria for Foods and Food ingredients. National Academy of Sciences National Research Council, National Academy Press, Washington, DC.

13. Foodborne Disease Surveillance

Charles A. Bartleson

Washington State Department of Social and Health Services

INTRODUCTION

Foodborne illness has always been with us; but reporting of incidents of foodborne illness only began in the United States during this century. In the early 1900s, summaries of outbreaks of gastrointestinal illness associated with milk were published. In 1938, summaries of outbreaks linked to other kinds of foods were added. The Centers for Disease Control (CDC) began publishing annual reports in 1961. The format and contents of these reports have evolved over the past 25 years. Annual summaries are furnished to interested state and local health agencies. To produce these annual summaries, CDC collects, collates, and analyzes data provided by the state, local, and territorial health agencies. Canada began publishing an annual summary of foodborne and waterborne illness in 1973.

Reporting, analyzing, and publishing information about foodborne disease clearly allows identification and investigation of trends. For example, the information provides answers to the following questions: Which organisms are causing foodborne illness? What are the trends in illness frequency? Which foods are involved? Where are these reports of outbreaks of foodborne illness occurring? Is the reporting uniform between states? Which factors are responsible for or associated with outbreaks of foodborne illness?

When food protection professionals understand some of this basic information about foodborne illness, then still more difficult questions can be addressed. For instance, what are the economic costs of foodborne illness? How can we develop programs which will prevent these illnesses? How can we most effectively justify our programs politically so they can not only survive but also grow to meet future needs? How can we encourage enough flexibility to meet the demands of newly recognized pathogens, new food products and procedures, and the threats of tampering and terrorism?

When discussing questions and issues regarding foodborne illness reporting, the following definitions apply. The definition of an outbreak of foodborne disease in the United States is an incident in which

(1) two or more persons experience a similar illness after ingestion of a common food, and (2) epidemiologic analysis implicates the food as the source of the illness. A few exceptions exist; for example, one case of botulism or chemical poisoning constitutes an outbreak. In Canada the definition for an outbreak of foodborne disease is almost exactly the same as in the United States but additional information on single cases of foodborne illness are collected and analyzed as well.

IMPACTS OF FOODBORNE AGENTS

Foodborne illness-causing agents affect health and finances, either directly as in medical care costs or indirectly through higher food prices at the grocery store. Health risks increase with today's apparent change in diet and eating practices. Recent trends suggest for instance, that currently an increasingly large percentage of meals are eaten in public food establishments as opposed to meals eaten at home. Additionally, increasing amounts of processed foods are consumed. As a result of these changes in diet, exposure to various newly recognized causes of foodborne disease such as carcinogenic, acutely toxic, or pathogenic agents has increased. These include: pesticide residues, toxicants, viral, bacterial, fungal, parasitic pathogens and toxic plants and animals, and food additives (see Chapters 1–15). The impact of these newly recognized agents, as well as some historically well-known but unresolved foodborne disease problems, are addressed below.

Pesticides

Pesticide residues have caused much concern in the food protection community during the last five years. The recognition of ethylene dibromide (EDB) as a carcinogen prompted the U.S. Environmental Protection Agency (EPA) to ban its use as a fumigant in grain and fruit in 1984. This resulted in massive national publicity when certain states began testing foods and finding EDB residues. A tremendous amount of money was spent analyzing food samples for EDB. Many food producers were adversely affected economically when their products were identified as contaminated, and as many as ten states required the removal of products from display in retail food stores. While it seems unlikely that any cases of cancer will be reported as foodborne from EDB exposures, the public was sensitized by the media to the issue of pesticide residues in foods.

During 1985, two outbreaks of acute illness were linked to systemic pesticide residues. In each outbreak, the systemic pesticide aldicarb (TEMIK), which can build up to acutely toxic levels if it is applied to certain crops too close to harvest, was implicated as the causative agent. In the first outbreak, during June of 1985, over 100 individual reports of illness associated with ingestion of English cucumbers occurred in British Columbia, Canada, and Washington state. Samples of leftover cucumbers showed aldicarb sulfoxide residues present at high enough levels to produce acute illnesses. All cucumbers were recalled and destroyed although only one of 20 growers was involved. Substantial losses resulted from lawsuits because of illness or loss of product by the growers not linked to the outbreak.

The second and much larger outbreak, in July 1985, involved aldicarb contamination of watermelons on the west coast. Watermelons grown in California were linked to over 1,000 reports of acute illness in five western states and Canada. All watermelon sales were banned two days before the Fourth of July holiday weekend and several million pounds of watermelons were destroyed. The watermelon growers in California had their businesses decimated by the loss of product and lack of markets for their remaining crop caused by negative publicity. Even watermelons grown in areas not associated with the contamination problem were shunned by consumers. Although the health costs to ill individuals were substantial, the financial losses were staggering. Additionally, cost in staff time to the regulatory agencies was great.

Environmental Contaminants

Toxic chemicals and other environmental contaminants make the news almost daily. While many of these chemicals adversely affect groundwater and air quality, many also threaten the food supply. Luckily, problems in the United States pale when compared with those of Japan where mercury and cadmium from chemical plants entered the food chain and struck the local populace with both acute symptoms and permanent neurologic damage.

However, several incidents in the United States have shown us that diligence is needed here as well if injury and illness from environmental contaminants in the food supply are to be prevented. In Michigan in 1973 and 1974, polybrominated biphenyls (PBBs) were mistakenly mixed with animal feed. As a result, thousands of farm residents were exposed to PBB in varying amounts. Other Michigan residents were generally exposed to lower levels in milk, eggs, meat, and other foods. The health effects of this exposure are unknown and still being studied, but the economic losses of farmers and the food industry, as well as increased government regulatory costs, were staggering.

In 1975, contamination of trout, salmon, and other freshwater fish in the Hudson River and Lake Ontario with polychlorinated biphenyls

(PCBs) resulted in a total ban on commercial fishing in those areas and health advisories being issued to sport fishermen to severely limit their dietary consumption of these fish. PCBs again made the news during 1979 when a leaking transformer at a packing house in Montana contaminated animal feed. As a result, nearly 500,000 eggs, 300,000 chickens, 6,000 swine, and 700,000 pounds of animal feed and meal were destroyed as contaminated. The pending lawsuits exceed $40,000,000. Even though no adverse health effects have been proven in this incident, the economic consequences of these food contamination situations are significant.

Bacteria

Bacteria continue to be the major reported cause of foodborne illness in the United States, Canada, and other countries that summarize these illnesses. *Salmonella spp.*, *Staphylococcus aureus*, and *Clostridium perfringens* are most often the major causative organisms in confirmed outbreaks. However, over half of the outbreaks of foodborne illness lack a known etiologic agent. Although during the last few years several newly recognized bacterial pathogens have been identified as causing outbreaks of foodborne illness, for example, *Listeria monocytogenes*, *Vibrio spp.*, *Campylobacter jejuni*, *Yersinia enterocolitica*, and *Escherichia coli 0157:H7*, it is clear that these newly recognized bacterial pathogens will not account for the majority of the "unknown etiology" foodborne outbreaks. In 1985, bacterial pathogens were responsible for both the largest reported outbreak of foodborne illness (over 16,000 cases of salmonellosis epidemiologically linked to an Illinois dairy) and the deadliest outbreak (63 deaths from listeriosis linked to Jalisco brand soft cheese in California) in this century.

Fungal Toxins

Contamination of our food supply with fungal toxins continues to be a problem. Aflatoxin in peanuts and corn used as animal feed sometimes causes illness in livestock and is identified when milk samples are analyzed. Often potentially mold-contaminated foods are tested, rejected, and destroyed before entering the food supply. While this is a protective measure to food safety, the economic impacts may be substantial. Public health authorities must, therefore, remain vigilant for mold contamination problems.

For example, in Washington state in 1985, a trucker dumped a load of rye, which had been rejected for milling because of fungal contamination, in a parking lot behind a supermarket. Soon local people discovered the pile of rye and collected it for animal feed. In addition, several hundred pounds were donated to a local food bank for grinding into bread flour. Luckily, the food bank test-baked the flour before distributing it. Four food bank bread testers became violently ill with symptoms of ergotism when they ate rye bread baked with the salvaged grain. The local health department investigated, discovered the fungal contamination, and notified the public by radio and newspapers to destroy the grain.

Parasites

Parasites also continue to cause occasional reported outbreaks of foodborne illness. Occasional cases of trichinosis and toxoplasmosis, for example, occur every year. Moreover, amoebic dysentery remains a problem in certain areas of the United States and among certain ethnic groups. The first report of *Giardia* causing a foodborne outbreak occurred recently in Minnesota. Cryptosporidiosis has also been recently suggested as a new pathogen to look for in unexplained outbreaks of illness. Finally, changes in food preference for certain raw fish "delicacies" have resulted in case reports of nematode-associated illness and an increase in certain tapeworm infections.

Viral Agents

Foodborne outbreaks caused by viral agents are being reported more frequently. Since the laboratory methodology for confirming illness is well understood and widely available, hepatitis A remains the most frequently reported cause of viral foodborne outbreaks. However, many outbreaks are classified each year as "unknown" or "suspect." A sizable percentage of these outbreaks appears to be caused by Norwalk agent or related viruses. In addition to Norwalk virus, other probable causes of foodborne viral illnesses are nonpolio enteroviruses (NPEV), ECHO, and Coxsackievirus groups. Recent experiences in New York and New Jersey with thousands of persons ill from virally contaminated hardshell clams pose control problems for the regulatory agencies involved. In New York, a ban on the practice of eating raw clams has been proposed to halt these illnesses. Increases in population and nonpoint pollution sources near many estuarine environments where these shellfish grow has raised concerns about the continued approval of these beds. Questions about the ability to predict whether molluscan shellfish are safe to eat continue to arise.

Poisonous Plants and Animals

Poisonous, naturally occurring plants and animals remain a continued problem to health authorities. Because these outbreaks occur sporadically and relatively infrequently, they are easy to overlook. However, since many of these situations result in hospitalization and even death, the consequences of these outbreaks can be severe. Trends that encourage eating of native plants, gathering wild mushrooms, and use of herbal remedies all contribute to this problem. Several deaths have resulted from these practices. People have died, for instance, in the Pacific Northwest from water hemlock poisoning while on a wilderness adventure raft trip, from foxglove poisoning when the plant was mistaken for comfrey and brewed as tea, and from mushroom poisoning. In addition, paralytic shellfish poisoning problems continue to occur sporadically on both east and west coasts. Ciguatera poisonings also occur sporadically in tropical waters. Recently in New York City, a Japanese restaurant tried to import the tiger puffer fish, fugu, a natural source of tetratoxin. Tetratoxin is responsible for dozens of deaths each year in Japan among fugu eaters. Luckily, the FDA did not allow the fugu to be imported.

Food Additives

Food additives continue to cause concerns as increasing amounts of the food supply receive processing, and humans are subjected to an ever increasing array of food additives. Allergies to these additives attack many people. The recently identified problem with sulfiting agents threatening up to 1,000,000 Americans with potentially life-threatening reactions is one example. Though these allergies are quite common, since the symptoms initiated by sulfites may vary and may manifest themselves as breathing difficulty syndrome, which is exactly like an asthma attack, they may be very difficult to diagnose. The addition of too much niacin or monosodium glutamate (MSG) to foods has also been responsible for outbreaks of foodborne illness. Additionally, Yellow Dye No. 5 continues to require labeling since many thousands of people are allergic to it. The questionable safety of other food colors has resulted in the removal of many from the G.R.A.S. list approved by the FDA. Likewise, the safety of artificial sweeteners like cyclamates, saccharin, and aspartame are debated between persons who believe they are individually sensitive to these sweeteners and the food processing industry, and the FDA who believes they are safe to consume.

WHAT IMPROVEMENTS HAVE OCCURRED SINCE THE 1971 CONFERENCE FOR FOOD PROTECTION?

The First Conference for Food Protection recommended that the quality and quantity of foodborne disease reporting be improved and that these data be assembled and distributed to all interested parties. Specifically, the 1971 conference suggested the following steps to implement better surveillance:

1. Each state health department should appoint a "Food Poisoning Surveillance Team" to be headed by the state epidemiologist.
2. Regular lines of communication between the "Food Poisoning Surveillance Team" and other agencies need to be established with each part of the surveillance system familiar with its own duties. Each party needs to keep the others informed and to understand who is responsible for reporting incidents and outbreaks of foodborne disease to CDC.
3. Specialized training in foodborne illness reporting and investigation need to be provided to local health department personnel by the state health department.
4. The public health laboratory needs to be accessible by phone to local and state health department personnel for consultation and advice.
5. Local health departments and the food poisoning surveillance team need to contact private physicians and hospitals to encourage reporting of outbreaks of foodborne illness.
6. To provide uniformity in investigation and reporting, protocols need to be established.
7. State health departments should adopt regulations requiring reporting of possible foodborne illness outbreaks and positive laboratory isolates to state and local health departments.
8. Where staff and laboratory resources are lacking, the federal government should provide funding and CDC should provide staff to implement adequate investigation and reporting of outbreaks.

While many of these recommendations have been addressed and slow progress has occurred, there is still much to do. The number of reports of foodborne illness has doubled in the past 15 years, but, all too frequently, the outbreak summaries shed little light on what really happened. A variety of confounding factors may cause these poor reports, such as: untrained or poorly qualified staff, unavailable food samples, reporting delays, implicated food collecting and/or sample shipping errors, failure to test implicated foods for all probable pathogens involved, poor report forms which fail to facilitate adequate data gathering, lack of followup in filing and preparing reports to notify state health agencies and CDC of outbreaks, and lack of cooperation between various governmental agencies.

Although some state and territorial health agencies have admirably collected and reported foodborne illness, other jurisdictions have exhibited an abysmal, noncooperative, lack of concern for foodborne disease investigation and reporting. During 1980, 1981, and 1982 (the most recent data available from CDC) reporting rates for foodborne illness have varied from a low of one outbreak reported per ten million population per year to a high of over 21 outbreaks per million population per year. This means that some jurisdictions are reporting foodborne illness at over 200 times the rate of others (Table 1). Unfortunately, many more states are reporting poorly than are reporting well. For instance, 14 states report less than one outbreak of foodborne illness per million population per year. They are: Indiana 0.1, Missouri 0.1, Mississippi 0.2, Idaho 0.3, West Virginia 0.3, Nevada 0.4, South Dakota 0.4, Arkansas 0.5, Iowa 0.5, North Carolina 0.5, Louisiana 0.6, Michigan 0.7, Kentucky 0.8, and Oklahoma 0.8. An additional eighteen states report between one and two outbreaks per million population per year, whereas only six states report more than five outbreaks per million population per year.

In addition to the improvement in foodborne illness reporting that has occurred in some states, the quality of some reports has also slowly improved over the last ten years. These reports have more complete data. As laboratories and investigators began testing for more pathogens, a few of the previously "unknown etiology" reports began to confirm pathogenic causes. Errors in food preparation responsible for outbreaks have also been reported more completely.

Since the 1971 conference, reports of specific information about these foodborne illness outbreaks have been well distributed most quickly by Morbidity and Mortality Weekly Reports published by CDC, and also in a series of articles published in different journals which summarize and analyze data collected by CDC and collected in other countries. Both the United States and Canada now publish and distribute yearly summaries of foodborne illness reports.

1984 CONFERENCE RECOMMENDATIONS

The recommendations of the Second National Conference for Food Protection also identified needs for improvement in surveillance of foodborne illness. However, the more recent recommendations have not met with as much success as would be hoped for. For example, one conference recommendation suggests that during investigations of foodborne illness outbreaks, hazard analysis should be conducted rather than inspections for violations of the sanitary code. These hazard analy-

Table 1. Foodborne Disease Reporting for the U.S. by State, 1980–1982

State	Total Number of Outbreaks	Rate—Outbreaks per M./yr.
Alabama	13	1.1
Alaska	10	8.3
Arizona	18	2.2
Arkansas	4	0.5
California	118	1.6
Colorado	14	1.6
Connecticut	66	7.0
Delaware	3	1.6
Florida	43	1.4
Georgia	21	1.2
Hawaii	52	17.3
Idaho	1	0.3
Illinois	36	1.0
Indiana	3	0.1
Iowa	5	0.5
Kansas	17	2.3
Kentucky	9	0.8
Louisiana	8	0.6
Maine	8	2.4
Maryland	17	1.3
Massachusetts	34	1.9
Michigan	20	0.7
Minnesota	45	3.6
Mississippi	2	0.2
Missouri	2	0.1
Montana	6	2.5
Nebraska	10	2.0
Nevada	1	0.4
New Hampshire	10	3.7
New Jersey	30	1.3
New Mexico	31	7.9
New York City	451	21.1
New York State	264	8.3
North Carolina	10	0.5
North Dakota	5	2.3
Ohio	34	1.0
Oklahoma	8	0.8
Oregon	12	1.5
Pennsylvania	74	2.0
Rhode Island	11	4.0
South Carolina	11	1.1
South Dakota	1	0.4
Tennessee	16	1.1
Texas	43	1.0
Utah	6	1.3
Vermont	7	4.6
Virginia	24	1.5
Washington	139	11.3
West Virginia	2	0.3
Wisconsin	22	1.5
Wyoming	5	3.3

ses should evaluate the entire food flow and determine points of potential contamination; evaluate critical heat processes during cooking and reheating which would kill pathogens and denature heat labile toxins; and identify time-temperature situations which would allow growth of foodborne illness-causing bacteria if they were present in the food. Contrary to recommendations, only a handful of jurisdictions have instituted hazard analysis. Needed support for hazard analysis is lacking. The CDC form available for reporting foodborne illness outbreaks does not, for example, ask for food flows, or much of the other information which might be collected during such an investigation. However, more support may be forthcoming. The Food and Drug Administration (FDA) Current Concepts in Food Protection Course which is offered as training to many state and local health agency personnel, first covered hazard analysis this year. In addition, many state and local conferences have presented HACCP, and it will be covered during task force meetings at the Third National Conference. The IAMFES is including HACCP in its most recent additions of Procedures to Investigate Foodborne Illness. This may foster future gathering of this information.

Another recommendation suggested that surveillance data should be more effectively used in providing administrative guidance. The data compiled should annually be reviewed by the food industry as well as each state and local health department so that health agencies can determine which disease detection and control measures have been effective and which areas need improvement. Regulatory inspections need to focus on factors identified by surveillance data as those frequently contributing to foodborne illness outbreaks. Again, most health agencies do not use surveillance data in the planning process. Often existing programs are continued whether or not they are effective in reducing or preventing foodborne illness, and lack of funding kills new approaches. For example, many times health agencies sample soft serve ice cream, do swab-rinse tests on silverware or spend time meticulously evaluating plans for chain restaurants while they complain bitterly that they have no staff to investigate foodborne illness complaints or to try the HACCP approach. In addition, data compiled on foodborne illness is seldom shared with the food industry and often not even with local health department staff. Too often inspection forms track how equipment is designed, how clean the facilities are, and if routine maintenance of the floors, walls, and ceilings occurs. At the same time these forms largely ignore time-temperature factors most often identified as linked to foodborne illness outbreaks. The current FDA 44-item inspection form, which is in use by the vast majority of the nation's food protection programs, demonstrates these deficiencies and is, therefore, woefully inadequate in addressing time-temperature concerns. In contrast, the forms in use in Washington and New York states clearly focus more attention on the most important areas in preventing foodborne illness. (Figure 1).

RECOMMENDATIONS

1. Enhance investigation of outbreaks of foodborne illness by improving the education and knowledge of the medical and public health commmunities. This can be accomplished by cooperative efforts by CDC, state health departments, and local health agencies to provide training to the medical personnel in hospitals, clinics, and laboratories, as well as private physicians, about the need to properly report and investigate possible cases of foodborne illness. Obviously, for this process to function effectively, the public health agencies responsible for providing this training must be both knowledgeable and motivate to do the job. If the public health agencies need training, either CDC or state health department staff must be available to provide the training.
2. Improve communication between public health agencies and from public health agencies to the medical community and the public. Communication between public health agencies can be improved by: (1) developing computer networks for rapid sharing of current data and information, (2) regular telephone and written communications from CDC to states about new information, (3) an aggressive approach in providing accurate and timely public health information to the news media for dissemination to the public, (4) simple, complete report forms provided to state and local health agencies and hospitals and reporting physicians and laboratories so that needed information gets to the investigative and reporting agencies. Regular followup and feedback needs to be provided to those providing reports and effort must be spent to convince those not reporting to begin, and (5) newsletters can be effective in providing current information to all interested persons and agencies.
3. Outbreak investigation should first focus on known pathogens and causes of outbreaks. The information which has been learned about foodborne illness needs to be applied to all investigations. The factors associated with outbreaks of foodborne illness are well-known. Yet, all too frequently, this information is not used when an illness is investigated. Forms used when these investigations are carried out must stress these factors. HACCP and hazard analyses should be in use at the local and state health department level during selected outbreak investigations. The FDA and CDC should certainly enhance the availability and dissemination of revised forms for use when foodborne illness is suspected and is investigated.
4. Local health agencies need to focus their food protection programs

specifically on foodborne illness prevention and investigation. They can do this by: (1) developing better inspection forms which focus on the "critical items" responsible for causing foodborne illness outbreaks, (2) identifying "high-hazard establishments" and using the HACCP approach on a selected number of these each fiscal period, (3) applying information discovered during HACCP evaluations to other similar establishments to lessen foodborne illness potential there, and (4) working nonstandard work schedules when necessary to ensure that all hazardous operations are evaluated. Since many temporary food service establishments, special menu items such as prime rib, or types of service like Sunday brunches only occur on weekends or at night, staff must be available to work during these times, and (5) trained personnel must be available to investigate possible foodborne illness complaints when they occur after normal working hours or on weekends. Having staff coverage at these times improves the chance of success in investigations since leftover foods may be destroyed or ill persons disperse rapidly following an illness outbreak and are then unavailable for questioning or obtaining laboratory specimens.

5. A program should be developed, possibly in conjunction with the state and territorial epidemiologists or CDC staff, to address the problem of states that continually fail to report foodborne outbreaks at reasonable levels. Methods to enhance reporting could include incentives of additional funding based on improvements in reporting. A target level of reporting of five outbreaks of foodborne illness per million population per year should be a goal for every state in the country.

Figure 1. Food Service Establishment Inspection Report (Front)

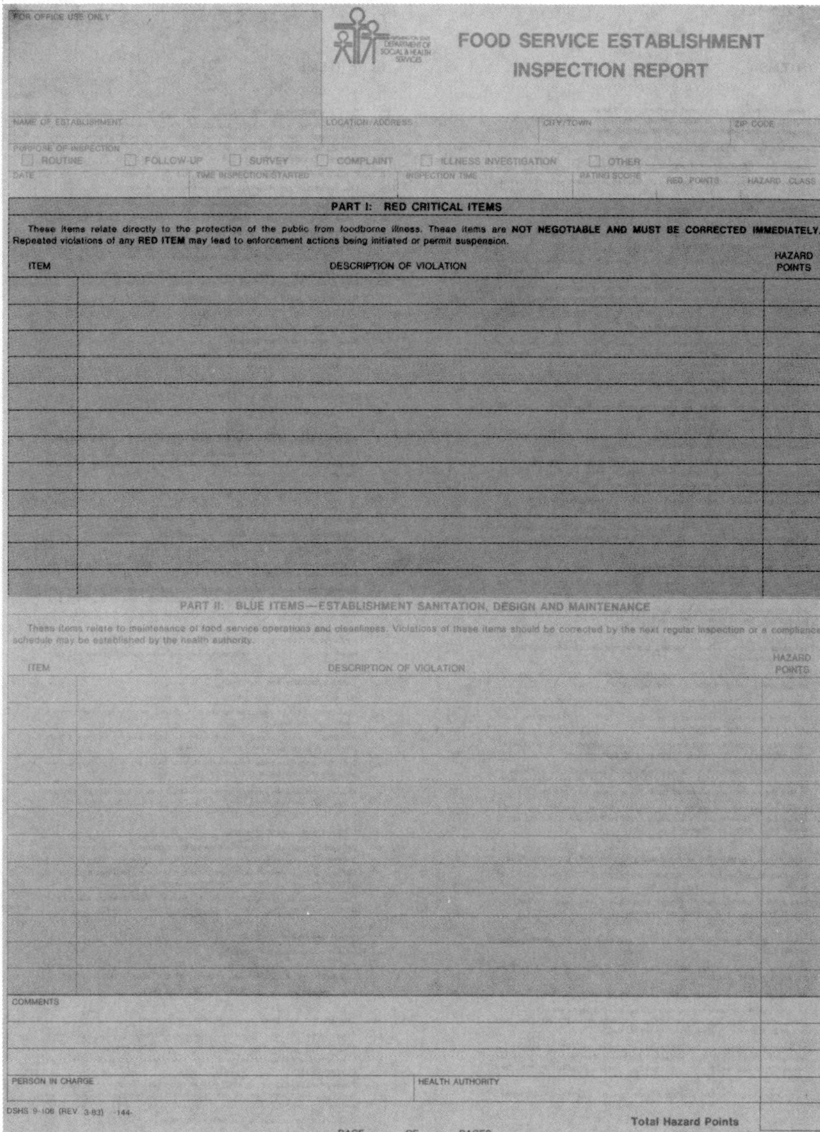

FOR OFFICE USE ONLY

WASHINGTON STATE DEPARTMENT OF SOCIAL & HEALTH SERVICES

FOOD SERVICE ESTABLISHMENT INSPECTION REPORT

NAME OF ESTABLISHMENT	LOCATION ADDRESS	CITY/TOWN	ZIP CODE

PURPOSE OF INSPECTION
☐ ROUTINE ☐ FOLLOW-UP ☐ SURVEY ☐ COMPLAINT ☐ ILLNESS INVESTIGATION ☐ OTHER ______

DATE	TIME INSPECTION STARTED	INSPECTION TIME	RATING SCORE	RED POINTS	HAZARD CLASS

PART I: RED CRITICAL ITEMS

These items relate directly to the protection of the public from foodborne illness. These items are **NOT NEGOTIABLE AND MUST BE CORRECTED IMMEDIATELY.** Repeated violations of any **RED ITEM** may lead to enforcement actions being initiated or permit suspension.

ITEM	DESCRIPTION OF VIOLATION	HAZARD POINTS

PART II: BLUE ITEMS—ESTABLISHMENT SANITATION, DESIGN AND MAINTENANCE

These items relate to maintenance of food service operations and cleanliness. Violations of these items should be corrected by the next regular inspection or a compliance schedule may be established by the health authority.

ITEM	DESCRIPTION OF VIOLATION	HAZARD POINTS

COMMENTS

PERSON IN CHARGE	HEALTH AUTHORITY

DSHS 9-106 (REV. 3-83) -144-

PAGE ______ OF ______ PAGES

Total Hazard Points

Food Service Establishment Inspection Report (Back)

RED ITEMS

		VALUES
1.	**Foods received from approved sources, not adulterated.**	
a.	Pasteurized milk and milk products used and served (except Grade A raw milk for off premises consumption and approved raw milk cheeses)	10
b.	Water and ice from safe and approved sources	10
c.	Shellfish from DSHS approved sources, properly labeled	5
d.	Canned goods obtained from approved commercial processor (no home canned items)	10
e.	Canned goods are in sound condition (no leakers, severe dents, rusty or swollen cans)	5
f.	Meat and meat products from USDA approved plants [no uninspected (marked "not for sale") meat on premises]	5
g.	Eggs clean and uncracked, if fresh. Liquid or frozen eggs and powdered eggs must be pasteurized	5
h.	All foods are wholesome and free from spoilage or adulteration [no rodent or insect filth, unapproved chemicals or mold (except cultured products)]	5
2.	**Foods protected from cross-contamination.**	
a.	Prepared food products do not contact equipment or work surfaces which have had prior contact with raw foods unless sanitizing of the food contact surfaces has occurred to prevent contamination	10
b.	Food workers do not handle raw and cooked or prepared food products without thorough handwashing in between	10
c.	Cooked or prepared foods are not subjected to contamination from juices dripping from raw foods	5
3.	**Foods protected from contamination by workers.**	
a.	Food workers do not handle foods if they are ill with a disease transmissible by foods such as: diarrhea ("stomach flu"), hepatitis, acute respiratory infections ("colds"), or have infected cuts or burns on their hands	15
b.	Food workers wash hands thoroughly (generate lather) after visiting the toilet, coughing, sneezing or otherwise contaminating their hands	15
c.	Food workers use proper utensils (tongs, spoons, spatulas, plastic gloves) to eliminate unnecessary hand contact with cooked and prepared foods	5
4.	**Foods protected from contamination by other sources.**	
a.	Toxic or poisonous chemicals are properly labeled, stored, used so that contamination of food will not occur	10

BLUE ITEMS

		VALUES
8.	**Food protection—general.**	
a.	Food protected during storage, preparation, display, transportation and service from potential sources of contamination, original containers, properly covered, labelling as required, no double stacking of pans, sneeze guards as needed. Storage of foods off the floor	5
b.	Perishable foods refrigerated as needed	3
c.	Unwrapped and potentially hazardous foods not reserved	5
d.	In use food including (ice) dispensing utensils properly stored	3
e.	Proper handling and storage of clean, sanitized equipment and utensils	3
f.	Single service items not reused, properly stored and dispensed	3
g.	Accurate thermometers available and used to evaluate hot holding, cooking, reheating and refrigerated storage temperatures	3
h.	Proper thawing procedures used, and frozen foods kept at required temperatures. Enough freezer space provided, freezers properly maintained	3
i.	Raw foods properly washed prior to serving	3
9.	**Hygiene and activities of food workers.**	
a.	Clean clothes, neat appearance	2
b.	Use of tobacco in food preparation, dishwashing and food storage areas is prohibited	2
c.	Hair properly restrained	2
d.	Dressing rooms provided and properly located	2
e.	Food and beverage workers permits current for all food workers	2
10.	**Sanitary design, construction and installation of equipment and utensils.**	
a.	Food (ice) contact surfaces are non-toxic; properly designed, constructed, installed and located [no cracks, open seams (no # 10 cans), pits, uncleanable or corroded food contact surfaces]	5
b.	Non-food contact surfaces are properly designed, constructed, and installed	3
11.	**Cleaning, washing and sanitizing of equipment and utensils.**	
a.	Tableware and utensils prewashed as needed, scraped and soaked	2
b.	Wash and rinse water clean and at proper temperatures	2
c.	Sanitizing rinse solutions at proper temperature, or concentration for proper time; hand dishwashing technique correct	5
d.	Accurate thermometers, chemical test kits and pressure gauges are present and functional to monitor dishwashing and sanitizing operations	2

b. Acid foods are not stored in containers or pipes that consist of toxic metals (antimony, cadmium, copper, zinc, lead) 10

5. **Temperature Control—Proper cooling and refrigerated storage of potentially hazardous foods.**

a. All solid or semisolid cooked, potentially hazardous foods are stored in refrigerators in containers with a food depth of 4" or less ....

b. Liquid potentially hazardous foods cooled by an approved method .... 25 (a–c)

c. All cooked, potentially hazardous foods cooled to 70°F. or less within two hours and to 45°F. or less within four hours ....

d. All potentially hazardous foods are stored under refrigeration except during necessary preparation procedures (no room temperature storage) 25

e. All potentially hazardous foods such as salads prepared from potatoes or macaroni are prepared using prechilled ingredients and are prechilled to 45°F. or less before they are stored on buffet lines; proper iced storage 5

f. Enough refrigerated storage equipment present, properly designed, maintained, operated so that all potentially hazardous foods are cooled properly and can be stored below 45°F. as required 10

6. **Temperature Control—Proper hot holding of potentially hazardous foods.**

a. Potentially hazardous foods stored hot are kept at 140°F. or above during hot holding 15

b. Enough hot holding equipment is present, properly designed, maintained and operated to keep hot foods at required temperatures 10

7. **Temperature Control—Adequate cooking and reheating of potentially hazardous foods.**

a. All poultry, poultry stuffings, stuffed meats and stuffings containing meat are heated to 165°F. or above 10

b. All pork or any food containing pork is heated to 150°F. or above 10

c. All other potentially hazardous foods requiring cooking are heated to 140°F. or above before they are placed in hot holding equipment except that rare roast beef or rare beef may be served at 130°F. or above 10

d. All potentially hazardous foods that have been cooked and then refrigerated, are rapidly reheated to 165°F. or above; steam tables, bain-maries, warmers and similar hot holding facilities are not used for reheating of potentially hazardous foods 15

e. Frozen foods larger than three pounds are not cooked unless they are first thawed 5

DSHS 9-106 BACK (REV. 3-83) QX A-68

e. Wiping cloths are clean, restricted in use and stored in approved sanitizing solution 5

f. Food contact surfaces of equipment and utensils are maintained, clean, free from abrasives and detergents, sanitized when needed 5

g. Non-food contact surfaces or equipment and utensils are maintained and clean 3

12. **Sanitary facilities and controls.**

a. Hot and cold running water, under pressure, provided 5

b. Sewage disposal by municipal system or approved, properly functioning on-site waste disposal system 5

c. Plumbing and sinks properly sized, installed and maintained, equipment and floors properly drained, no cross-connections present 5

d. Toilet facilities—adequate, conveniently located, clean, properly functioning and in good repair, toilet paper present 3

e. Handwashing facilities—lavatories accessible, properly located, clean and in good repair, soap and approved towels or hand drying devices are present 3

13. **Garbage and rubbish disposal.**

a. Adequate, leakproof, non-absorbent, vermin-proof, covered containers provided as needed 3

b. Garbage storage areas properly constructed, incinerators, food waste grinders or compactors clean and do not create a nuisance 2

c. Pick-up frequency, cleaning schedule of dumpsters and containers adequate to prevent a nuisance 3

14. **Insect and rodent control.**

a. Insects and rodents are controlled 5

b. Effective measures used to control entrance (rodent and insect proof construction) and harborage of rodents, flies, cockroaches and other vermin 3

c. Pesticide use as per label directions and state law 5

15. **Construction and maintenance of physical facilities.**

a. Floors, walls, ceilings properly constructed and installed, in good repair, clean surfaces 2

b. Lighting and ventilation provided as required, ventilation hoods, ductwork, filters cleaned properly 3

c. Premises maintained free of litter, unnecessary equipment or articles, living and sleeping quarters completely separated from food service operations, live animals and birds excluded 3

d. Proper storage of cleaning equipment and linens 2

14. Concerns about Foodborne Viral Gastroenteritis and *Listeria Monocytogenes*

Dean O. Cliver
Michael P. Doyle
University of Wisconsin-Madison

FOODBORNE VIRAL GASTROENTERITIS

Statement of the Problem

For many years, approximately half of the outbreaks of foodborne disease reported in the United States has been called "gastroenteritis of undetermined etiology." Where treatment of such illnesses with antibiotics was unsuccessful, or when known bacterial pathogens could not be cultivated from the stools of ill persons, the outbreaks were often classified as "acute infectious nonbacterial gastroenteritis" (AING). Such deductive diagnoses have, properly, been greeted with considerable skepticism. However, there are now three "kinds" of viruses that have been associated specifically with gastroenteritis and, to varying degrees, with foodborne outbreaks.

Norwalk-like viruses are called "small, round, structured viruses" (SRSV) in the United Kingdom. The name "Norwalk" derives from a town in Ohio that was the site of a gastroenteritis outbreak attributed to the virus. The causative agents, of which there are now known to be more than one serotype, are described as being 27 to 35 nm in diameter and, on the basis of characteristic depressions seen on the surface of the viral particle (the "structure" that affords the second "S" in "SRSV"), are thought to resemble the caliciviruses which contain RNA. The total number of serotypes in the group is unknown. One complication is that specific (antibody-mediated) immunity appears to be transient, so that those who have been infected by the Norwalk virus itself may become susceptible to reinfection after a year or two. Illness characterized by nausea, vomiting, diarrhea, and a few other protean symptoms occurs 24 to 48 hours after exposure (range of 10 to 51 hours) and typically lasts less than two days. No natural host, other than humans, is known for these viruses; they evidently do not replicate in any known cell culture, either. The Norwalk-like viruses are frequent causes of outbreaks of shellfish-associated gastroenteritis reported from Australia, the United Kingdom, and the United States. With the exception of some minimally cooked cockles in the United Kingdom and some lightly steamed clams in the United States, the shellfish seem generally to have been eaten raw, which is to say that the heat resistance of the viruses is not at issue. Elsewhere in the United States, food handlers have been sources of contamination, principally of salads. An exception is an outbreak of gastroenteritis affecting 3,000 persons in Minnesota, in which a food handler apparently contaminated butter cream frosting that was served on various baked goods. Water-associated outbreaks are also known. The apparent distribution of these outbreaks is constrained by the availability of means to diagnose the infections on the basis of immune electron microscopy, radioimmunoassay, or enzyme immunoassay, which are rare.

"Small, round, featureless viruses" (SRFV) are reported in the United Kingdom as causes of shellfish-associated gastroenteritis. These particles are perhaps 25 nm in diameter and have no perceptible surface features. They are thought to resemble parvoviruses, which contain single-stranded DNA. They evidently have not been cultivated in cell cultures and have not been reported to cause foodborne or waterborne disease elsewhere.

Rotavirus particles are about 70 nm in diameter and contain double-stranded RNA covered by a double-layered protein coat. They are a frequent cause of juvenile diarrhea and are also known to affect the elderly and to cause occasional community outbreaks of water-associated gastroenteritis. Rotaviruses that affect humans are cultivated with difficulty in cell culture. No food-associated outbreak of rotavirus illness has yet been recorded.

Evaluation of the Problem

Knowledge of the true incidence of food-associated viral gastroenteritis is limited by inability to cultivate the causative agents in laboratory cell cultures. The agents cannot generally be detected in samples of food or water, and diagnosis of the illness depends on detection of the virus in patients' stools by immune electron microscopy, radioimmunoassay, and, lately, by enzyme immunoassay. Such methods are costly and exacting. No government (state or federal) laboratory in the United States apparently has the necessary capability. Research to improve detection and diagnostic methods for these viruses is badly needed, as is the establishment of government diagnostic laboratories that can apply the methods that are now available. The incentive to investigate outbreaks is severely limited at this time by the lack of laboratory support.

Solutions to the Problem

Prevention of food contamination by gastroenteritis viruses might begin with a focus on shellfish. In the short run. this might best be done by paying closer attention to wastewater discharges by coastal communities and to other sources of fecal contamination of water. The depuration process, by which shellfish are induced to rid themselves of viral contaminants, has proven difficult to monitor for virologic effectiveness: human volunteers were used in Australia to test the shellfish, and some of them were made ill by samples that one would have thought should be free of virus. To prevent other foods from being contaminated by infected food handlers, there is need for better training and supervision of the workers and for provision of paid sick leave so that workers are not induced to come to work and handle food while they are ill.

LISTERIA MONOCYTOGENES

Statement of the Problem

Although not a common cause of human infection, *Listeria monocytogenes* is responsible for many deaths in certain high risk populations. The primary manifestations of listeric infections are meningitis, abortion, and perinatal septicemia, with immunocompromised individuals, pregnant women, and infants primarily at risk. Typically, the mortality rate for patients with listeriosis is about 30%, which coincides with the death rate among individuals involved in three recent food-associated outbreaks. The foods identified as vehicles in transmitting the organism in these outbreaks were coleslaw, pasteurized milk, and Mexican-style cheese. *L. monocytogenes* has also been recently isolated from several lots of soft, surface-ripened cheeses. Although all of these foods are refrigerated products, cold storage does not impede the growth of *L. monocytogenes* because the organism is a psychrotroph.

Because the organism has been associated with dairy products, several surveys have been done to determine the incidence of *L. monocytogenes* in raw milk in different parts of the United States. Results revealed that generally about 5 to 10% of the samples contained the organism. Excretion of *L. monocytogenes* in cows' milk is often intermittent but may persist for periods exceeding a single lactation. In many instances, the cows do not have obvious mastitis, and the milk they excrete appears normal.

The organism is known to be more heat tolerant than vegetative cells of most bacterial pathogens, and some reports indicate that *L. monocytogenes* can survive the pasteurization treatment that milk receives. Some investigators suggest the organism may be more heat tolerant when it is present in phagocytes, which is a natural condition in *Listeria*-contaminated milk. However, based on thermal inactivation studies in milk containing either artificially inoculated *L. monocytogenes* or *Listeria* within mouse macrophages, others have concluded that the organism will not survive the time-temperature treatment required for pasteurized milk in the United States.

Although *L. monocytogenes* is infrequently associated with human infection, the organism is widely distributed in nature. Studies indicate that a fecal carriage rate of about 5% among healthy humans is typical. Cattle, poultry, swine, and sheep have also been identified as apparently healthy intestinal carriers. Recovery of *L. monocytogenes* from decaying moist vegetation prompted some investigators to suggest that the organism has a saprophytic existence wherein the plant-soil environment may serve as a reservoir for *Listeria*.

Evaluation of the Problem

The brain-damaging and often fatal consequences of listeric infections in high-risk groups are of serious concern. However, we need to know much more about *L. monocytogenes* before its significance as a foodborne pathogen can be fully understood. It is important to know if the virulence potential of the organism is strain dependent, and what is the infectious dose of virulent strains in high-risk populations. We need to know the source and fate of the organism in foods, particularly in ready-to-eat foods such as soft, surface-ripened cheeses. We need to know what time-temperature treatment is needed to pasteurize milk to assure destruction of the organism. We need to know how prevalent virulent strains of *L. monocytogenes* are in our food supply. Hence, substantially more research is required to obtain the information needed to fully evaluate the problem of *L. monocytogenes* in foods.

Solutions to the Problem

Substantial research is needed to elucidate the association of virulent types of *L. monocytogenes* with foods. Developing measures to control the foodborne transmission of *Listeria* will depend largely on results of these studies. Furthermore, there is an immediate need for an improved, rapid but sensitive method for detecting the organism in foods. Existing isolation procedures are too long (e.g., cold enrichment requires several weeks of incubation) or are not sufficiently sensitive in recovering injured *Listeria* from foods.

GENERAL RECOMMENDATIONS

1. Develop better (cheaper, more sensitive) methods for detecting the human gastroenteritis viruses in patients' stools and in foods.
2. Establish a government reference laboratory that can do tests for gastroenteritis viruses and will provide laboratory support for epidemiologic studies.
3. Reduce viral contamination of shellfish by applying more stringent standards for wastewater discharged by coastal communities and for other potential sources of fecal discharge to estuarine waters.
4. Reduce viral contamination of other foods by more effective training and supervision of food handlers and by providing sick leave for food handlers so they are not induced to handle food while ill.
5. Research is needed to identify means of controlling the foodborne transmission of *Listeria monocytogenes*. This includes development of more rapid and sensitive methods for detecting the organism in foods, identifying virulence factors to determine if pathogenic and avirulent strains exist, identifying reservoirs and foods that are common carriers of *L. monocytogenes*, and determining the fate of the organism in foods during processing and storage.

15. Microbiology Committee Conclusions and Recommendations

On Hazard Analysis Critical Control Point (HACCP) Methods

RECOMMENDATIONS

1. Implementation of the HACCP approach must be a top priority in the food industry and among food regulators. Industry and regulatory representatives must work together in its implementation. The Conference for Food Protection should endorse and promote the use of the HACCP concept in food protection programs.

2. Government, academia, and industry should work together to develop training and resource materials for government and industry to use in implementing HACCP programs. Regulators must receive training, developed and provided cooperatively by regulators and industry in food microbiology, epidemiology of foodborne disease, and HACCP. Food manager training should prepare managers to recognize hazards associated with their operation's critical control points and to implement appropriate monitoring procedures.

   Steps toward implementation should include the following:

   The Conference for Food Protection should promote the use of HACCP concepts through the use of allied professional organizations.

   The Conference should promote HACCP through correspondence, media contacts, and presentations at professional symposia and other national conferences.

   The Conference should write to the Commissioner of the U.S. Food and Drug Administration requesting the following actions: (1) that the FDA endorse the HACCP concepts and incorporate them into all of its food programs and (2) that the FDA be a clearinghouse for

HACCP, providing information to state and local governments and to industry.

The Conference should promote the incorporation of HACCP principles into existing training programs conducted by government, industry, and academia.

The Conference should promote HACCP certification courses for inspectors and food service managers.

The Conference shold consider the feasibility of developing a task force tht can serve as a resource to groups developing HACCP training and certification programs and eventually become an HACCP certification and accreditation body.

3. Additional research must be carried out by regulators, industry, and academia to provide greater flexibility in the criteria for cooling potentially hazardous foods; to develop equipment and procedures needed to meet these criteria; and, where applicable, to define the practical use of pH and water activity, either alone or in combination, as monitoring procedures for critical control points.

On Rapid and Automated Methods

The following set of recommendations are in reference to rapid and automated methods for the enumeration and identification of foodborne pathogens and spoilage and indicator organisms, and for monitoring critical control points.

RECOMMENDATIONS

1. Strong efforts should continue in the development of more sensitive and specific test methods based on current and new concepts and technologies. Basic research invariably precedes practical applications and must be properly funded.

2. Research efforts should be aimed at the development of procedures to detect virulent food pathogens or a virulence determinant, with acceptable levels of accuracy, within eight hours or less. Assays should be developed for all important microbial toxins and metabolites. They should be simple, rapid (1 hour or less), sensitive, and quantitative. An ideal kit would analyze for a variety of toxins simultaneously, using food homogenates or extracts.

3. Research should be directed to simplify or automate new technologies that are inexpensive and not labor intensive. These methods should not disrupt the work flow in the food microbiology laboratory; they should be part of the computer reporting system.

4. Research is needed to improve enrichment procedures for microorganisms. Filtration and other techniques need to be improved for concentration of organisms.

5. We need to take a closer look at total plate count and at the relevance of bacterial numbers in measuring food quality.

6. The BAM procedures, to which all new rapid methodology must be compared to receive AOAC approval, are not foolproof. The possibility that a new procedure might be more sensitive and produce positive results even when the BAM is negative must be considered. Alternative procedures for certifying the accuracy of new assays should be considered.

On the Applicability of Microbiological Criteria

The safety and wholesomeness of food is the responsibility of the food industry. It is the task of regulatory agencies to assure that the various segments of the industry fulfill their responsibility. The primary purpose of the use of microbiological criteria by regulatory agencies is to supplement other means available to them (see HACCP recommendations above) to assure that foods produced under their jurisdiction are safe and wholesome.

To develop effective and uniform microbiological criteria for foods and food ingredients in the United States, the Microbiology Committee recommends:

RECOMMENDATIONS

1. Establishment and implementation of a microbiological criterion for a food or food ingredient in the United States only when there is a need for it and when it can be shown to be effective and practical. The criterion must accomplish its objective, i.e., adequately measure the contaminants of concern, be technically attainable under commercial conditions by Good Manufacturing Practices, and be administratively feasible.

2. Endorsement of the National Research Council's plan for a national

program to identify foods for which microbiological criteria are needed, develop criteria for those foods, and appropriately implement such criteria in regulatory programs.

Specific recommendations embodied in the NRC plan include the following:

a. The Center for Food Safety and Applied Nutrition of the Food and Drug Administration, the Food Safety and Inspection Service of the Department of Agriculture, the National Marine Fisheries Service, and the U.S. Army Natick Research and Development Center, being the primary federal agencies having the responsibility of assuring that foods available to their constituencies are safe and of acceptable quality, jointly should establish an ad hoc commission on microbiological criteria for foods.

b. The objective of the commission should be to develop microbiological criteria for foods that will be responsive to need.

c. The Food and Drug Administration, having the broadest responsibility for food protection among the federal agencies, should take the initiative to arrange, with the cooperation of the other three agencies identified above, for appointment of members to the commission.

d. Membership of the commission should consist of appropriate personnel selected from each of the four above-mentioned federal agencies, the food industry, and academia. The number of members on the commission should be kept to a minimum in the interests of expediency in reaching decisions and economy of operations.

e. Initially, the commission should address two areas: (1) the selection of foods and food ingredients on the basis of need for microbiological criteria that can effectively supplement and be incorporated in food protection programs of federal, state, and municipal regulatory agencies; and (2) the development of criteria for each food for which it is determined that microbiological criteria would serve a useful purpose. (The selection process described above would include, but not be limited to, a review of foods for which criteria currently exist and an evaluation of the suitability of these criteria.)

f. The commission should appoint expert working groups to carry out the task of selecting foods and of developing criteria. The expert working groups should present their recommendations to the commission.

g. The commission should present its recommendations to the federal agency that has primary jurisdiction for the safety and quality of the food or food ingredient involved. The federal agency should take necessary steps to promulgate regulations embodying the recommended criteria and to promote uniform adoption of the criteria by state and municipal regulatory agencies, if it deems it appropriate to do so. Insofar as possible, the Department of Defense should utilize microbiological criteria recommended by the commission in developing military specifications for foods and food ingredients.

h. The development, review, and modification of microbiological criteria for foods is a continuing task. Therefore, as soon as the initial work of the commission has been set in motion, it should take the steps necessary to organize itself as a continuing body elected by vote or other suitable means of the participating groups indicated above, except that the representative of each federal agency should be appointed by an appropriate official of the respective agency. Finally, the commission should address the continuing need for research to understand better the effects of current and emerging processing and distribution practices that may affect the microbial safety and quality of foods, and to improve and develop methods for the detection and enumeration of microorganisms and their toxic agents that are important to the safety and quality of foods.

On Pathogens and Toxins

RECOMMENDATIONS

1. The processing of dairy foods, including the pasteurization of milk and the aging of cheese made with unpasteurized milk, needs to be evaluated for its efficacy in preventing the survival of pathogens,

e.g., *Listeria monocytogenes* and *Salmonella spp.* Implementing this recommendation can be approached in three areas:

a. Research is required to identify the time-temperature treatments to destroy *Listeria* in milk and other dairy products
b. Pasteurization plants need to be engineered to prevent post-pasteurization contamination
c. The regulation for the 60-day holding period for cheese made with unpasteurized milk should be removed or changed because of its ineffectiveness in eliminating pathogens and toxins

2. Research is needed to identify means of controlling foodborne transmission of emerging pathogens, e.g., *Listeria*, *E. Coli 0157:H7*, *Norwalk* agent, and certain parasites. This research includes development of sensitive and rapid methods for detecting these organisms in foods, identifying virulent strains, determining reservoirs, routes of transmission, foods that carry these organisms, and their fate during processing and storage. Implementation of this research can be helped by the establishment of interdisciplinary committees to exchange ideas and results of current studies, as has been done with seafood toxins and *Clostridium botulinum*.

3. Alternate preservation systems, processes, and practices (such as water and energy conservation) for newly formulated products should be evaluated for microbiological safety before the products are introduced into the market. Specific criteria need to be developed for each evaluation.

4. Present technology does not exist for industry to ensure that raw foods of animal origin, including meat, poultry, and seafood, are free of pathogenic bacteria, viruses, and parasites. Through training courses, professional publications, and the popular press, the public and food industry employees should be made aware of the potential hazards of animal-derived raw foods. Through professional publications and the popular press, consumers should be made aware of the risk of eating raw or insufficiently cooked meat, poultry, and seafood. Consumers should also be informed of the risk of cross-contamination during food preparation and the need for rapid cooling or hot holding of foods following cooking.

5. Innovative and effective techniques should be instituted for the education of primary and secondary students on the safe preparation, handling, and storage of foods.

6. Meaningful differences in nutritional value between pasteurized and unpasteurized milk have not been demonstrated and other purported benefits of raw milk have not been substantiated by scientific studies. Epidemiologic studies have shown that consumption of raw milk has been responsible for hundreds of cases of campylobacteriosis and salmonellosis within the last five years. Appropriate regulations should be promulgated to prohibit the distribution of milk and milk products for consumption before pasteurization. Because of the increased awareness of the need to prohibit the sale of such milk, the Conference recommends this concern be given priority action by state regulatory authorities without further delay.

7. The Conference calls for the development of better (cheaper, more sensitive) methods for detecting human gastroenteritis viruses in clinical specimens and foods. The establishment of a reference laboratory is recommended to carry out specific tests for these viruses to support epidemiologic studies and confirm outbreaks.

8. The Conference urges the reduction of spills and untreated wastewater discharged into rivers and the sea to prevent contamination of fish and shellfish by bacterial and viral pathogens and hazardous chemicals.

9. Seafood toxins, such as paralytic shellfish poison, diarrhitic shellfish poison, ciguatoxin and tetrodotoxin, require research into better methodology for rapid methods, identification of dinoflagellates that produce them, surveillance of marine products for these toxins, control methods for the prevention of illness by consumers, and development of specific antidotes.

10. Water used in food processing should be of adequate microbial quality (e.g., safe from *Giardia*).

On Surveillance

RECOMMENDATIONS

1. National foodborne disease data should be presented in a different fashion than in present annual summaries. Instead of the existing categories of "confirmed" and "unknown," reported outbreaks should be placed into "confirmed," "presumptive," and "suspect"

categories. This will require greater emphasis at the state and local levels in identifying vehicles, etiologies, and causes for reported outbreaks. In addition, information on outbreaks of foodborne illness available from the FDA, USDA, and other federal agencies should be included in the annual summary. The Centers for Disease Control should make a concerted effort to adequately analyze the data and present it in a more timely and more usable fashion. Implementation of this recommendation can require the following:

a. To assure continuity and improve quality control of the data, CDC should assign responsibility for foodborne disease surveillance to a staff epidemiologist. This epidemiologist should communicate with reporting agencies to clarify and complete data provided on the report forms.
b. To improve coordinate reporting among the various local, state, and federal health agencies, and to improve the utility of the information in the HACCP systems, the responsible agencies should agree on standard reporting forms, e.g., those in *Procedures to Investigate Foodborne Illness*, fourth edition, published by the International Association of Milk, Food and Environmental Sanitarians.
c. To improve the timeliness and ease of reporting, CDC should consider automating the reporting system using existing computer networks.

2. Enhance investigation of outbreaks of foodborne illness by improving the education and knowledge of the medical and public health communities. This can be accomplished by cooperative efforts by federal and state health departments and local health agencies to provide training to medical personnel in hospitals, clinics, and laboratories as well as private physicians about the need to properly report and investigate possible cases of foodborne illness. Obviously, for this process to function effectively, the public health agencies responsible for providing this training must be both knowledgeable and motivated to do the job. If the public health agencies need training, staff from either state health departments or the Public Health Service must be available to provide the training.

3. Improve communication between public health agencies and from public health agencies to the medical community and the public. Communication between public health agencies can be improved by:

a. Developing computer networks for rapid sharing of current data and information
b. Regular telephone and improved written communication from CDC to states about new information
c. An aggressive approach in providing accurate and timely public health information to the news media for dissemination to the public
d. Simple, complete report forms provided to state and local health agencies, hospitals, reporting physicians, and laboratories, so that needed information gets to the investigative and reporting agencies
e. Regular followup and feedback to originators of reports and also to agencies not providing reports to convince them to report
f. Newsletters to provide current information to all interested persons and agencies

4. Local health agencies need to focus their food protection programs specifically on foodborne illness investigations, evaluation of data gathered, and prevention. Investigations should start by using information learned from previous outbreaks, e.g., check known causes and pathogens. HACCP evaluations to identify factors that contribute to outbreaks other than violations of the local sanitary code should be stressed. Outbreak investigation reports should include recommendations for prevention and control. These data can then be included in surveillance reports which can be shared with industry and other regulatory agencies.

5. Local health agencies have to be flexible enough in their working hours to ensure that all food operations are evaluated, e.g., temporary food service establishments, Sunday brunches, fairs, and catered events, during evenings, weekends, and holidays. Having inspection personnel and, in an outbreak situation, laboratory staff available at those times improves the chance of success in investigations, since leftover foods may be destroyed, or those ill may be dispersed following an outbreak and be unavailable for questioning or providing laboratory specimens.

6. The Conference recognizes the importance of animal traceback during epidemiologic investigation of foodborne disease, and endorses the activities of other relevant organizations to encourage systems for tracing animals through the food chain.

On Measurement of Economic Loss

RECOMMENDATIONS

1. Cost-benefit studies should be conducted to assist government and industry in evaluating options to solve substantial and expensive food contamination problems (e.g., reducing contamination of poultry by *Salmonella*).

2. Collection of data depends on the cooperation of health agencies, industry, scientific organizations, industry associations, insurance companies, and legal firms. An organization should promote interest in this by setting up a study with representatives of these agencies to clearly define food losses.

GOOD MANUFACTURING PRACTICES AND QUALITY CONTROL

16. The Hazard Analysis Critical Control Point Concept

Howard E. Bauman

The Pillsbury Company

The HACCP concept was introduced at the National Food Protection Conference in 1971. Since then HACCP has become familiar to and used by many. Lately, however, the way in which the HACCP concept has been interpreted and utilized is somewhat disturbing. Many people have developed a very narrow approach to the HACCP system. This appears to be dependent to a large degree on the profession of the individual or the particular position of the person within the company.

Perhaps it is time to review the HACCP concept and system. To do this, a short history on how and why it was developed might be helpful, followed by a broad view of what HACCP is, how it should be used, and what it covers.

The HACCP system came about within the Pillsbury Company because of the production of food for use in the space program in the 1960s. It became obvious during this effort that the old procedures of quality control were not sufficient to produce food for the astronauts that could be assured to be safe. The amount of testing required by standard quality control techniques used at that time was so large that meeting the stringent criteria for space use resulted in destructive testing of most of each batch of product.

It was felt there had to be a better way to assure safety, and the Pillsbury Company began to evaluate current methods, processes, and production methods to determine how to establish a preventive system, rather than a reactive system which discovers problems after they have occurred.

The Pillsbury Company worked in conjunction with NASA and the Natick Laboratories using a modes of failure concept that had been developed at Natick. The various facets of food production, from crop growing to harvesting, processing, packaging, and distribution through final use by the consumer were examined to determine whether the weak links in the system could be identified, i.e., which failures were likely to occur in the food system being used and, if they could occur, which controls might be installed in order to prevent them.

The first thing discovered, that should have been obvious, was that it is essential to treat the production of the food as a total system. The system was broken down into its various components and each examined by a mode of failure analysis. At the end of this exercise, it was found that the components then had to be linked together because it was conceivable that something done or recommended in the early stages might have an effect in the later stages of production.

We started by laying out a typical food production system then in use. The first component was the growing and harvesting of the raw material. It was determined what the conditions were under which the raw material was grown and harvested and whether or not these conditions could result in problems. From this alone, it was discovered that a list of sensitive raw materials that would most likely be carriers of pathogens, chemical contaminants, toxins, and other potential safety issues could be developed. It was also discovered that it is important to know how these raw materials are processed. This allowed an evaluation of the treatments (i.e., pesticides) received in storage and the conditions under which it was transported to an area for further processing. It was, of course, not always possible to instigate tight controls in these areas. However, a number of tests were found that could be run on these materials in order to determine whether they had been abused or rendered unsafe.

The next area of control found to be important was the supplier. It was necessary to ensure that the suppliers producing the raw materials or the further processed materials were operating under controls sufficient to produce a food product that would meet the criteria that had been established. This meant supplier inspections and agreements with suppliers that, when they were producing a particular product or ingredient, they do so under certain agreed upon conditions. Wherever possible, suppliers were encouraged to install their own HACCP systems. It is believed that if the total food system operated under HACCP criteria, many food safety issues would disappear.

As facilities were evaluated, several important areas were identified. Essential is the collection of samples for testing, proper storage, and appropriate inventory control of the materials. Within the plant the system was divided into components of storage, testing, the environment (which must be controlled and monitored), and the equipment. By equipment is meant the physical systems and the controls that are essential, such as diversion valves, temperature recording devices, etc. Additionally, the people factor is critical, particularly if there is the opportunity for cross-contamination between raw and finished product.

In a food plant, the environment is checked principally from a bacteriological standpoint. However, if there is any possibility of vapors, dusting, or other hazardous situations that might occur, then appropriate controls are set up. The environment is viewed from the standpoint of potential hazard to the food product, as well as potential hazards to personnel. Within the equipment area, it is possible for screens to break, knife blades to shatter, metal to be scraped from surfaces, and other things to happen to equipment that can cause safety hazards in products. It is essential to identify those areas and ensure that they are monitored at appropriate times and that the observations and times of observation are recorded. This can be of utmost importance if, for instance, a wire screen breaks. If it happens to be nonmagnetic stainless steel, unless the screens are checked frequently, it is possible to produce product with wire contamination for days without realizing it.

Thermometers and temperature recorders must be calibrated and checked frequently, and the dates and times recorded. Flow diversion valves must be checked frequently to make sure they are working properly and the dates and times of these checks recorded. The Pillsbury Company requires that the recording of date and time take place at every station where a hazard might occur. The person doing the checking also initials the report.

A critical people issue is the possibility of transmission of disease through the food products. The appropriate controls and education must be undertaken to ensure that good hand wash stations are available and that the employees are utilizing the facilities so that they are not bringing potential hazards into a plant. Another people factor, as mentioned before, is the possible cross-contamination between raw and finished goods. It is important that this be controlled.

Chemicals for use in plants on occasion have resulted in major problems. To prevent this, it is important that the chemicals be kept in locked storage, separate from food or food ingredients. Withdrawals of chemicals should be recorded as well as signed for by the individuals who are getting the material. Only individuals trained in the use of the chemicals should be permitted to withdraw them. The chemicals that are used must be appropriate for the food plant and authorized for use. It is possible, and it has happened, that the wrong chemicals are used. At the Pillsbury Company, it has been found that color coding can be used to ensure that the appropriate containers are connected to the appropriate lines. There are many controls that can be established with a little thought and foresight, and it is not too difficult to determine what they should be.

Packaging material is another factor. We must be assured that the packaging material is suitable for the purpose for which it is used and results in appropriate protection of the product. We now have the additional possibility of tampering so it is essential that tamper-evident packaging be utilized.

An area that a number of companies do not pay enough attention to is the transportation and distribution system. It is possible, for instance, in shipping refrigerated and frozen foods to incorporate a temperature recorder in a case of product so that the temperature throughout the transportation and distribution system up to the point that the case is opened will be recorded. The Pillsbury Company has found that those that receive the shipments are very cooperative in returning this information. It gives good insight into those distribution organizations that are capable of maintaining and controlling the appropriate temperatures. The transportation and distribution system is one of the weakest links because it is at this point that many companies seem to lose control of their product. In fact, once it is out of the plant, many believe it is no longer their responsibility. This is, of course, far from the truth because any problems that occur will reflect back on the product and the manufacturer of the product.

Another weak link is the consumer. It is important to determine how a product might be abused by the consumer and, if so, how safe would the product be under the conditions of abuse. It is conceivable and is known that companies have not marketed certain types of products because it was determined that a combination of transportation, distribution, and consumer abuse could cause the product to become a hazard under conditions that one might expect will occur. In most instances, labeling alone will not ensure the safe handling of the product.

It is worth elaborating on the physical systems hazard control portion which is critical because many things can happen in plant that can negate a carefully laid out HACCP program. A good way to start is to make a flow diagram for every line in the plant with all of the equipment and interfaces identified. At Pillsbury, this is done by numbers—thus, the identification number on any flow chart will relate to a specific brand of equipment. The number is also key to the equipment specification document. A flowsheet is essential to show the interrelationship of the equipment and the piping between all of the components of the process line. Other physical points critical to product safety are metal detectors, sifters and screens, magnets, thermometer calibration, and any other mechanical devices that may affect product safety. It is also essential that these documents be dated, since any time a change is made in the process line or equipment, the total process must be reevaluated as to the possibility of its contributing to a hazard. By dating the documents, one can be assured of using the right one. It is possible to make what might conceivably be considered minor changes in a process line that can undo or at least jeopardize a portion of the safety features that have been built into the system.

It is important, if a HACCP system is installed, to frequently educate and train plant people in the use of the system. With employee turnover, in a few years there may be very few people left in a plant who understand what the HACCP system is or understand the need for the various controls that have been established. Training should be done on a semiannual basis and must be an ongoing living process.

A word of caution – in most instances, the specification for producing a product will contain points critical to safety and some critical to quality. It is important that these not be blended together so that plant people will be unable to determine which are for safety and which for quality. They should be differentiated, since one may compromise on quality but never on safety.

In review, these are five hazardous areas involved in product safety:

- Physical
- Biological
- Environmental
- Chemical
- Distribution, transportation, and consumer abuses

In summary: The key to success of the HACCP concept is to:

1. Have in-depth knowledge of the specific manufacturing process, so critical points necessary for producing a safe product can be identified
2. Establish the critical points *and* determine the best way to control and document these points on an ongoing basis
3. Keep the number of critical control points to a minimum so the system does not become burdened with a complex cumbersome network that can become outdated when small line changes are made
4. Review the HACCP program with key plant personnel on a semiannual basis to ensure that the critical control points for product safety are understood and followed.

17. Process/CIP Engineering for Product Safety

Dale A. Seiberling

Seiberling Associates, Inc.

INTRODUCTION

Recent incidents in Massachusetts, Illinois, and California have caused some concern about the capability of the nation's dairies to produce a high quality product guaranteed safe from a public health point of view. The general media reporting of these incidents has not helped to improve the consumer image of the dairy processor or of the regulatory agencies involved. In addition, reported commentary from some industry sources expresses concern about the problems being created by the nation's new, complex, computer-controlled dairies.

It should be noted that a relatively small percentage of the total milk processed in this nation is handled via computer-controlled systems. Audits of product flow/Clean in Place (CIP) piping in many plants, completed during this past year, provide adequate evidence that the older processing facilities require substantially more consideration than the "complex new plants." The design of automated process/CIP systems 15 to 20 years ago was generally the work of "system-oriented specialists" in the form of consultants, equipment manufacturer engineering groups, or dairy equipment jobbers with engineering capability. During the intervening period of time, the welded pipeline, air-operated valve, and automatic CIP system have come to be considered "old hat" by people at many different levels of responsibility in most companies. Maintenance and operating personnel have figured out the means of accomplishing all sorts of operations and activities not included in the original design. Some of today's innovative solutions to current problems may lead to bigger problems of a different nature in the future.

The design, installation, and operation of an automated processing system requires a proper plan followed by proper installation. Then, personnel must be trained and supervised to assure proper operation and maintenance of the system. Finally, modifications of the original design must be given as much attention by management and regulatory agency personnel as was the original installation.

PLAN

A CIP piping system, whether simple or complex, provides many opportunities for mistakes of omission or commission. Long-term problems can be best prevented by starting a project with a proper plan. The *3-A Accepted Practices for Permanently Installed Sanitary Product-Pipelines and Cleaning Systems*[1] states under *Layout and Engineering Requirements* that, "Prior to installation a drawing or equivalent plan shall be made available to the controlling Regulatory Agency by the processor for each installation, or subsequent addition or modification, showing each permanent circuit to be cleaned, noting thereon the size and length of piping, fittings, pitch, drain points, access points, relative elevations, location and specifications of circulating unit, and other pertinent facts." This same section includes additional commentary about the circulating unit, minimum average solution velocities, suction side conditions of the primary circulating pump, maintenance of solution temperature, and other criteria. The *single most important point in this standard* is the requirement for keeping proper drawings on file in the plant and with the regulatory agency. Only on this basis can company personnel and the agency have the means for maintaining control of permanently installed piping systems, whether simple or complex.

Serious problems may arise well after completion of the original new construction or renovation project. All work of this nature is, of necessity, performed to a specified scope, i.e., the plant is designed to make certain products, at certain capacities, and package those products using specified machines. Some months or years later, market conditions may change and suddenly personnel at the operating level are asked to do things not provided for in the original plan. Occasionally, the original dairy engineer, equipment supplier, or dairy equipment jobber will be called in to assist with the modifications. More commonly, plant maintenance personnel make the necessary changes per the direction of a production or plant manager. As mentioned above, recent audits of existing installations have disclosed substantial discrepancies between the "As Installed" drawings prepared several years earlier, and the existing piping system or control system installation. In many instances, these audits have shown that no drawings were available, or that drawings were far from current, and obviously the responsible regulatory agency and operating management lacked proper documentation.

COMPONENTS OF AUTOMATED SYSTEMS

The components of the automated dairy, all of which are developments of the past 25 years, include:

1. Permanent piping systems of nearly "all-welded" construction, which reduce the amount of manual cleaning of tubing and fittings to an absolute minimum
2. Silo-type storage tanks and dome-top processors designed to be effectively cleaned by CIP procedures
3. Air-operated, CIP-cleaned sanitary valves, which eliminate the manual cleaning of plug-type valves, and provide for remote and/or automatic control of product and CIP solution flow
4. Automatically controlled CIP systems, which have provided a means of assuring uniformly successful cleaning of tanks, valves, and piping systems on a day-to-day basis
5. Control system based on relay logic, "dedicated" solid-state controllers, small computers and programmable logic controllers wired or programmed to control complex operating and cleaning sequences
6. Processing equipment designed either for susceptibility to CIP cleaning (homogenizers, plate heat exchangers, and certain types of fillers) or which *requires* CIP cleaning, such as the self desludging centrifugal machine

These components function most effectively when properly integrated into a complete processing system—designed and installed for automated control of all production operations, cleaning, and sanitizing. However, the greater percentage of this equipment has been purchased as a replacement for older machines through implementation of expansion and upgrading programs which are based on acquisition of the major machinery first and followed by application of automation at some subsequent date.

CONTROL

The word "control" for the purposes of this chapter is very broad in that it relates to (1) flow control via air-operated valves, (2) control of a process through proper initial piping system design, (3) control of the continued integrity of the process through proper updating of documentation, and (4) control of how the system is used via provision of proper instructions for operating and cleaning.

Control of Product Flow

Piping systems used for processing dairy products exist in three different forms: (1) takedown lines, (2) permanent piping incorporating manually operated plug-valves, and (3) permanent piping incorporating air-operated valves. The expression "permanent" may refer to welded lines or lines assembled with CIP joints and is intended to mean all piping with exception of those connections required to separate product containing lines from solution containing lines or to facilitate disassembly of the equipment for maintenance.

The evaluation of a "takedown" piping system or the "takedown" portion of any piping system can be made only when the connections are intact. The determination of the function of a permanently installed plug-valved system is relatively easy since the configuration is visible. Therefore, a concerned individual can, given sufficient time, determine how the system is applied to meet the daily production requirements, and how (or if) it can satisfactorily be cleaned using CIP procedures.

In comparison, the incorporation of air-operated valves to control flow of product or cleaning solutions makes it necessary to consider many factors when attempting to determine the suitability of the system from either a quality control or regulatory compliance point of view. These factors include:

1. Valve body and actuator combination
2. Installation position with respect to drainage of valve and associated piping
3. Valve actuation (air hose, toggle switch, machine tool switch, or integrated process/CIP control system).
4. Interlock against improper actuation (applicable only to integrated process/CIP control systems)
5. Condition (position) valve assumes on failure of air supply or power

The air-operated sanitary valve currently in use evolved through modification of a hand-cleaned compression valve used for automatic level control in evaporators in the early 1950s. The body (gasket area between the body and bonnet), plunger, and stem O-ring components were all modified in accordance with the generally accepted criteria for CIP cleanability at that time and two specially fabricated valves were evaluated using a radioactive isotope tracer technique to determine cleanability. Approximately 20 of these valves were installed in an Ohio plant in 1957, and 17 more were applied in a Florida plant in 1958;[2] these were the initial commercial applications of CIP systems. It was determined during this early work that effective cleaning was assured if the valve was actuated three to four times during the prerinse and postrinse portions of a cleaning cycle and a minimum of six times during the solution wash. This actuated movement became required to effectively rinse, clean, and postrinse the stem O-ring area. Further, the cleaning application within the piping system must provide for flow through all ports during the CIP program, i.e., it is not permissible to merely wash through the body with the valve remaining deenergized and the plunger remaining fixed.

Though widely applied during the last 25 years, the air-operated valve has not been recognized by the adoption of a 3-A standard. A program has been initiated to correct this situation. The three most commonly used compression-type air-operated valves are shown in profile view at the top of Figure 1. Other variations are available including kettle valves, tangential tank outlet valves, and upper and lower bodies for any design with either single or double ports. Actuators of the air-to-raise (A/R) or air-to-lower (A/L) type may be combined with any of these bodies. Many actuators can be field converted from A/L to A/R operation or vice versa.

A less familiar mid-position operator is often used on the 3-port divert valve. This is available in the air-to-lower configuration only. The length of stroke of the mid-position piston is mechanically adjustable. This operator may also be used on a HTST regenerator bypass (tee-type shutoff) valve to provide restricted flow through the bypass so as to cause the raw regenerator to be filled with liquid and brought to temperature before the unit achieves forward flow, thus assuring continued forward flow.

Any of these valve bodies will be free-draining when mounted in a vertical position. If the stem is in a horizontal plane, one body port of all types must be pointed downward at not less than a 45-degree angle to make the body cavity drainable.

Air-operated valves were initially used in the "automated system" approach. In the early 1960s, processors began purchasing and installing these valves on an individual basis as a replacement for manually positioned (and cleaned) plug valves, and many applications have been made using air hoses, toggle switches, or machine tool switches to control the air supply to the actuator. Some computer-based control systems include a "Manual-Off-Auto" switch in the output line to each valve solenoid. It should be noted that a valve subject to control by operating personnel in this manner cannot be properly integrated in a total system to prevent improper handling during processing operations or to assure effective cleaning.

Even though the operation and installation requirements for the valves and actuators are understood, the visual evaluation of the piping

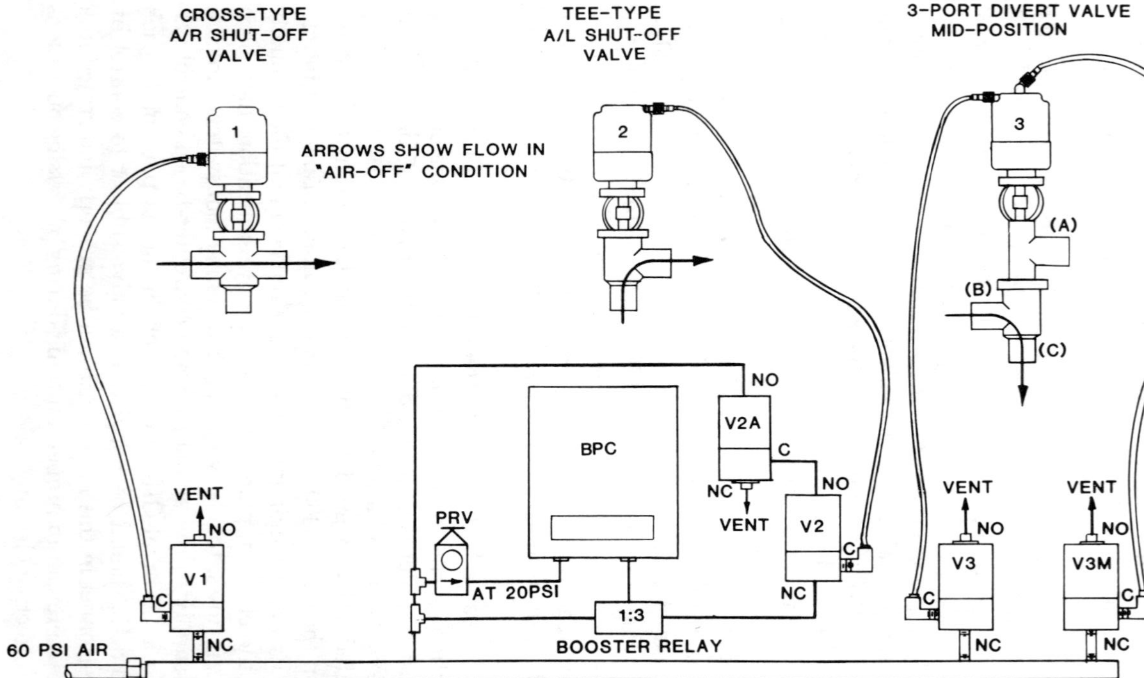

Figure 1. Typical air-operated valves and control solenoids. Valve 1 is a cross-type shutoff valve with an A/R (Air-to-Raise) actuator, commonly used in both valve manifold assemblies and fill headers. Valve 2 is a tee-type shutoff valve fitted with an A/L (Air-to-Lower) actuator for a throttling control application. Value 3 is a 3-port divert valve fitted with a mid-position actuator.

system alone does not assure understanding of the application. The lower portion of Figure 1 shows typical methods of controlling the air supply to both shutoff valves and divert valves. Three-port solenoid valves, operated by 110-volt or low-voltage power, are most commonly used for this purpose. These solenoids have common, normally closed, and normally open ports identified C, NC, and NO, respectively, on the air piping schematic beneath the valves. With reference to Figure 1, some typical valve applications would function as follows:

1. Valve 1 is a cross-type shutoff valve which may be used in valve manifold assemblies (also called valve groups or valve clusters) or in fill headers above a row of tanks or processors. Flow would be through the upper body only with the valve deenergized, and product would pass through the bottom port when the solenoid was energized, admitting air under the actuator piston, raising the plug from the seat, and permitting flow from the header or manifold through the valve to the connecting piping.
2. Valve 2 is a tee-type shutoff valve. If fitted with an A/R actuator, it could be used as a blocking valve in valve manifold assemblies or as a tank outlet valve. Figure 1 shows this valve fitted with an A/L actuator, and the air supply is controlled by two solenoids. With air available and no power on any solenoid, full-line air pressure is applied through V1A and V1 to keep the valve "closed." The NC port of the primary solenoid receives a signal from a back pressure control instrument using a 1:3 booster relay, which uses the 60-psi air supply pressure to increase output from 3 to 15 psi to 9 to 45 psi. The back pressure controller shown could be replaced by any other type of level, flow, or pressure control instrument, or by an I/P (current/pressure) transducer to interface the valve to a computer or programmable controller-based control system.

   With both valves deenergized, air is supplied by the NO/C ports of V2A (auxiliary solenoid) and the NO/C port of V2 (primary solenoid) to the actuator, closing the valve tightly as long as the air compressors are functioning. In the event of an air failure, the valve will open, but this would not normally be a problem as the same air failure would have caused all the other valves in the system to go to a "safe" condition.

   If Solenoid V2 is energized, the valve is placed under instrument control and would modulate to maintain the desired back pressure at the instrument sensing point. If

Solenoid V2A is energized (with V2 deenergized), the cylinder would be vented via C/NO of V2 and C/NC of V2A, opening the valve. This would be a required condition for cleaning the line in which the valve was installed. The valve will then be "pulsed" intermittently to assure proper cleaning of the stem O-ring.

3. Valve 3 is a 3-port divert valve fitted with a mid-position actuator. With both solenoids deenergized, flow through Valve 3 would be B/C. If Solenoid V3 is energized, the valve plunger will travel full stroke, changing the flow path from the B/C port to the B/A port. If Solenoid V3M is energized (with V3 deenergized), the plunger will move to a mechanically set position permitting flow through all three ports.

   If both solenoids are energized simultaneously, this valve will assume the full-stroke position providing flow from B to A. Full-stroke travel is essential during the CIP operation to properly clean the stem O-ring, but the mid position may be used to provide "split-flow" of solution through the circuit at other times.

Process and cleanup operating personnel can be trained to use toggle switches or machine tool switches to operate shutoff valves. The control of the more complex systems using two switches must be considered a much more difficult task and subject to occasional human error.

Figure 2 is a photograph of a typical air output box showing the actual physical arrangement of some of the components described above. From top to bottom on the left side, internal components include (1) pressure reducing valves with output gauges, (2) I/P (current-to-pressure) transducers for converting a 4 to 15 ma current output signal to a 3 to 15 psi air signal, and (3) solenoid valves, booster relays and pressure-reducing valves as shown for Valve 2 of Figure 1. The right-hand subpanel contains additional solenoids for controlling air-operated valves plus three air-blow control solenoids in the lower right-hand corner. A small instrument and gauge at the bottom of the right-hand door constitute an operator-adjustable set-point knob and gauge. Note that all components are labeled with identification legends. These components, applicable to the control of valves associated with the HTST pasteurizing system, are generally located in a stainless steel enclosure near the unit. This enclosure also houses the instrumentation required in the form of booster differential pressure controllers and back pressure regulators to keep these instruments close to the process and permit the use of standard capillary tubes. A major portion of the remaining solenoid valves may be placed in a single enclosure near the control panels as it is generally more economical to install Poly-

Figure 2. This typical air output box is used for all of the valves and instruments associated with an HTST pasteurizer and would be installed similar to Figure 1 in close proximity to the control devices and streams being monitored by pressure instruments.

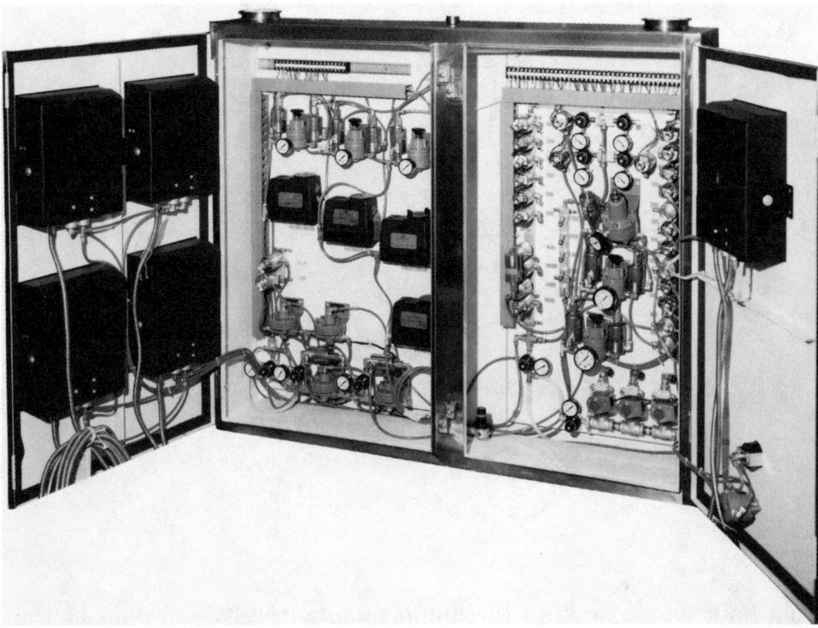

Flo tubing for air supply to the valves than to install the required wiring to a number of remote air output boxes.

The 3-port divert valve is often considered as a replacement for 3-port plug valves, whereas the preferred approach would be to use three shutoff valves at considerably greater cost. Figure 3 shows a commonly used application on the suction side of a transfer pump to select either of two tanks as the source of product to be moved. If the most common flow path is C/B, the indicated application is satisfactory. If, however, the most common flow is A/B, requiring the valve to be energized whenever flow is initiated, then that portion of the piping from the valve body to the left will constitute a "dead end" containing product which will warm up in the pipe, and a small portion of which will mix with the product in the A/B stream whenever the valve is energized or deenergized. Proper application would configure the valve and/or solenoid control to maintain the integrity of the most commonly used paths accompanied by positioning the valve to minimize the resultant "dead end" associated with the alternate port. The more astute innovator will ultimately recognize the use of mid-position actuators to achieve flow

Figure 3. This 3-port divert valve is frequently used to control flow from two sources to a common point or from a common point to two destinations either individually or together as shown.

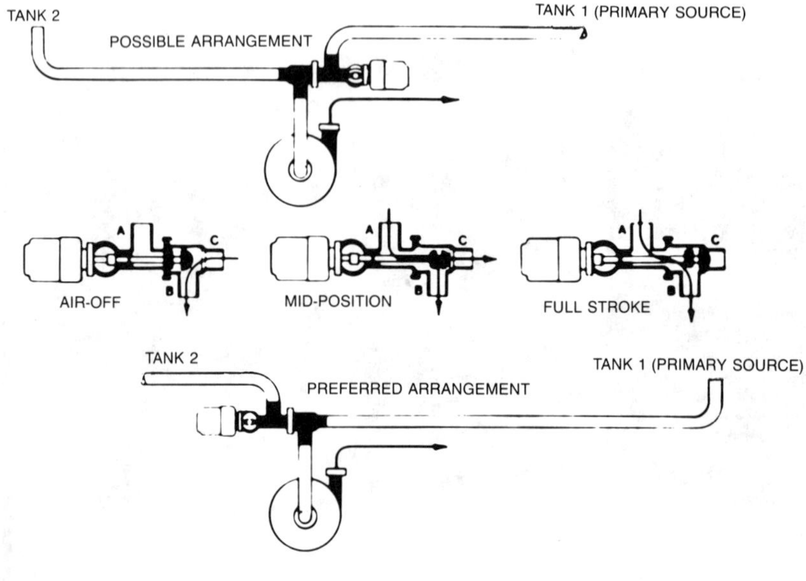

from both source tanks to the pump simultaneously, as would be the case during cleaning operations.

In some applications, one flow path may be of primary concern and the second of secondary interest, i.e., used only occasionally. If the valve is fitted with a mid-position actuator and installed as shown in the schematic view (Figure 3), each time this valve is actuated (normally when the pump is started), the plunger will move from top to bottom of the body cavity. When this action occurs, the product pathway from Tank 1 may now enter the pipeline that is connected to Tank 2. Some product can be expected to enter the Tank 2 leg. It will remain there, warming up, until the valve is subsequently deenergized. Then, if the same process operation is reinitiated without first moving product through the Tank 2 leg, this process will repeat, resulting in the mixing of some of the product which had been in the "dead end" with the fresh product going through the system. This would be of little concern on a two-tank by-product batching system where all product would be repasteurized, but could be troublesome if this piping configuration was used to supply a filler from two adjacent pasteurized surge tanks.

If a 3-port valve and mid-position actuator must be used for this flow control purpose, a preferred approach is to place the valve adjacent to that transfer path least frequently used, with Port A connected to that path. This has the combined effect of (1) preventing product from entering the "dead end" portion of the system except when in use and (2) reducing the volume of the "dead end."

Control of Design and Installation

As noted previously, the *3-A Accepted Practices for Permanently Installed Sanitary Product-Pipelines and Cleaning Systems*[1] requires that "a drawing or equivalent plan shall be made available to the Regulatory Agency by the processor for each installation, or subsequent addition or modification . . . ," but, fortunately, is not explicit in describing the type of drawings required. The design of a piping system for a new milk processing facility may require 40 to 100 sheets to provide all of the information necessary for installation purposes. These drawings will include schematic flow diagrams, plan and elevation views of the piping, isometric drawings of the product and CIP piping, large-scale isometric of complex parts of the system, shop drawings of special tanks or recirculating units, panel drawings, utility requirements, ladder diagrams for electrical and air piping, etc. However, two basic drawings can be prepared to provide all of the information necessary to production, maintenance, quality assurance, and regulatory personnel who require an understanding of how the system is to be used to complete processing operations and how it is to be cleaned.

The first drawing would be a small-scale, but accurate, floor plan showing the building and equipment arrangement for the areas which include process/CIP equipment and piping. Major equipment in the form of tanks, HTST pasteurizing systems, fillers and pumps should be identified, and for automated systems, the location of the various control panels, subpanels and air output boxes are also useful. Then, a schematic flow diagram developed to the same geography as the layout drawing should show in detail the piping, pumps, lines and valves, processing systems, and all connections required for establishing processing or cleaning circuits. For automated systems, the usefulness of that drawing will be increased through the addition of minimal information about switches, other controls, and instrumentation which interface the process to the operator and to the control system. Finally, sepia reproducibles of the schematic flow diagram can be heavy lined with numbered arrowheads to indicate all of the product flow paths and CIP circuits required for the proposed application. In some instances, color-coded versions of a new print are used for this purpose, but color-coded drawings cannot be easily reproduced or updated, so the sepia reproducible of the basic schematic flow diagram is the preferred method.

Properly detailed and complete drawings provide the *only* means of reviewing a system prior to installation, on paper, and then confirm that the installation is in accordance with the plan. These steps are necessary to evaluate the design with respect to:

1. Provision of all of the required connections to accomplish the proposed operation
2. Assurance that all of the piping and processing equipment is included in appropriate CIP circuits, or identified for manual cleaning
3. Assurance that the design incorporates no cross-connections between pasteurized product equipment and piping and raw product equipment and piping, whether direct, or indirect through a portion of the CIP return system

The PMO[3] requires that pasteurized milk products be protected from contamination (with raw milk) and that, during processing, all product containing piping or tanks shall be effectively separated from piping or tanks containing cleaning and/or sanitizing solutions. In this nation, the interpretation of this requirement is that a "physical break" exist between the two streams involved, via removable jumpers, key-pieces, swing-elbows, or U-bend connections. The intent is that these connections be changed to accomplish the required task, and that the connections and the ends of the permanent lines be properly cleaned and sanitized by manual procedures when making that change. These actions require people, and people require training and supervision, and there is a tendency to weld everything possible to reduce the number of removable connections.

The above mentioned audits of existing processing systems suggested that potential "cross-connections" between product and cleaning solution, or raw product and pasteurized product, could be found in most plants. The obvious examples include:[4]

1. Welded-in CIP supply connections, CIP return connections, or CIP tie lines incorporating hand-operated plug valves to establish CIP circuits
2. Permanent connections from flow diversion valves, divert and leak-detect lines to an HTST constant-level tank below the top of the tank. These "submerged" connections provide an opportunity for raw product to be drawn into the flow diversion valve body during momentary interruption of forward flow

3. Submerged termination of the HTST recycle line in the constant-level tank, generally provided to reduce foam buildup during long periods of recycle
4. The use of two 3-port divert valves comparable to the CIP-type flow diversion valve as the means of separating cleaning solution and product lines or as the means of returning a pasteurized product to the raw side of the system (standardization via mixing en route to the HTST constant-level tank). These concepts might be considered "safe" if the anticipated relative pressure were always maintained, but those applications observed failed to incorporate any method for assuring change in valve position on failure of relative pressures. Further, such an approach would require new regulations and procedures for testing and inspection of components involved
5. Less obvious "cross-connections" existed in the form of common CIP return systems which supported both raw and pasteurized product CIP circuits, installed in a manner which would permit the return line to become a tie-line between the raw and pasteurized side of the facility during production operations

Most fluid milk plants use common fillers for packaging the high-volume products such as homogenized milk, 2.0%, and low-fat (skim) milk, and by-products in the form of cream, buttermilk, shake mixes, juices, and drinks.[5] It is a fairly common practice to supply the high-volume products via air-operated valves through one line to a filler, and to supply the by-products through a separate by-product line, using normally capped tees or plug-valves in close proximity to, and generally above, the packaging machine. The final connections and the packaging machine may be only "rinsed" between products and the by-product supply header may receive only similar treatment, if any. If all products supplied to the filler are properly pasteurized, this arrangement does not represent a "cross-connection." However, it provides substantial opportunity for problems associated with dead ends and for contamination of the final connections when changing from high-volume fluid milk products to by-products and back to fluid milk products again. Some alternatives to physical "make-break" connections create opportunities for product mixing, i.e., the packaging of a buttermilk, homo, orange, or 2.0% blends. This may not be detrimental to product safety, but it has a substantial impact on the consumer and results in the production of substantial "returned milk" which, in itself, is another serious problem.

Many of the new plants constructed during the past decade have replaced the random plug valves and takedown connections described previously with U-bend transfer panels designed and installed to (1) simplify and improve control of the connection procedure, (2) provide an organized method of handling that procedure in a manner which permits improved sanitation, and (3) makes it impossible to connect a single filler to two different sources of product simultaneously. U-bend connections in each filler supply line permit filler to be connected to (a) pasteurized surge tanks containing high-volume products with flow controlled via air-operated valves, (b) individual by-product supply tanks, or (c) CIP supply header for automatically rinsing or cleaning filler during production run. A typical filler supply transfer panel is shown in Figure 4.

Control of CIP Procedures

Automatic control of the CIP (Clean-In-Place) procedure assures the uniformity and reliability of results essential to obtaining approval for the installation of silo-type tanks, welded piping systems and air-operated valving. The reliability of this procedure is dependent upon the control of four variables which include time, concentration, temperature, and physical action. The last variable, physical action, is established during the design and is related to the selection and installation of sprays in tanks, the sizing of pumps, and development of circuits to assure proper velocities in the piping and the elimination of "dead ends" in cleaning circuits. A constant effort must be made during the design and installation phase to assure that CIP means clean-in-place, rather than "clean-in-part." Any design effort which reduces the number of make-break connections, especially on the pasteurized side of the system, will contribute to improved product quality and shelf life. Beginning in 1978, every major fluid milk processing project handled by our firm, whether a new plant or a renovation program, has included the procedure of cleaning all of the pasteurized tanks, lines, valves, and fillers in a single circuit.[6] In several instances this has provided the opportunity to clean 5 to 7 fillers, 5 to 6 silo tanks of up to 15,000 gallon capacity, as many as 100 valves and 2,500 to 3,000 feet of piping, in a single circuit, via a single program, generally in the duration of 2-1/2 to 2-3/4 hours. The required physical connections include (1) swing-elbows at the beginning of the HTST discharge lines, (2) appropriate setup of the jug or paper fillers included in the system for CIP cleaning, and (3) opening manhole doors and installing CIP manifolds at the silo-type pasteurized surge tanks. The remainder of the system is cleaned in the process configuration, assuring maximum freedom of recontamination when preparing for the next day's production.

Figure 4. Filler supply transfer panel installed in large milk processing facility.

Benefits not so obvious in this approach include (1) substantial reduction in time required for the single program as compared to a multiplicity of programs for cleaning the individual system components and/or circuits and (2) a reduction in capital cost through elimination of CIP system-related piping, as only a return line from the fillers is required. This concept has been applied with consistent and acceptable results in plants which pump to the fillers, plants which enjoy the benefits of gravity supply to the fillers, and plants which have a little of each. The engineering effort is substantial, but the results have proven this approach to be worthwhile.

Assuming proper engineering design will assure the application of the proper physical action, the next most important matter of concern is control of cleaning operations, i.e., control of the people running the systems and control of the program. The 3-A Accepted Practices[1] require the installation of a recording thermometer in the return line. However, a return line temperature record is not an effective means of evaluating system performance, for a circuit can be filled with hot solution, which may then remain stationary for a substantial part of the total cycle time, and still produce an acceptable record. Evidence of this situation was obtained during a project which involved the evaluation of two different types of cleaning systems for washing bulk milk tankers.[7] Though the experimental procedure included in this report did not so state, personal knowledge of the work involved to make an existing CIP system mechanically and hydraulically reliable required considerable time and effort. Observation of the poor cleaning performance would have been unrecognized except for the installation of a dual-pen temperature/pressure recorder on each of the two systems. The CIP pump discharge pressure record as an adjunct to the temperature record is valuable in the following manner:

1. If a spray device plugs, flow will be restricted, and the pump discharge pressure will be higher than normal.
2. If a supply pump seal fails and the pump loses efficiency, the supply pump discharge pressure will be lower than normal.
3. If return flow to the solution tank on reuse systems is inadequate or interrupted for any reason, the supply pressure will drop to zero as the tank empties or will become low and erratic as the tank level approaches the bottom.
4. On a single-tank systems which provide for the return pump to discharge directly to the inlet of the supply pump, the wash and rinse pressure records will show a light increase in pressure during the wash and postrinse, due to the 3–5 psi stuffing pressure normally recommended. This confirms proper flow rates and adequate water in the system and proper return pump performance.
5. When cleaning line circuits which require the sequencing of a large number of valves to establish subcircuits, a repetitive pressure profile will be developed which should be consistent from one day to another. Any inconsistency may be the

result of failure to complete a manual connection, an air hose disconnected from a valve, or a failed control system component.

The return line temperature pen provides an adequate record of time and temperature. The method of controlling concentration is dependent upon the type of system involved. We would recommend that quality assurance supervisory personnel require the checking and recording of concentration for at least one program involving pasteurized equipment (tanks or lines) on a daily basis.

Automated dairies with complex cleaning circuits which include a large number of air-operated valves create a special problem for operating, management, and regulatory personnel alike. The documentation which describes how the system is to operate is often ambiguous, nonexistent, or difficult to understand. Matrix drawings provide an excellent solution to this problem by documenting operation of air-operated valves to achieve cleaning solution flow through all portions of complex air-valved CIP circuits. The simplified schematic of the specific circuit involved makes it easier for responsible personnel to understand the relationship of valve operation to flow through the various subcircuits of the total program. This approach worked well during the earlier years when valve sequencing was done by process planners or drum stepping switches, and it works equally well with today's programmable controller-based systems. As shown in Figure 5, the matrix includes, on the left side, the identification of every valve or pump included in the circuit. The sequence steps are shown across the top, though the description was not included in this example. An X in the box means that the particular valve or pump should be operative on that step. Most control systems provide the means of "single-stepping" the sequencer through its program, and this permits the concerned person to check every controlled device to assure compliance with the program. It is necessary also to look at all other controlled devices to make certain that some are not operating improperly, creating short circuits, parallel paths, or otherwise affecting the expected hydraulic performance of the system.

The next problem is to relate the information on the matrix drawing to the physical installation of the pipes, pumps, and valves. Though the sepia reproducible of the schematic flow diagram referred to previously may be available, it is a much larger and more difficult task to interpret this drawing. A preferred practice is to place small-scale, simplified versions of each CIP circuit on the same drawing with the matrix drawing.

Figure 5. Valve sequencing documentation.

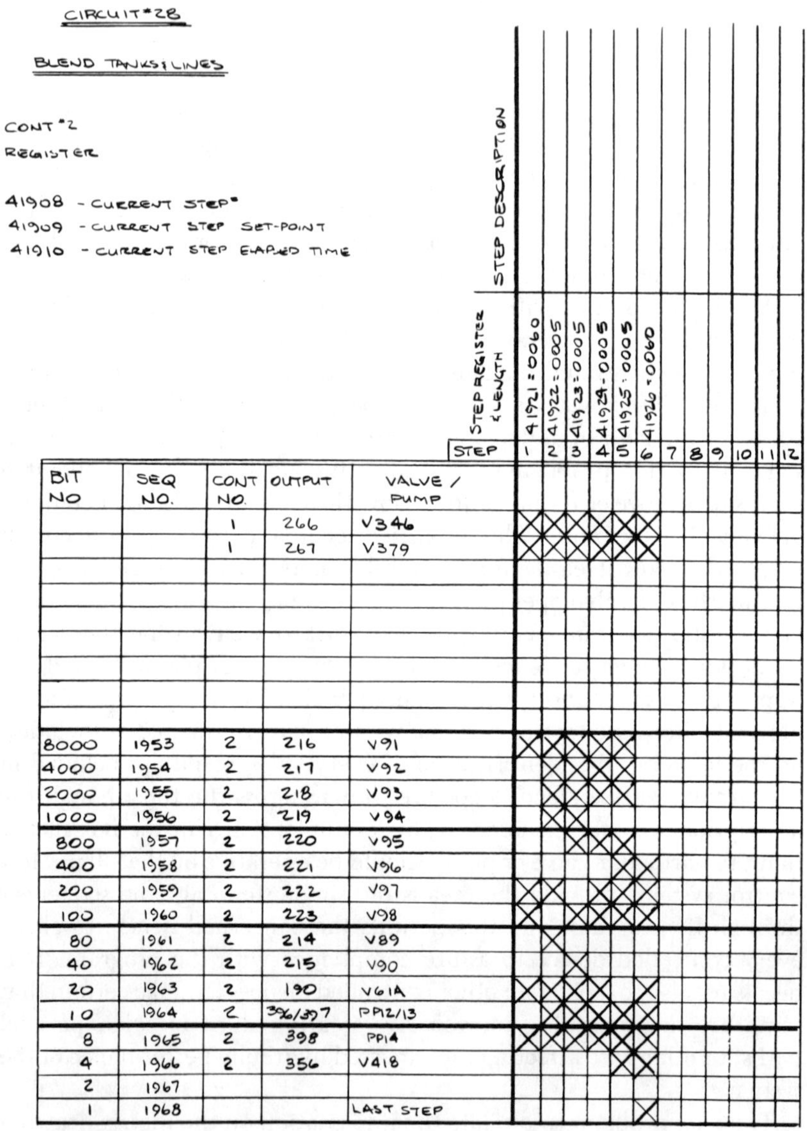

CONCLUSION

This chapter will conclude with a paragraph from an operating manual written in 1961, used continuously since, and equally applicable in today's complex computer-controlled dairies.[8]

"Automated product flow/CIP cleaning systems provide highly sophisticated approaches to controlling processing and cleaning operations in milk and food processing plants. By comparison with many of other forms of automation, however, they are extremely simple. Unfortunately, upon installation of such systems, many people tend to 'forget all they ever knew about milk plant operation.' When operational or sanitation problems occur, the concern is always with the control system . . . whereas the problem is generally with a simple component such as a pump, a valve, or a motor starter. A thorough understanding of the system and its intended operation will make it possible for most people to become highly competent with respect to both day-to-day operation and with respect to finding and eliminating operational problems which will occasionally occur."

REFERENCES

1. 3-A Standards. 3-A Accepted Practices for Permanently Installed Sanitary Product Pipe Lines and Cleaning Systems. International Association of Milk, Food and Environmental Sanitarians, P. O. Box 701, Ames, IA 50010 (Continuously revised; year not applicable).
2. Seiberling, D. A. and Harper, W. J. 1957b. Automatic Sanitary Valves for Dairy Plants, *Am. Milk Rev.* 19 (1):30–31, 34.
3. Grade A Pasteurized Milk Ordinance—1983 Revision. U. S. Department of Health and Human Services, Public Health Service, U. S. Food and Drug Administration.
4. Seiberling, D. A. September, 1972. Second Generation Approaches to Automated Production. *Dairy and Ice Cream Field.*
5. Seiberling, D. A. January and February 1974. Computer-Controlled Digital Separating/Standardizing/Blending of Fluid Milk Products. *Am. Dairy Rev.*
6. Seiberling, D. A. 1979. Process/CIP Engineering for Shelf-Life Improvement. *Am. Dairy Rev.* October, 1979, 41 (10):14–22, 64–67.
7. Richter, R. L., Bailey, J., and Frye, D. D. 1975. A Field Study of Bulk Milk Transport Washing Systems. *J. Milk and Food Technol.* 38 (9):527–531.
8. Seiberling, D. A. 1961. Custom system operating manual for confidential party.

18. Aseptic Processing of Low-Acid Heterogeneous Foods in Relation to Current Good Manufacturing Practices

Dilip I. Chandarana
Austin Gavin III
Dane T. Bernard

National Food Processors Association

INTRODUCTION

Aseptic processing technology has gained wide acceptance among food processors, particularly for fluid foods, since the 1981 approval of hydrogen peroxide (H_2O_2) as a container sterilant by the U.S. Food and Drug Administration (FDA). The various advantages of aseptic processing over traditional in-can processing include:

1. Superior product quality is possible because the bactericidal effect of a thermal process increases at a greater rate than chemical changes (nutrient destruction) as temperature is increased.
2. There are potential energy savings.
3. Processing is continuous, resulting in higher efficiency.
4. Heat transfer to the food is faster and more uniform.
5. High level of automation can reduce variations in process parameters.
6. Containers can be less expensive than metal cans.

There are also some possible disadvantages that must be considered such as:

1. Initial capital investment is high.
2. Cost of training personnel may be substantial.
3. Instrumentation and controls are more complex.
4. Destruction of certain enzymes may be more difficult.

The objective of this chapter is to analyze various process establishment procedures for aseptic processing of heterogeneous foods (food containing discrete, particulate matter) in relation to the current regulatory framework. As with other low-acid products, those aseptic products containing discrete particulate matter must be processed to a condition of commercial sterility. If the product contains more than 2% poultry or 3% raw red meat, it will fall within the regulatory jurisdiction of the U.S.D.A. and the process to be used for that product must be approved by the agency before that product can be manufactured. If the product is a low-acid food but contains little or no meat, then the FDA has regulatory responsibility, and a process must be filed with and accepted by the agency before product can be shipped in interstate commerce. Regardless of which agency has regulatory responsibility, the steps in assuring a commercial sterility process for the product and providing appropriate documentation will be the same.

The problem posed by establishment of aseptic processes for this type of product is the accurate prediction of the amount of heat which penetrates to the center of a food particle while that particle is continuously moving through an aseptic system. One solution to the problem would include a model for predicting heat penetration into the particulate matter, a model for prediction of the residence time of the fastest moving particle, and a well-tested method to validate the above models (biological testing).

PRODUCT DESCRIPTION

Aseptic particulate products now under development utilize a viscous carrier medium. The difference in the specific gravity of the particulate matter and that of the carrier medium prevents particulate matter from settling out in the system and aids in producing a uniform distribution of particles throughout the product during processing. The product must be pumpable and well within the physical limits set for available pumping equipment. These products are normally non-Newtonian and would heat by conduction in a conventional retort process.

SYSTEM DESCRIPTION

Heterogeneous foods could potentially be aseptically processed in a variety of systems, all with the ultimate aim of consistent delivery of commercially sterile product.

Current system designs involve the use of a positive displacement pump to move product through heating scraped-surface heat exchangers, a hold tube, cooling scraped-surface heat exchangers, and a lobe-type pump to produce a "back-pressure" throughout the system. Other possible system designs are being investigated.

Flow Control

The sterilization time for heterogeneous products is related to the rate at which the product is moving through the system, the physical and rheological properties of the carrier medium and particle, and the design of the system components. A positive displacement pump is used to produce a uniform, controllable flow rate through the system. As noted earlier, also included as part of a flow control system is a means to pressurize the system. The back pressure must be greater than the vapor pressure associated with the highest product temperature in the system. This eliminates the possibility of "flashing" or boiling of the product as it is being sterilized. This is usually accomplished by a positive displacement pump which is operating at a slightly lower flow rate than the feed pump.

Other possible designs involve the use of a pressurized surge tank to produce back pressure. If this option is used, the surge tank may need to be equipped with agitation to maintain a uniform particle distribution within the product.

Heating and Cooling

Scraped-surface heat exchangers are currently used in aseptic systems to heat and cool heterogeneous foods because of their ability to process viscous foods with suspended particulate matter. The scraper blades are designed both to prevent the buildup of product on the heat exchanger wall which would hamper heat transfer, and to induce agitation, thereby enhancing heat transfer. Certain scraped-surface design changes have been introduced which minimize particle damage and result in a more uniform distribution of particulate matter. The flow rate, mutator RPM, and blade design all have an effect on product flow patterns and particle distribution in the heat exchanger.

Hold Tube

The hold tube is used to hold product for a specific period of time at a specific temperature to produce a desired lethality. Other than the relative length, hold tubes currently used for experimental particulate systems are essentially the same as those used for fluid processing.

PROCESS ESTABLISHMENT AND CONTROL

Traditionally, process establishment for low-acid foods involves the marriage of the heating characteristics of the product with the bacterial resistance of certain target microorganisms. A mathematical model is used to develop thermal processes which are confirmed by bacterial inoculated pack studies. The use of a mathematical model allows maximum flexibility concerning process alternatives while minimizing the amount of process testing necessary. The Ball formula method essentially serves this purpose for traditional in-can retort processing.

The conventional approach for aseptic processing of homogeneous products is that the thermal process includes only that lethality accumulated in the product as it passes through the hold tube. A mathematical model is used to determine the minimum residence time and minimum product temperature necessary to produce a desired lethality. The flow rate, physical dimensions of the hold tube, and the rheological properties of the fluid are used to calculate a residence time for the fastest fluid particle. The minimum temperature is verified by physically measuring the temperature at the end of the hold tube. Since no heat is applied to the hold tube, this represents the minimum temperature for the product flowing through the hold tube.

When considering processes for particulate-containing foods, however, the situation becomes much more complex, due to the increase in the number of variables which must be considered. The process design must be such that the lethality imparted to the center of the largest, fastest moving particle is adequate to achieve commercial sterility within that particle, while at the same time minimizing the overprocessing of the fluid carrier medium.

Several possible approaches could be considered when designing aseptic particulate processes. A few examples are selected for discussion. An approach which considers only the lethality incurred in the hold tube is one possibility. Within the system, the carrier medium serves to transport the particle through the system while also providing the vehicle for transferring heat to the particle. This approach assumes the center of the particle could be brought to a minimum processing temperature in the heat exchanger, transported through the hold tube and cooled. One drawback associated with this approach is that a means would have to be developed to assure that the slowest heating region of the particle has reached processing temperature prior to entering the hold tube. In order to accomplish this, research must not only involve heat penetration and heat transfer rates for individual particles, but factors which affect particle residence time within the swept-surface heat exchangers and temperature distribution profiles as well.

Since the carrier medium is used to heat the particle, the thermal process received by the carrier medium is substantially greater than that received by the center of the particle. This differential will increase as particle size increases. The amount of heat used to bring the particle center to processing temperature may degrade the carrier medium and particle surfaces. Therefore, processes which consider only the lethality in the hold tube may prove to be impractical for some heat-sensitive products.

A second possible approach was introduced by Dail of the Dial Corporation.[1] This approach considers only the heat received by the particle as it moves through the hold tube. The assumption made is that the temperature of the slowest heating region of the particle is the same when it enters the hold tube as the temperature of the particle when it entered the system. The thermal contribution of the heat exchangers is, therefore, essentially ignored. A second assumption states that the temperature of the carrier medium in the hold tube is constant and equals the minimum temperature at the end of the hold tube. A mathematical model is used to calculate the lethality at the center of the particle given the above conditions. This approach should be very conservative and would result in a more drastic "overprocessing" of the carrier medium than the first approach discussed.

Both the first and second approaches involve a determination of the residence time for the fastest particle as it moves through the hold tube. Assumptions concerning rate of particle movement in relation to the carrier medium will need to be clarified. The effect of solid to liquid ratios and particle density on the flow behavior of the particulate slurry will also need to be investigated.

A third possible approach, and by far the most complex, involves using the thermal contribution of the heat exchangers, plus that accumulated in the hold tube. The contribution during cooling would not be considered since particle integrity cannot be controlled, and a fractured particle may cool at a faster rate than predicted. In order for this approach to work, minimum temperature profiles would have to be determined for the carrier medium and combined with fastest particle residence times for both the heat exchangers and the hold tube.

All three possible approaches involve a determination of the heat transfer conditions at the carrier medium/particle surface interface. These heat transfer conditions will be influenced greatly by the relative velocity of the particle compared to the surrounding medium. The relative velocity of a particle may be minimal within a hold tube but be radically different when considering the same particle moving through a scraped surface heat exchanger. The rheological properties at this critical interface between particle and carrier medium will also need to be defined in order to reliably predict the amount of heat transferred to the particle surface and ultimately through conduction into the slowest heating area of the particle.

Whatever approach is selected, a mechanism would have to be established to verify that the model accurately predicts conditions within the aseptic system. Process simulators could be used which would allow direct measurement of particle temperatures under controlled conditions. Coupled with bacterial inoculated pack studies under controlled and commercial conditions, these mechanisms would provide the strongest evidence that the predicted process reliably produces the intended lethality and can be controlled within predefined operating conditions.

PRESENT WORK AND FUTURE OUTLOOK

If aseptic processing of particulate-containing low-acid products is to achieve its potential of producing a high-quality finished product, every effort must be made to collect the information needed to establish the commercial sterility process, based on lethality delivered in the heat exchangers and the hold tube rather than on the hold tube alone. While this approach is unconventional, both the FDA and USDA regulations or guidelines are written in language that is inclusive rather than exclusive. Thus, implementation of new technology is not excluded within the current regulatory structure, but, as always, appropriate documentation must be available for regulatory review.

Of necessity, however, initial process designs for aseptically processed heterogeneous foods will probably be conservative in order to provide a relatively large margin of safety, thereby sacrificing some product quality. In view of the existing gaps in our knowledge, the margin for error may be less defined than traditional thermal processing. As experience with this processing method is gained, further steps can be taken toward optimizing processes.

Process design, by competent process authorities who are knowledgeable in the area of aseptic technology, is of primary concern. There is also a need for a united effort on the part of the various groups involved in research on these products and processing systems because of the great needs in terms of personnel, equipment, and funds. Working with regulatory officials during these stages should help bring about the acceptance of thermal processes involving particulate-containing products without unnecessary delays.

REFERENCE

1. Dail, R. 1985. Calculation of required hold time of aseptically processed low-acid foods containing particulates utilizing the Ball Method. *J. Food Sci.* 50:1703–1706.

19. Quality Control and Quality Assurance Progress in the Food Safety and Inspection Service, USDA

Ralph W. Johnston

U.S. Department of Agriculture

Quality control and assurance are not new concepts; they have been practiced in this country for decades. However, the adoption of, commitment to, and formalization of the concepts have grown dramatically in the last decade and in particular since the last conference on food protection. A renewed interest in quality control began vigorously in the 1960s, particularly in the automotive and electronic industries.

This decade marked a worldwide shift from shortages in almost all commodities to adequate supplies. Consumers became knowledgeable of quality and aware of the lack of it. Exporting countries foresaw the need for quality to compete in the international market and consumer brand loyalty began to disappear. These changes have continued through today. There is now excessive production of just about everything, and it's a buyer's market. While the United States in general was a little slow to see this trend for quality coming, it has now adopted strong quality control programs. These are no longer options of elegance; rather, they are essential for survival.

With several notable exceptions, the food industry also lagged in instituting strong quality control programs. Again, however, it is now well-recognized that strong quality control programs in food processing are essential to corporate vitality and growth.

Food regulatory agencies such as the Food Safety and Inspection Service (FSIS) have been even slower in developing quality control concepts, partly because they believed their numerous regulations to be adequate quality control mechanisms. This may be true to some extent, but it is no longer recognized as the best approach to quality control. For many years, U.S. Department of Agrigulture (USDA) inspectors have done much of the quality control work of smaller processors and perhaps caused these firms to become dependent instead of progressive.

This too has changed. The FSIS is committed to developing ways to stimulate quality control programs in the industry it regulates. Industry should be responsible for developing and operating its own quality control programs designed to meet, and preferably exceed, FSIS regulations. Such a concept has benefits for both the industry and the FSIS. During the last decade alone, the number of federally inspected meat and poultry plants increased 60% while the cost of processing inspection has approximately tripled. Clearly, with the implications of the Gramm-Rudman-Hollings Law, we must continue to develop new concepts.

The agency is well underway in streamlining and modernizing its inspection program, and this effort contains a heavy emphasis on quality control systems which are primarily operated by the inspected firm.

It is likely that this redirection of manpower and modernization will continue to expand. In 1985, the National Academy of Science, National Research Council produced two reports entitled "Meat and Poultry Inspection: The Scientific Basis of the Nation's Program" and "An Evaluation of the Role of Microbiological Criteria for Foods and Food Ingredients." Both of these comprehensive documents, prepared by separate expert committees, strongly recommend that the FSIS utilize the HACCP system in carrying out its responsibilities.

Quality control and/or quality assurance programs now utilized by the FSIS cover most facets of USDA responsibilities, including laboratory operations, inspection of meat and poultry slaughter, inspection of meat and poultry cooking and fabrication, and others. Some of these are as follows:

Total Quality Control (TQC). Under TQC, the meat and poultry processing inspector becomes a quality control inspector who monitors a firm's TQC program. TQC is voluntary and applies primarily to firms that further process. Details of the program are submitted to the FSIS and reviewed prior to approval. TQC plants are somewhat limited due to the extensive level of commitment required of the firm.

Partial Quality Control. This system is very similar to TQC except that the controls are applied to a portion but not all of a firm's operations. More firms participate in PQC programs and we hope that many of these will eventually progress to TQC status.

Cooked Beef Regulations—1983. This is a revised version of an earlier regulation and incorporates HACCP principles. A number of plants have TQC or PQC programs which address their individual controls and thereby amplify the basic controls called for in the regulation.

Canning Regulations. This regulation has been proposed and will soon be finalized. It is comprehensive for canned meat and poultry products, fully compatible with the FDA regulations, and fully embraces the HACCP principles. In-plant QC programs again further enhance this regulation.

Laboratory Operations. Each of the four USDA laboratories has its own quality control program for both chemistry and microbiology. In addition, quality assurance samples for each major analytical determination are prepared and analyzed by each laboratory, and reviewed by division directors. Quality assurance samples are conducted quarterly or more often so that, in total, a sizable portion of laboratory resources is committed to quality control and laboratory safety.

Slaughter Operations. Extensive and very comprehensive quality control procedures have been written or are nearly completed for chicken, turkey, beef, and pork slaughter and dressing operations. The majority of chicken processors have adopted these procedures, and there is great interest by processors of the other species. The prospect for similar participation in the newer programs is encouraging.

In summary, the FSIS has been active in developing quality control and assurance programs for the industry and in accepting existing programs of individual corporations. By law, these must be voluntary. But, the advantages for both the FSIS and the firms inspected by the USDA are large enough that the rate of participation is encouraging. The USDA is also beginning to incorporate more HACCP principles into its regulations. Although these changes are accelerating, they are not easily adopted. They create some confusion, there is some resistance, and naturally some details have to be modified. Nevertheless, the overall perception is that the changes are beneficial to both government and industry and that more QC activities are both needed and under development.

20. Good Manufacturing Practice Regulations, Guidelines, and Voluntary Programs

Thomas R. Mulvaney

U.S. Food and Drug Administration

INTRODUCTION

In 1969, the U.S. Food and Drug Administration (FDA) promulgated its first current good manufacturing practice (CGMP) regulation for the manufacturing, processing, packing, or holding of human food. This regulation has often been referred to as the umbrella CGMP because it applies to the entire food industry.

The FDA has also promulgated CGMPs for frozen raw breaded shrimp, smoked and smoke-flavored fish, thermally processed low-acid canned foods, acidified foods, bottled drinking water, and cacao products and confectionery. In addition to these final regulations, the FDA proposed but has not finalized CGMPs for bakery foods and for tree nuts and peanuts.

The FDA considered for a period of time issuing separate CGMP regulations to cover requirements for each major segment of the food industry. However, after preparation of and experience with a number of specific CGMP regulations it became apparent that many of them were based on common or similar requirements. As a result, it was determined that the most appropriate course would be to revise the original umbrella CGMP regulation.

At about the same time, and in recent administrations, emphasis has been placed on reduction of the number of new regulations that may result in unnecessary controls. Often the amount of time, money, and other resources used by government and industry in proposing and finalizing new regulations is not realized. Regulations in the past have been used for government to let companies know what an agency expects of the regulated industry for the protection of consumers and industry. However, fewer, more inclusive regulations can save everyone time, money, and resources.

Some may question why the Food and Drug Administration has

CGMP regulations at all. The opinion may be stated that food manufacturers know what practices in their manufacturing operations are critical, and there is no need for FDA to specify to them which items are critical control points in manufacturing operations. It is true, of course, that the larger manufacturing companies have food scientists, technologists, and engineers on their staff who know what steps in a manufacturing operation are critical, and which ones should be carefully monitored. However, many of the small manufacturing companies have few, if any, technical people on their payroll, and they do not necessarily know what type of equipment to use or what controls should be utilized. Therefore, the agency provides an outline of current good manufacturing practices for all food manufacturers, both small and large. It is the responsibility of the food industry to produce safe, nutritious and wholesome foods.

An important point of current good manufacturing practice regulations is that they contain the criteria and definitions which apply in determining whether a food is adulterated within the meaning of section 402(a)(3) of the act in that the food has been manufactured under such conditions that it is unfit to be used as food; or within the meaning of section 402(a)(4) of the act in that the food has been prepared, packed, or held under unsanitary conditions whereby it may have become contaminated with filth, or whereby it may have been rendered injurious to health. The criteria and definitions also apply in determining whether a food is in violation of section 361 of the Public Health Service Act (42 U.S.C. 264).

REVISED UMBRELLA CGMP

The FDA published, on June 8, 1979, a proposal to revise the current good manufacturing practice (CGMP) regulations for the manufacturing, processing, packing, and holding of human food (the umbrella CGMP regulations, 21 CFR Part 110). The primary purpose of the proposed revision was to establish new, updated, and more detailed CGMP provisions concerning food industry personnel; plants and grounds; sanitary facilities, controls, and operations; equipment and utensils; processes and controls; product coding; warehousing and distribution; recordkeeping; and natural or unavoidable defect levels. FDA designed the proposed revision of the umbrella CGMP regulations to address the problems associated with foods for which specific CGMP regulations had not been promulgated and thus to better ensure the production of safe and sanitary foods. FDA provided a period of 204 days for the filing of comments.

To gather information and opinions on the impact of the proposed revision, FDA also held hearings in Chicago, San Francisco, and in Atlanta. Approximately 250 persons attended the three hearings. Of the roughly 50 persons who made presentations at the hearings, nearly two-thirds represented small businesses. Because FDA was particularly interested in the impact that the revised regulations would have on small businesses, the agency solicited comments from small businesses and trade associations representing small businesses. The hearings resulted in a voluminous hearing record.

In addition to the comments received at the hearings, FDA received 132 written communications reflecting the comments of suppliers, manufacturers and processors, trade associations, operators of small businesses, consumers, and other interested persons.

The comments on the 1979 proposal, including those received at the hearings, suggested revisions of practically every section of the proposed rule. In response, the FDA adopted many of these suggestions in the regulation.

The most significant revision responds to comments concerned with the proposed requirements for coding and recordkeeping. The vast majority of these comments questioned the need for these requirements for segments of the food industry that contend that (1) their products pose little or no risk to the public health, and (2) their products, in the rare event that a risk arose, could be removed from the market expeditiously and effectively without the need to comply with the costly proposed requirements. Most comments also pointed out that the cost of coding and recordkeeping would be excessive, especially to small businesses.

To evaluate more fully the validity of industry's comments concerning the cost of the proposed regulations, the FDA contracted for a study of compliance costs associated with the 1979 proposal as the FDA then had considered modifying the proposal in response to comments. The study was conducted by ICF, Inc., of Washington, DC. ICF concluded that total compliance costs would be \$81 million. The costs were primarily attributable to the proposed recordkeeping (\$76 million) and coding (\$4.5 million) provisions. (Costs are adjusted to represent 1985 dollars.) ICF also found that 95% of the large manufacturers sampled and 93% of the small manufacturers sampled were already coding their products sufficiently to be in compliance with the proposed regulation. The recordkeeping costs would have been high because, although all manufacturers have detailed records, few have recordkeeping systems based on lot numbers, which the proposed regulation would have required.

The purpose of proposing coding and recordkeeping was to facilitate a manufacturer's recall of suspect products in case such a recall was recommended by the FDA. Although such information is potentially useful in determining the production time period affected by a recall, thereby limiting manufacturers' risk exposure, it is not needed to protect consumers from products that have been purchased but not ingested. Furthermore, all manufacturers either currently code all their products and/or keep shipping records in the ordinary course of business. As these sources can provide most of the information that would have been required in the proposed rule, and all of the information needed for a recall, it is not necessary to impose other economically burdensome recordkeeping requirements. This decision will save manufacturers and consumers approximately $80.5 million annually in costs which would have been incurred if the regulation had gone forward as proposed in the notice of proposed rulemaking.

For consumer protection, the most effective safeguard is product, not lot, identification and swift dissemination of such information by mass media. These mechanisms will in no way be compromised by the deletion of coding and recordkeeping requirements.

In addition, the products most likely to involve risk of recall (low-acid canned food) are already subject to coding and recordkeeping requirements.

Accordingly, because industry voluntarily codes and keeps records adequate for consumer protection, the FDA decided not to require coding or recordkeeping. FDA, after reviewing comments and the ICF study, concluded that an industry confronted with little likelihood of recalls of products subject to the proposed rule could decide that removal of all offending products from the market in the presence of a recall would protect the public health and would be more cost effective than maintaining records and coding products. On the other hand, an industry confronted with a high frequency of recalls or with the apparent potential for infrequent but serious contamination of a limited quantity of product could decide that coding and recordkeeping are essential to accomplishing a recall. Under either option, the public is protected and industry has the opportunity to decide which recall strategy is appropriate.

FINAL RULE

In the *Federal Register* of June 19, 1986 (Docket No. 78N-0296), the FDA published a final rule revising the current good manufacturing practice regulations for the manufacturing, packing, or holding of human food. The final rule became effective on December 16, 1986. Although the the FDA published a final rule, it provided a comment period. Comments were to be received by August 18, 1986. If the FDA decides on the basis of the comments received that any changes in the

Table 1. Part 110—Current Good Manufacturing Practice in Manufacturing, Packing, or Holding Human Food[a]

Section	
Subpart A—General Provisions	
110.3	Definitions.
110.5	Current good manufacturing practice.
110.10	Personnel.
110.19	Exclusions.
Subpart B—Buildings and Facilities	
110.20	Plant and grounds.
110.35	Sanitary operations.
110.37	Sanitary facilities and controls.
Subpart C—Equipment	
110.40	Equipment and utensils.
Subpart D—[Reserved]	
Subpart E—Production and Process Controls	
110.80	Processes and Controls.
110.93	Warehousing and distribution.
Subpart F—[Reserved]	
Subpart G—Defect Action Levels	
110.110	Natural or unavoidable defects in food for human use that present no health hazard.

[a]AUTHORITY: Secs. 302, 303, 304, 402(a), 701(a), 52 Stat. 1043–1046 as amended, 1055 (21 U. S. C. 332, 333, 334, 342(a), 371 (a)); sec. 361, 58 Stat. 703, (42 U. S. C. 264); 21 CFR 5.10, 5.11.

final rule are necessary, it will publish those changes in the *Federal Register*. Industry-wide compliance costs associated with this final rule are between $272,000 and $623,000 per year. These costs are for the installation and maintenance of thermometers in industries where food products or processing techniques might allow the growth of microorganisms. No coding or recordkeeping requirements are included in the final rule.

The primary purpose of the revision is to establish new, updated, or more detailed provisions for the food industry to help ensure a safe and sanitary food supply. The revision includes providing for foods for which specific CGMP regulations have not been promulgated.

Contents of the revised umbrella CGMP regulation are listed in Table 1. The specific provisions of this final rule are too numerous and involved to adequately describe here. They are available by reference to the *Federal Register* of June 19, 1986 (Docket No. 78N-0296).

In addition to the final rule revising the umbrella CGMP regulation, actions are being taken to revoke the frozen raw breaded shrimp and the cacao products and confectionery CGMP regulations, as well as to withdraw the bakery foods and the tree nuts and peanuts CGMP proposals. The smoked and smoke-flavored fish CGMP regulation was revoked previously. Continuing in effect will be the thermally processed low-acid canned foods, acidified foods, and bottled drinking water CGMP regulations, in addition to the revised umbrella CGMP regulation.

GUIDELINES AND VOLUNTARY PROGRAMS

To address important needs, efforts have been intensified to identify and increase utilization of efficient, effective, and flexible alternatives to detailed mandatory government regulations. While the current degree of emphasis on voluntary and cooperative programs may be relatively recent, their use is not new.

Two relatively recent successful activities of this type are the reduction of lead in canned foods and improved metal can integrity programs between industry and FDA.

The program to reduce lead in canned food had to do with lead migrating from the solder of metal cans. FDA first worked cooperatively with the Evaporated Milk Association and then with the National Food Processors Association and the Can Manufacturers Institute. Part of the success of the program resulted from industry's efforts in cleaning up and improving soldering lines. A near ultimate solution appears to be the change by industry from soldered to nonsoldered (e.g., welded or seamless sidewall type) cans.

The metal can integrity program developed because of incidents of botulism that resulted from cans that were damaged. The most significant of these incidents occurred in 1978 and 1982 when people died of botulism from canned salmon. The first instance was caused by apparent cable type damage, and the second instance resulted from a torn can flange when the can was reformed in Alaska. As a result of FDA and industry concerns, the National Food Processors Association and the Can Manufacturers Institute formed a joint container integrity task force to investigate container defects, make microbiological assessments, examine quality assurance and review leak detection methods. A result of this program was the classification and definition of can defects. The FDA obtained photographs chiefly from the can manufacturers and the National Food Processors Association and developed the AOAC Can Defect Chart after consultation with representatives of Canadian and British health agencies, and American industry. Thousands of copies have been distributed. A separate program is in progress under the Codex Alimentarius Commission Food Hygiene Committee to make this or a similar classification and manual on can defects available as an international guide.

In addition to this program to further develop and make more internationally available guidelines on can defects, other voluntary or cooperative programs in the area of food packaging and processing in which progress is being made include the following.

1. *Flexible package integrity.* Industry and the FDA have recognized for some time the need for guidelines on flexible package integrity. As a natural extension of the metal can integrity program, industry has been proceeding with the development of flexible packaging guidelines. The National Food Processors Association, Flexible Package Integrity Group Committee, has four subcommittees to address the various package types: pouches, paperboard packages, plastic packages with heat-sealable lids, and plastic packages with metal lids. Each subcommittee is working on developing documents on defect definitions, good manufacturing practices, and test methods. The NFPA recently reported completion of three draft documents for each type of package, and current projections are for possible publication in 1987.

2. *Aseptic processing and packaging systems.* Aseptic systems are controlled by the low-acid canned food regulations. In conjunction with these regulations, FDA and industry have conducted a voluntary program of review of new systems before commercial operation which appears to have worked well in ensuring consumer protection and compliance with the requirements and intent of these regulations. Plans are for this program to continue. FDA's guidelines for inspection of aseptic systems have been updated based on technical information and cooperation from industry (Inspection Operations Manual No. 637.6, June 17, 1985). The Food Processors Institute in 1985 updated and supplemented training materials on aseptic systems for the Better Process Control Schools. In addition, specialized courses on these systems have been developed and are presented by several institutions. The Codex Alimentarius Commission Food Hygiene Committee has begun an activity directed at updating its low-acid canned food guidelines on aseptic systems.

3. *Computer process control.* The use of computerized systems to perform process control, quality assurance, and documentation activities is increasing in aseptic processing and packaging operations, retort thermal processing operations, and the food processing industry overall. As these systems become instrumental in ensuring the safety, quality, and integrity of FDA regulated products, it becomes important for the agency to verify that proper controls were used to ensure the correct performance of the computer system prior to its implementation and for the maintenance and monitoring of the system once it has been installed. The FDA is currently developing "A Guide To The Inspection

Of Software Development Activities" which it expects will be applicable to the industries it regulates. Consideration also is being given to the possible development of a guide to the inspection of computerized systems in food processing. This guide would provide an overview on inspection of a complete computer processing system, in addition to the guide on software. It is important that industry continue to exercise its role and participation in the development of such guidelines if this program is to be successful in meeting the needs and concerns of the consumer, industry, and regulator.

Few would dispute the many important advantages of voluntary programs and self-regulation. The problem, it seems, is the few manufacturers that do not voluntarily comply and adversely affect consumer protection and proper competition, thereby undermining the credibility of industry regulatory programs to the point that its innovative efforts and consumer choice fall prey to declining confidence or required government regulation. The need continues, it appears, to identify and implement effective means to achieve voluntary compliance with generally accepted standards of performance. It is important that voluntary efforts continue, for as there are improvements and sufficiently successful programs, there are decreased needs for government regulations.

21. Good Manufacturing Practices and Quality Control Committee Recommendations

1. The Board should establish a task force to develop methods for cooperation between industry and food regulators for training and education programs to support the HACCP concept.
2. The Board should develop a means to provide a list of rapid methods that are used for in-process HACCP analyses and provide a clearinghouse for new methods.
3. The Committee recommends to the industry that processes that are carried out for microbial control in the food safety area be validated (installation qualification as well as operational qualification) before they are put into use, after changes are made in the system, and subsequently on an annual basis.
4. The Committee recommends to the Education and Training committee that basic food safety be reinforced in the educational system and to the general public.
5. The Committee recommends that the Education and Training Committee encourage the initiation or expansion of internships for food science/engineering and related science students.
6. The Board should compile a list of texts and training programs covering HACCP principles. Any need for additional materials should be brought to the attention of NSF and/or other organizations for development.
7. USDA grading standards for grains should be reflective of current FDA defect action levels.
8. The Board should form a task force to evaluate the potential hazards in the food transportation system. This should include incoming raw products as well as finished product distribution.
9. The Board should establish a task force to study a means for traceback to point of origin of all food ingredients.

STANDARDS AND REGULATIONS

22. A Model Food Establishment Unicode

Darrell J. Schwalm

U.S. Food and Drug Administration

INTRODUCTION

The old adage "variety is the spice of life" may well have been written with the food industry in mind. Regulating the food service industry and retail food stores involves a highly diverse system of food and public health laws and programs. More than 2,700 state and local regulatory agencies assume primary responsibility for monitoring retail food operations and ensuring that industry is adequately protecting the consumer. This has led to a myriad of problems, including a substantial lack of uniformity in inspectional and enforcement programs between jurisdictions.

The FDA is charged under Part B, Section 311(a) of the Public Health Service Act (PHS Act)(42 U.S.C. 243) to:

> . . .assist states and their political subdivisions in the prevention and suppression of communicable diseases . . . (and) cooperate with and aid state and local authorities in the enforcement of . . . health regulations. . . .

To provide this assistance, FDA directs its activities toward establishing an effective system of state public health control programs by recommending uniform technical guidelines, administrative procedures, and *model regulatory codes* and standards.

The history of model sanitation code development goes back over 60 years. In 1934, the Public Health Service (PHS) began developing an ordinance relating to the sanitary control of food service establishments. At that time, in cooperation with the Conference of State and Territorial Health Officials and the National Restaurant Code Authority, PHS drafted suggested minimum food sanitation standards. At the request of many state and local health authorities, the proposed standards prepared by PHS became the basis for a model ordinance published in tentative form in 1935. It was expanded in 1938 and revised in 1943 and again in

1962 with the publishing of the model Food Service Sanitation Ordinance and Code (1962 Recommendations of the Public Health Service).

In 1968, the FDA and the PHS assumed responsibility for the retail food protection program and the development of model codes. The FDA updated the Food Service Sanitation Code in 1976. In 1978, the FDA published an updated model code on "The Vending of Food and Beverages." Finally, in 1982, the FDA and the Association of Food and Drug Officials (AFDO) jointly published a model Retail Food Store Sanitation Code. This latter code was the result of an eight-year cooperative effort by the FDA and AFDO.

Thirty-six states and 157 local jurisdictions have adopted the 1976 edition of the model food service code. Many other state and local jurisdictions have adopted earlier editions. About half of the states and some local jurisdictions are known to have adopted the provisions of the model food vending code. Nine states and six local jurisdictions are reported to have adopted the model retail food store code.

PREVIOUS NATIONAL CONFERENCE ACTIONS

At the 1984 Second National Conference for Food Protection (SNCFP) the status of model food codes was addressed by the Standards and Regulation Committee. The following pertinent recommendations came out of the committee and were adopted by the Conference.

1. The FDA should develop and implement a plan to promote the adoption of model uniform laws and regulations in those states that have not done so. This effort should involve appropriate associations of industry and regulatory officials as well as representatives of appropriate governmental agencies and industry within affected states.
2. The FDA should determine what segments of the food industry are not covered by uniform model laws and regulations and proceed to develop such laws and regulations and assure that they are kept current. Appropriate associations, industry representatives, and state and local officials should be involved in the development process.
3. The FDA should coordinate the development and promotion of uniform interpretations among all state and local regulatory agencies.

UNMET PROGRAM NEEDS

While the level of adoption of the food service model code by states has been high, adoption of the food vending model code has been only moderate. Adoption of the newer food store code has just begun and is still low. Developing a new model code must be carefully considered to ensure that (a) the most current scientific evidence for food and consumer protection is obtained, (b) important public health program needs will be met, (c) opportunities to improve upon present codes and procedures will be acted upon, and (d) existing problems with the level of adoption by states and local jurisdictions will be addressed.

The FDA's consultation with state program officials and evaluations of the effectiveness of state programs has revealed significant control program problems that could either be solved or started to be solved by developing a model unicode. These unmet needs include the following:

(a) Removing inconsistencies between existing model codes in scope, definitions, sanitation requirements, grandfather clauses and compliance procedures;

(b) Removing inconsistencies between existing model codes in plan review provisions which help to discourage reciprocity in accepting plans, structures, and operations between jurisdictions;

(c) Reducing duplicative and unnecessary administrative time and cost at the state and local level to adopt, update and publish three or more very similar but separate codes to cover the retail food industry;

(d) Changing the form of the model code to a loose leaf format similar to FDA's model Pasteurized Milk Ordinance (PMO) in order to facilitate easy updating;

(e) Updating the inspection forms and the scoring system in order to provide:

- an improved means of reflecting the degree of seriousness of a violation
- a format which accommodates HACCP-type inspections
- a format which accommodates state-of-the-art electronic data processing and records management
- a reference to specific code provisions for each observed violation; and

(f) Recognizing that many food establishments in the retail segment now perform one or more food processing or service functions within the same establishment which can subject them to multiple permits and differing and often inconsistent regulations enforced by different agencies.

Developing a new unicode may provide an opportunity for states to improve program controls and existing codes and procedures. These opportunities include the following:

(a) Initiating a new and improved effort to involve state and local regulatory officials and the industry into the code development process, including specifically focused approaches that will increase input and foster greater commitment to the process and the end product;

(b) Devising a new and improved mechanism to revise the codes to provide a more formal approach that better reflects joint concurrence for changes by the FDA and state and local regulatory officials with appropriate industry input;

(c) Working toward eliminating duplication and conflicts in states where the retail food and food service responsibilities are in separate agencies by encouraging the agencies to work together on joint adoption of the unicode and interpretations; and

(d) Developing new and improved code language and provisions that can be applied to existing codes irrespective of whether the unicode is adopted by a particular jurisdiction.

ALTERNATIVE UNICODE DEVELOPMENT STRATEGIES

National Public Health Organizations

National public health organizations such as the Association of Food and Drug Officials (AFDO) have developed model codes in the past. It is also possible that the Conference for Food Protection could undertake this project. Having a national organization develop the unicode has the advantage of getting a high level of involvement and commitment from members of the organization to adopt the code. The disadvantages are that (1) most national organizations do not have a paid staff available to work on a project, or committees that meet regularly that could develop a code on a timely basis, (2) the same problems with lack of staff can prevent the code from being interpreted or otherwise maintained on a timely basis, and (3) having one national organization take the lead in developing the unicode might discourage other national organizations and other concerned parties not members of the lead organization from taking an active role.

Private Standards Setting Organizations

A unicode could be developed and maintained by a nationally recognized private public health standards setting organization such as the National Sanitation Foundation (NSF) or the 3-A Standards Committee. These organizations usually have the staff and public health screening processes needed to develop and maintain standards on a timely basis. They also usually have current scientific evidence and can produce high quality documents. The disadvantages are that (1) private third party organizations have not been involved in retail food model code development in the past and are not generally recognized by federal and state regulators as an authority on codes, (2) the source of funding to pay for code development, promotion and maintenance would most likely be the regulated industry itself, (3) it is unlikely that a national health organization or FDA could be a co-signer on the code or interpretations, and (4) a code provision or an interpretation which takes a position contrary to an established federal or state regulatory position could be adopted.

Food and Drug Administration

The FDA has had a long history of model code development, as discussed earlier. FDA has considerable scientific and regulatory resources that can be used to help provide a sound scientific basis for a code and interpretations. It has program resources necessary to develop and maintain the unicode and procedures in place to coordinate development with national public health organizations, industry associations, academia, and private third party organizations. The FDA is recognized as the authority on model food codes by most federal, state, and local regulatory agencies. In addition, all codes and interpretations issued by the FDA would be consistent with any relevant federal food laws and regulations. The disadvantages of the FDA taking the lead are that FDA resources are limited, and that some states and local jurisdictions do not welcome federally authored codes that do not provide adequately for state and local consensus input.

PROPOSED UNICODE DEVELOPMENT STRATEGY

The most viable strategy for developing a unicode and ensuring that it will be effectively maintained is to have the FDA assume the leadership role. The process to be followed by the FDA, however, *must* be carefully structured to maximize state and local regulatory, consumer, and industry input. There should also be procedures to ensure that when a controversial issue arises, no one party can veto or otherwise detract from the unicode as a consensus document.

Two examples of mechanisms which have been effectively used in developing food regulations and related standards based on federal, state, and industry cooperation and consensus are (1) the procedures followed by the National Conference on Interstate Milk Shipments (NCIMS) and the Interstate Shellfish Shippers Conference (ISSC) and (2) the process used by the National Sanitation Foundation (NSF). These procedures are used as the basis for the strategy proposed.

Key Participants

Persons involved in the development of a unicode and the periodic revision process should be limited to representatives from key organizations who are willing and able to commit the resources necessary for participating in order to keep the process manageable. Key organizations include the following.

1. Food and Drug Administration (FDA)
2. Association of Food and Drug Officials (AFDO)
3. American Public Health Association (APHA)
4. Conference of State Health and Environmental Managers (COSHEM)
5. International Association of Milk, Food, and Environmental Sanitarians (IAMFES)
6. Institute of Food Technologists (IFT)
7. National Conference of Local Environmental Health Administrators (NCLEHA)
8. National Environmental Health Association (NEHA)
9. National Society of Professional Sanitarians (NSPS)
10. Food Marketing Institute (FMI)
11. National Automatic Merchandising Association (NAMA)
12. National Restaurant Association (NRA)
13. National Sanitation Foundation (NSF)
14. A national consumer group

Although the joint involvement of regulatory, consumer, industry, and third party representatives is important, their roles should be sufficiently different to avoid any possible conflict of interest. This can be accomplished if the industry and third party representatives have an advisory rather than a concurring status. Such an advisory status would allow industry and third party representatives to take an active leadership role in developing scientific and technical information and to fully participate in committee deliberations, but not vote on any regulatory provisions of the unicode. This approach works well in the Interstate Milk Shippers program, which is considered by many public health officials to be the nation's most effective federal-state cooperative program.

Unicode Development Task Force

The old saying that a camel is a horse put together by a committee can provide valuable guidance in deciding upon a strategy to develop a unicode in a timely and efficient manner, but with as wide a base of involvement as possible. Mechanisms recently used by the FDA in developing the bulk foods guidelines in cooperation with AFDO and the updates to the National Shellfish Sanitation Program (NSSP) Manuals of Operation in cooperation with the Interstate Shellfish Sanitation Conference (ISSC) provide worthy models to consider. These documents were successfully developed in a timely manner because key participants were involved at critical phases of the development process. Based upon this experience, the following development process is proposed.

1. Key associations would be approached by the FDA and asked to participate in the development of a unicode and to designate a representative to serve on a Unicode Development Task Force. Funding for their participation would be provided by the organization being represented. Each organization would advise its members of the cooperative unicode development project and elicit their support.
2. The task force would meet for the purpose of agreeing upon an overall philosophy, scope, and structure for a unicode.
3. The FDA would complete a first draft of the unicode based upon the guidance provided by the task force.
4. The first draft would be sent out for concurrent task force and public comment, including a notice of availability in the *Federal Register*. Copies of the draft would be sent to as large a number of representatives of the key organizations and other known interested persons as is practical. Comments would be provided to FDA in writing.
5. The comments would be compiled by the FDA and provided to the task force along with a second draft of the unicode based upon the FDA's consideration of the comments.
6. The task force would meet a second time to consider the

public comment, the FDA's second draft, and to agree upon a final redraft as necessary.

7. The final task force redraft would be cleared by the FDA and presented to each key organization for endorsement.
8. A final unicode would be published by the FDA in concert with one or more of the key organizations, a notice of availability would be provided in the *Federal Register*, and copies of the code would be sent to all participants and made available to all interested persons.
9. Each key organization would be requested to develop a program to encourage the adoption of the unicode by appropriate local, state, and federal jurisdictions. The FDA would monitor the level of adoption and provide periodic feedback to the key organizations as to the success of the program.

PROPOSED UNICODE MAINTENANCE STRATEGY

There will be a need to develop interpretations and revisions to the unicode on a continuing basis. These code maintenance activities should be accomplished in a timely and efficient manner with wide input from key organizations and interested individuals. In addition, there should be a formal mechanism for key regulatory organizations, with industry consultation, to concur with any unicode interpretations and revisions. The following unicode maintenance process is proposed.

1. A unicode maintenance committee composed of representatives of the key organizations would be established.
2. Proposed interpretations and unicode revisions would be requested or presented in draft form to the FDA by an interested party.
3. If the FDA concurs with the need for an interpretation or revision, a priority and time table for its development would be established, and a draft interpretation or revision would be prepared.
4. The FDA would periodically report to the committee as to requests received which did or did not receive the FDA concurrence, the FDA's time table for drafting the interpretation or revision, and the status of those under development.
5. A majority of the committee could request that the FDA reconsider and develop an interpretation or revision on any request originally rejected by FDA.
6. The FDA would develop a draft of the interpretation or revision, send it out for concurrent review and comment by

the committee and other key interested individuals, and redraft the document on the basis of the comments received, as necessary.

7. The redraft would be resubmitted to the committee for majority concurrence by the regulatory representatives *if* there were any controversial issues and the committee had requested the opportunity to review and concur with the final document.
8. Any interpretations or revisions would be printed and distributed to all interested parties and would be reported upon by each key organization in its newsletters and periodic meetings.

In order for the unicode to be truly a consensus document, procedures are also needed to handle situations when regulatory representatives cannot agree upon an interpretation. It is proposed that when this situation occurs the committee prepare an *Issue Paper* which objectively discusses the problem and the different opinions held by committee representatives on how to address the problem. Issue papers would be distributed to all interested parties the same as with interpretations but with the clear recognition that they are advisories and not interpretations.

UNICODE ADOPTION STRATEGY

In order to maximize the level of adoption of the unicode once it has been developed, specific strategies should be established. These strategies should also be based upon the premise that involvement fosters commitment, which leads to effective implementation. It is proposed that the strategies include the following:

1. The unicode maintenance committee also function to promote the adoption of the unicode by working through the representatives' respective organizations to encourage and assist states where necessary in adopting the unicode.
2. The FDA regional offices program time to encourage and assist states with adopting the unicode.
3. States be encouraged to establish a food program advisory committee made up of representatives of the public and private sectors whose function is to (a) foster communications and improve working relationships, (b) assure consistency and uniformity of code implementation and enforce-

ment, and (c) address problems and needs for formal code interpretations and revisions.

4. The unicode adoption and maintenance committee work to have some organizations develop complete packages of audiovisual training materials, including prepared texts and lectures to accompany the code and aid the states in their process of implementing the new code.

ASSESSMENT OF UNICODE DEVELOPMENT PROPOSAL

The coordinated development of a unicode by federal, state, and local regulatory officials, the industry, and other interested parties may be feasible. However, developing and maintaining a unicode would require considerable FDA resources as well as a commitment of time, money, and people by several key national organizations. There are also several important factors that might limit the interest of some state and local jurisdictions in a unicode. These potential adverse factors should be carefully weighed against the potential benefits before a commitment to developing a unicode is made. The potential adverse factors include the following:

1. Public health officials who are now promoting the adoption of the individual codes might oppose the adoption of another code which would replace the code they had been promoting.
2. Since state laws are often slow to change, a proposal for another change might not be well received.
3. A unicode might not be appropriate for a state which has retail food protection responsibilities split in separate laws and administered by more than one state agency.
4. Some states may not support making changes in code format and the reporting of inspectional findings in order to accommodate HACCP-type inspections and computer use.
5. A single code which covers commercial and institutional food service, including mobile and temporary operations, all types of retail food stores, including bulk merchandising, and food vending, might prove to be unwieldy.
6. It may not be possible for several national organizations with different agendas and priorities to work together and with FDA to agree upon, in a timely manner, the various provisions, revisions, and interpretations necessary to have a viable unicode document.

FINDINGS AND RECOMMENDATIONS

On balance, the FDA believes the potential benefits to be derived from having a consensus document with a broad base of support by state, local, consumer, and industry officials that is compatible with up-to-date inspectional techniques and records management systems outweighs the potential adverse consequences. Therefore, it is recommended that:

1. The *National Conference on Food Protection* endorse the development of a model retail food protection unicode as a priority project by the FDA and other key national organizations; and
2. The *National Conference on Food Protection* endorse the proposed cooperative, consensus approach to developing and maintaining a unicode that involves participation by the FDA and several key national organizations.

23. Development of Standards and Regulations by a Third Party Consensus Process

George A. Kupfer

Milwaukee, Wisconsin, Health Department

INTRODUCTION

There has been continuing discussion as to the most effective and meaningful way to promulgate food sanitation standards and regulations. It is generally accepted that standards and regulations should be reasonable, uniform, and result in the protection of the public's health. Standards need to be technically viable and responsive to change. They need to bring about feasible solutions to potential problems which affect the public's health. A long-standing controversy has been the manner in which these characteristics of standards should be attained and the process by which they evolve.

BACKGROUND

Food sanitation standards have been promulgated since Biblical times. Throughout the years, as technical knowledge increased, as food technology became more sophisticated and as food processing and handling became more complicated, increasingly complex standards and regulations have been promulgated. Basically, such standards relate to equipment, facilities, and food handling. Interested and involved parties in the field of food sanitation include the federal, state, and local levels of government, manufacturers of food processing equipment, industrial processors of food, transporters of food, operators of food establishments, and consumers. Each of these participants has a role and is affected by food sanitation standards and regulations. Each, therefore, is keenly aware of the importance of the process of setting regulations and standards and the ultimate impact that they have.

In the 1984 "Proceedings of the Second National Conference for Food Protection,"[1] it is noteworthy that 7 of the 13 recommendations of the Standards and Regulation Committee related to the standards development and implementation process. There is increasing concern as standards and regulations become more complex and more restrictive due to the huge potential impact on the regulators, the industry, and consumers. In the past there have been a number of processes through which standards have been created. Regulatory agencies at all levels of government, in an effort to protect the public's health and minimize food sanitation problems, have promulgated numerous regulations governing equipment, facilities, and food handling practices. Various industries have also developed standards in an attempt to respond to government regulations, to provide some uniformity within the industry, and to guide manufacturers and operators in their efforts to achieve an acceptable level of performance as it relates to proper food sanitation.

Some standards and regulations have been developed by organizations that specialize in developing standards. These organizations generally develop model codes or standards which can be adopted by regulators as well as industry. Their goal also is to promote uniformity.

A more recent development has been the occasional use of a third party which develops standards and regulations through a consensus process. In the consensus process the various participants are brought together by a "neutral" party and asked to discuss their various concerns and problems. It is an experience and expertise "exchange" mechanism that has as a goal a "bargained" standard that achieves the protection of the public's health and is generally acceptable to regulators, industry, and consumers. The federal government has recently supported a "regulatory negotiations" process for environmental safety standards. In the consensus process regulations and standards may be developed which serve as a model only or may be adopted with an accompanying certification program. The certification program is one in which equipment, facility, or process is evaluated in light of the consensus standard on an ongoing basis. Where compliance is obtained, the product, facility, or process is certified to signify compliance with these standards.

DISCUSSION

The discussion and controversy surround the benefits and disadvantages of the various methods utilized to promulgate standards and regulations. On one end of the spectrum are the government agencies, which may feel that the most effective manner to protect the public's health is to utilize their experience and technical knowledge to promulgate rules and regulations. This process may be accomplished with little input from the industry or the public except for public hearings on the final product. It can be done rather quickly, especially at the local level.

On the other end of the spectrum are those industries, which may feel that they can regulate themselves and do not need government regulations. They may point to their experience, reputation, and the effect of competition as adequate "regulators" to assure proper food sanitation in their final product. Obvious problems can be cited in each of these extremes. While it is true that the government agency has traditionally been the regulator, it is also true that various government agencies have varying experiences, expertise, and support. The ultimate result of "government only" promulgation has been a myriad of varying regulations and standards which result in varying levels of food sanitation. It is true that there are special circumstances and conditions which can require differences in regulations to accommodate those special needs. Geographic differences may result in the need to vary regulations to take into account weather extremes. Certain ethnic and religious factors may also result in the need to vary regulations in order to accommodate such sensitive issues and needs.

Water supply variances can result in special problems which need to be addressed. While there are reasons to have variances, those reasons do not justify the vast differences that exist between like regulations promulgated by regulatory agencies across the country.

On the other hand, industry's self-regulation has also resulted in varying degrees of effectiveness in protecting the public's health. There are numerous reasons that could be cited for this observation. The ultimate commitment of a specific organization toward the protection of the public health can greatly affect the regulations it self-imposes. The balance between economic considerations and the protection of the public's health in industry has varied considerably. The commitment has also tended to vary due to other influencing factors, such as design and aesthetic concerns. Many industries have exhibited a major commitment to protecting the public's health. To be sure, this is to their advantage, since the buyers and consumers of their goods, facilities or processes normally will react more favorably over time where strict adherence to good food sanitation standards is followed.

Nevertheless, experience has shown that while some, if not most, industries are extremely competent and committed to good food sanitation, there are also a large number which have exhibited little or no commitment, thus affecting the usefulness of self-regulation.

If one accepts that the problems multiply as one approaches either end of the spectrum, it then becomes a meaningful discussion to determine how one might develop an alternative process which would combine important aspects of each in the most effective way.

There have been attempts by federal agencies, such as the Food and Drug Administration, to develop model standards by soliciting input through questionnaires and other means in order to involve the various participants in the promulgation of the model standards. This has been partially successful in producing model standards. A problem with this process has been that it is a very lengthy process, and while it solicits input, it does not result in a forum in which ideas and alternatives can be discussed face-to-face and an acceptable solution jointly developed. It is more a polling process whereby each one sends in their suggestions based on their own knowledge and experience but with little exchange between the participants. The results are reviewed and a decision made by the agency on the issues. The final product often is unacceptable to many and thus experiences varying degrees of acceptance.

As one comes down the scale from the industry side of the spectrum, one approaches the work of organizations which represent broad industrial areas such as the baking or restaurant industries. Then there are those organizations which have attempted to be neutral and have convened internal technical groups to develop regulations. As the process approaches the midpoint of the spectrum there tends to be more trust and wide general acceptance by the participants. A main deterrent to wide acceptance of food standards developed by these methods is the feeling by one group or the other that their problems have not been thoroughly understood or discussed. Another deterrent is a standard which is often developed under either a "best" or a "worst" case scenario with expectations that everyone should adopt such regulations. Since special needs are not accommodated, nor is there any mechanism to accommodate them, regulators often develop variances to such efforts. At the time of the 1984 Second National Conference for Food Protection, a major concern was expressed about the nonuniformity of laws and regulations, the lack of uniform interpretations of standards and the ultimate need to develop a more meaningful and acceptable process for standards development to resolve these problems.

The answer to several of the issues and expressed concerns may be the development of regulations and standards through an effective consensus process. An example to study is the National Sanitation Foundation (NSF), which has so successfully implemented this consensus process in promulgating standards for food equipment. This organization acts as a third party to bring together the participants to jointly develop standards. By the very nature of the purpose of the standards, the NSF must be biased to assure an outcome of reasonable protection of the public's health. In the process of doing this, however, this third party assumes the role of convening the best technical and knowledgeable sources who represent, as closely as possible, the major involved parties affected by the proposed standards. Thus begins a lengthy process to develop the needed standard. The value of this third party process is that it allows face-to-face debate by experts representing the involved parties and an appreciation for each other's positions, concerns, and problems. The involved process allows substantial input. This results ultimately in proposals which are voted upon. There are several checks and balances, and exhaustive attempts to resolve differences.

While the final product may not be what was originally conceived by any of the participants, it most often results in a standard which is acceptable to most, and which is adopted by regulatory agencies and is subscribed to by the industry. It would appear that this process of third party development of standards and regulations through a consensus process is perhaps as close as one can get to balance the interests of the various participants, to obtain maximum uniformity, and to reasonably protect the public's health. Perhaps there is no process which can ultimately result in absolute acceptance by all or which will solve all problems. There will always be differences of opinion in the interpretation of the word "reasonable" and the phrase "reasonable protection of the public's health."

There are other problems which are inherent in this and perhaps most other efforts to develop joint standards and regulations to protect public health. One such problem is the length of time it takes to respond to new technology or new research findings which show critical flaws or gaps in the existing standard. The process, by its very nature, requires substantial time, since full input is the key to its success. Emergency problems or conditions which occur result in the need for a regulatory agency to be able to react quickly to correct a problem which is not addressed in an existing standard. If a third party consensus approach is the choice, mechanisms must be built in to allow a quick interim resolution to resolve critical problems immediately. Such emergency procedures should be followed by quick review and change of the standard through the regular process.

Another problem is the funding of a third party consensus process. Local governmental agencies, by their very political nature, are seldom budgeted adequately to travel out of state and spend time with other organizations and industries to develop national uniform regulations. With increasing budgetary and resource restrictions, this problem will probably not improve. Industries have in the past been more willing to spend the funds necessary when it was apparent that their voice and concerns could be heard and could impact upon regulations being adopted by regulators. While industries dedicated monies to this process in the past, they, also, may now face restricted budgets and a hesitance by their companies to devote funds necessary to support this type of process. While a third party consensus standard development process may be the most acceptable approach, this problem must be addressed for it to be a success.

Recently the Environmental Protection Agency has entered into a grant project arrangement for the third party development of consensus standards relating to additives, both direct and indirect, to drinking water. The Agency, faced with decreasing resources and increased environmental problems, recognized the value of a third party consensus process to develop the complex standards needed to regulate direct and indirect additives to drinking water. The federal grant is an experiment to determine the value of the third party consensus process in developing complex standards to protect the public's health. The program is being funded by the federal grant and by industry monies. The goal is to bring together the expertise of the participants to formulate standards which are reasonable and yet effective. If successful, this project may serve as a model for future efforts. While this process has just begun, it holds great promise as an effective method to promulgate complex standards and regulations which affect the nation.

PROPOSAL

It is recommended, on the basis of the previous review, that this National Food Conference support and promote the development of a standards promulgation process which involves the various levels of regulatory agencies, applicable industry, and the consuming public. A model that has already demonstrated success and is being further tested is the third party consensus process as developed by the National Sanitation Foundation. It is believed that this process holds great promise in the development of uniform national model standards for food sanitation.

The ultimate test must be the reasonable protection of the public's health through proper food sanitation. The standards should have built into them a process by which new problems that create immediate threats to the public health can be handled immediately while the revision process is quickly initiated. The process must provide regular review and updating to assure relevance of the standards to current technology and knowledge. To be successful the process must assure full participation of each of the affected groups. Funding mechanisms, perhaps similar to that recently employed for the drinking water additives standard project, must be developed. It is essential to success that any third party that is involved in consensus standards development must be neutral, reputable, and respected. It must be dedicated to resolving issues through a fair and equitable process, with the ultimate goal to provide reasonable protection to the public health. It is recognized that the cost and length of this process may limit its usefulness to major standards of national significance. However, this principle of standards development has great value for more limited regulations also. There will be some specialized conditions which exist locally which may result in some variance from a national consensus standard. However, with proper development those variances should be minimal.

There will need to be an all-out educational effort by industry, by consumers, and by the regulatory agencies to promote maximum acceptance of the third party consensus standards.

The perceived result of the development of standards in this manner is a more effective, feasible, and more widely accepted standard. It would also promote better trust and a cooperative attitude between regulators, industry, and consumers, with more uniformity and better compliance than now exists.

REFERENCE

1. U.S. Department of Health and Human Services, Food and Drug Administration. 1984. "Proceedings of the Second National Conference for Food Protection." Washington, DC.

24. Forces That Initiate Change in Standards and Regulation

Sherwin Gardner

Grocery Manufacturers of America

INTRODUCTION

To some people, Harvey Wiley's reign at the Bureau of Chemistry is the source and continuing supply of standards for regulating the safety, labeling, and wholesomeness of the U.S. food supply. Without taking anything away from the very significant contributions made by Dr. Wiley to American food law, this representation is too simple and far from the truth.

Clearly, Wiley's work and what must have been a strong personality had a marked influence on the development of the Food and Drugs Act of 1906 and the regulations that flowed from it. There were other influential personalities and events at the time that also contributed to the development of food safety standards. For example, Upton Sinclair's book *The Jungle*, is generally recognized as the inspiration for the Federal Meat Inspection Act. Another personality, of course, was President Theodore Roosevelt, who urged enactment of the Meat Inspection Act. President Roosevelt also personally contributed to the safety evaluation of food additives when he told Wiley that, "Anybody who says saccharin is injurious to health is an idiot." Perhaps the more significant act of Roosevelt, in addition to signing the Food and Drugs Act and the Meat Inspection Act, was to establish the Referee Board of Consulting Scientific Experts, chaired by Ira Remsen of Johns Hopkins University, to rule on controversial issues of food safety.

Indeed, a study of history will show that the origins of a national food law can be traced to the mid 19th century. Great Britain adopted a national food and drug law in 1875. Strong personalities, usually together with adverse events arising from food adulteration, have been responsible for changes in standards and regulations governing foods. Several books and papers have been written on the subject; a comprehensive paper summarizing the history of government regulation of the food supply from biblical times to the present was presented at the Food and Drug Law Institute "Food Update" Meeting in 1983.

It is relatively easy to see from a historical perspective what were the forces of change in food standards and regulation. Perhaps it is more difficult to see these forces when you are living what will become history in the next century. Certainly, there is a very complex set of interdependent factors at work, the taxonomy of which has not changed throughout history:

- Social and cultural factors
- Scientific and technical factors
- Economic factors

SOCIAL AND CULTURAL FACTORS

Social factors generally refer to those conditions under which we live and work and conduct our daily lives. For example, an important social factor today is the increasing number of women in the work force. Today, 60% of women between the ages of 20 and 44 are employed outside of the home. As a result, more than half of all households now include two wage earners. The trend of female employment outside the home is likely to continue and is strongly suggested in the employment patterns of young mothers. Thirty years ago, only 10% of women with preschool-aged children worked outside the home; today, more than 50% of women with preschool-aged children are employed.

Some regard this changing role of women as the most significant social shift of the past generation. Certainly, it has changed the image we have of the so-called average household. Time is at a premium, and the results are that (1) people eat out more often and (2) when they do eat at home, they spend less time shopping and preparing their food.

Today's society is also considered more affluent than past generations. This is due in part to the large increase in the number of working women, and in part to the general expansion of our economy. We now have more people working in this country than ever before. The overall result, however, is increased affluence, and this has resulted in a shift in people's eating habits.

Some of the effects of these social factors on foods are found in the areas of packaging, preparation, and storage. Precooked dinners or main courses, including meat, fish, and poultry are increasingly found on supermarket shelves. The availability of plastic wraps that display food and that also serve as cooking utensils have enhanced the marketability of precooked ready-to-serve products. The microwave oven has also contributed greatly to consumer acceptance, providing a quick and easy heating method while retaining palatability and appeal of freshly prepared foods.

It seems obvious that these factors increase the importance of safe packaging materials and safe manufacturing processes. However, we will have to wait and see the impact of these kinds of products on food manufacturing facilities, i.e., whether there will be an increase in the number of smaller, highly specialized facilities; whether new entrepreneurs are attracted to food manufacturing; and what this means for state and federal inspection capabilities.

Other important social considerations would include the increased foreign travel by U.S. consumers and the emphasis of many consumers on physical fitness. The increase in tourism has exposed people to food habits of other cultures and broadened their interest in other foods. One need only observe the weekly food section in newspapers published in most parts of the country to see the astounding array of recipes featuring ethnic dishes, i.e., dishes representative of the cuisine of countries other than the United States. The exposure to other foods has also increased the demand for imported products or for domestically produced products in the same styles as imported products. To the extent that these products compete with other products for which standards have been established, they could create pressure for changes in regulations. Similarly, to the extent these imported products increase the inspection workload of the FDA and USDA and raise questions about the safety and quality of imports, they could create pressure for changes in laws and regulations.

The interest in physical fitness in the United States has grown steadily throughout the past ten years. Diet and exercise are "in," thin is "in," and people seem to be responding to the stimuli from scientific studies of diet and physical activity and their effects on personal health. In this regard, social customs have been strongly influenced by the science and technology that have produced these findings and that have also helped to disseminate it through the visual impact of television as well as the miniaturized and ubiquitous radio and cassette players.

SCIENTIFIC AND TECHNICAL FACTORS

Clearly, science and technology have had a major impact on the foods people eat, both directly and indirectly. Virtually everything from communications systems, the widespread use of computer electronics, and advances in packaging materials, to the explosion in the use of microwave ovens and other hardware products, can be seen at work in homes, restaurants, and manufacturing facilities. It would not be possible to forecast what specific changes lie ahead as a result of these products, but it is certain from our current experience that they will have an influence.

One of the greatest influences of science derives from the findings concerning nutrition and health that are reported almost daily in scientific journals as well as newspapers and television reports. Hardly a day goes by, it seems, that some aspect of nutritional research is not reported. Fat, cholesterol, sodium, calcium, fiber: these and other components of our daily food supply have been subjected to a great deal of research and people appear to be reacting to the reports of this research by changing their consumption habits. A few statistics will illustrate the point:

- Meat consumption has decreased 18% in the last decade. Some of this reduction probably is due to nutritional research indicating that the consumption of fat ought to be reduced in order to prevent cardiovascular disease and other illnesses. Beef and poultry producers are selecting strains of animals that will yield less fat and are promoting their products with this in mind. (One major poultry producer has run double-page advertisements in major newspapers in the northeast claiming lower-fat poultry.) Indeed, the USDA has changed the grading factors for beef to account for people's preference for lower fat content products.
- The consumption of distilled spirits is down approximately 5% in the five-year period 1980 to 1984. Some of this may be attributable to stiffened drunk driving laws; some of it to concerns about the health effects of alcohol, particularly in the case of expectant women. Some of it may be due simply to changes in preference and substitution of other competing beverages, such as wine and soft drinks.
- The sales of calcium supplements have increased dramatically in the last year, up by 40% over 1984 and growing at the rate of 5% a month (*The Wall Street Journal*, March 19, 1986). This change is probably attributable to the concerns of women that calcium deficiency is linked to osteoporosis.
- The per capita consumption of fish in the United States has increased by approximately 10% since 1980. This, too, may be attributable to reports by nutritional researchers indicating the beneficial health effects that may result from consuming more fish and poultry as opposed to red meat.

Some of the implications of these trends have already occurred in the way foods are marketed for their beneficial nutritional characteristics and in increased nutrition labeling interests by both retailers and manufacturers. Manufacturer interest in providing information relating food and health recently caused the FDA to modify its policy in this area.

Science and technology also have influenced attitudes about food safety. It is generally recognized that advances in analytical chemistry have dramatically increased the ability to identify very small amounts of unwanted materials in foods. Further, changes in toxicological testing practices have increased the ability to identify potentially hazardous effects of food use materials. The combined result of these advances in hazard analysis is to increasingly call into question the use of food materials that were formerly believed to be safe.

In addition to affecting specific substances, these increased analytical capabilities have generated a number of proposals for change in the laws governing food safety. One consideration for change is based on the concept of de minimis risk, and would cause the prohibition of use of food substances only when the risk exceeds a negligible or insignificant level. Indeed, this approach has already been put into practice for some food additives and pesticides. This approach provides balance to the enormously increased ability to identify potential risks.

Other proposals for change would increase the restrictions for marketing approval of products. Societal experience with market withdrawals because of safety questions caused by after-the-fact risk analysis has created the desire to avoid such problems by increasing the level of evidence before granting marketing approval. Pesticide approval is an example of one area affected by this concern; biotechnology is another. The food production pipeline is complex and has both time and space dimensions. The challenge for our society is to build balance and flexibility into the standards for safety of food products, both new and existing, so that beneficial effects of technology can be realized without unduly delaying their introduction to the market, and so that sudden banning actions do not create chaos in the market.

ECONOMIC FACTORS

Changes in consumer preferences arising from social factors and the introduction of new products arising from scientific and technological factors quickly translate into economics. Those with established products seek to protect their market from the introduction of new products. Established standards and regulations are frequently used for this purpose. The parties seeking to protect market interests are both private and public. Manufacturers of traditional products with a long history of use and for which standards may exist, such as hot dogs, pasta, and cheese, clearly want to distinguish their products in the marketplace from new products that will compete with them. Because food products are also agricultural products, states which have a substantial part of their economic base in the affected agricultural area will also become parties seeking protection through restrictive regulations and standards; for example, through marketing barriers, such as imitation labeling.

Those seeking to introduce new products face the obstacle first of demonstrating their safety, and second of satisfying the demands of regulations and standards that govern their identification. This situation can also bring about changes in law and regulation—to relieve existing restrictions or to carve out clear ground rules under which new products can enter the market with more certainty and less delay. For example, although manufacturers of substitute protein products or products derived from recombinant DNA techniques may not seek additional regulation, it may be helpful to establish regulations clarifying the circumstances under which they can be approved for marketing.

CONCLUSION

Social and cultural factors, scientific and technical factors, and economic factors are strong driving forces that will change and shape standards and regulations for foods.

Of the three, the social and cultural factors are likely to be the strongest, given the increased public participation in legislative and rulemaking activities at both state and federal levels. It is, after all, the reactions of society that ultimately determine the way science, technology, and economics are applied to the laws governing food products.

In the past 20 years, public participation in regulatory activities has become much more intense, organized, and flexible. Consumer groups have increased their range of interest, working at the state levels, in Congress, the federal regulatory agencies, and in international agencies as well. Industry groups have generally worked at these same levels and have increased in number to provide representation in new areas of prospective regulation or to focus on specialized issues.

The impact of the media is also an important element in the public process of changing standards and regulations. Television, radio, and newspapers carry the information about new scientific findings to the public. Given the abbreviated format of television and radio, coupled with intentional attention-getting newspaper headlines, both "good" information as well as "bad" information is frequently absorbed by people without a full understanding of the subject. Thus, we have consumers vigorously supporting bans of pesticides even when risks are negligible and hypothetical. We also see consumers enthusiastically embracing new dietary regimens that have not been fully evaluated but which hold out the prospect of improving personal health.

The purposes of food standards and regulations are to protect the public against unsafe and unwholesome products and against economic fraud and deception. Each of the groups affected by these rules—consumers, state and federal regulatory agencies, and manufacturers—has a shared interest in seeking a reasoned and equitable approach to protection. Overprotection with unduly strict requirements will delay, deny, or increase the costs of the advantages that new products can bring. Underprotection with inadequate requirements will undermine confidence in products and the regulatory process, and may cause harm. Despite the complexity of forces that bring about changes in requirements, it is essential to balance potential risks against benefits. It is also essential to seek a uniform regulatory approach across the country, and to inform and bring understanding to those who are involved in the regulatory process.

25. The Forthcoming Merger in Quantitative Risk Assessment

Robert L. Sielken, Jr.

Sielken, Inc.

INTRODUCTION

Quantitative dose-response modeling and risk quantification comprise two important inputs to cancer risk assessment which, in turn, can be a major ingredient in cancer risk management and the regulatory process. However, the dose-response models currently used in the regulatory process are generally overly simplistic probabilistic representations of highly complex biological phenomena; these models are not biological models. Furthermore, the risk quantifications currently used to summarize the dose-response relationships are usually inadequate reflections of several risk dimensions and are generally poorly understood by the public. The discussion and procedures presented herein are intended to promote improvements in dose-response modeling and risk quantification. Additional research will hopefully provide further improvements.

Researchers have been developing new dose-response modeling techniques which attempt to utilize more of the available scientific knowledge and expertise. These techniques involve

1. Representations of the exposure in terms of dose scales based on cell turnover rates, repair processes, immune system responses, and physiological and pharmacokinetic models of the absorption, delivery, metabolism, and elimination of chemicals
2. A new family of dose-response models reflecting the distribution of individual background exposure levels and individual susceptibilities to low levels of a chemical
3. Inclusion of age-dependent and time-dependent changes in the number of target cells and/or intermediate cells in a multistage process
4. Methods for exposures where the dose levels, susceptibilities, and background exposures are not constant over time

5. Inclusion of competing risks which can precede and, hence, censor or "prevent" a specified response

The new modeling components are biologically meaningful and physically identifiable. Hence, their functional form and dependence on dose, time, and age can frequently be determined through scientific studies before they are incorporated into the estimation of the cancer dose-response relationship from chronic bioassays or epidemiological studies.

New statistical procedures have been and are being developed to keep pace with the modeling advances. Maximum likelihood estimation procedures have been extended to include:

- The time at which the subject was observed
- Censoring due to competing risks
- Uncertainty in a subject's cause of death

The statistical and practical differences between an estimate of risk and a bound on risk have been emphasized and illustrated. New advanced bootstrap procedures which reflect statistical variability instead of worst-case analyses and which provide both upper and lower bounds on risk characteristics have also been developed.

Risk quantifications have been broadened to reflect more of the time-to-response component and consideration of not only lifetime probabilities but also how the probability of a specified response changes with the time period being considered. In addition, procedures have been developed to put the appearance of mathematical certainty that surrounds quantitative risk assessments into a more realistic perspective. Techniques for representing the quantitative impacts of the assumptions, value judgments, policy decisions, choices, etc., made in a risk assessment have also been demonstrated.

The merger of these many advances in each of the areas of modeling, estimation, and quantification should be forthcoming in the near future.

MODELING ON MORE RELEVANT DOSE SCALES

The dose level should be expressed on a biologically relevant scale. For example, the dose level corresponding to an inhalation exposure can be expressed in terms of (1) the concentration of the chemical in the air inhaled, (2) the total amount of the chemical inhaled in a particular period, (3) the amount of the chemical or its active metabolite reaching the target cells in a particular tissue, (4) the amount of interaction with

DNA, (5) the amount of cell damage that escapes repair, etc. That is, the dose scale can be defined in a progressively better manner as either (a) the administered dose, (b) an intermediate dose, (c) the dose delivered to the target site, or (d) the biologically effective dose (reflecting not only the amount delivered to the target site but also the amount actively participating in the carcinogenic process). Using a more biologically reflective dose instead of the administered dose

1. Allows the modeler to incorporate known biology, physiology, pharmacokinetics, etc.
2. Reduces the uncertainty and the number of unknowns in the dose-response model
3. Aids extrapolations from high doses to low doses, from species to species, from route to route, and from time frame to time frame

If there is insufficient scientific knowledge to establish the complete transformation from administered dose to a biologically effective dose, then using what is known and even a partial transformation or best approximation should lead to a better risk assessment than using only the administered dose. When there is scientific uncertainty about the ultimate biologically effective dose, dose-response modeling should be evaluated using alternative expressions for the effective dose.[1]

When a cancer dose-response relationship has been modeled on an administered dose scale, there has frequently been little success in accurately extrapolating that dose-response relationship from one sex to the other, from one strain to another, or from one species to another using equivalences based solely on the administered dose or the administered dose with a body weight or surface area adjustment. Equivalent doses need to be equivalent in terms of the amount of cancer-related activity induced, and not simply equivalent in terms of intake or levels at unrelated sites. While carcinogenic mechanisms are certainly not fully understood, a lot is known or discoverable about the differences in absorption, distribution, metabolism, and elimination between species, strains, etc., and this information should be sought out and utilized when human cancer risk is estimated using a species, route, or time-frame extrapolation. For example, when the only dose-response data are experimental data in a rodent species and that data is believed to be relevant to humans, then

1. The rodent's carcinogenic risk should be modeled using the best representation of the rodent's effective dose.
2. The transformation in humans from administered dose to effective dose should be modeled.

3. The human cancer risk for a particular administered dose and exposure scenario should be estimated by, first, transforming the human administered dose to a human effective dose; second, estimating the rodent's cancer risk when the rodent's effective dose level equals the human effective dose level; and, third, estimating the human cancer risk to be the same as this estimated rodent cancer risk (Figure 1).

Figure 1. Using an estimated rodent dose-response relationship on the effective dose scale and the estimated relationship between the human administered dose and human effective dose to obtain an estimate of the human cancer risk for a particular administered dose.

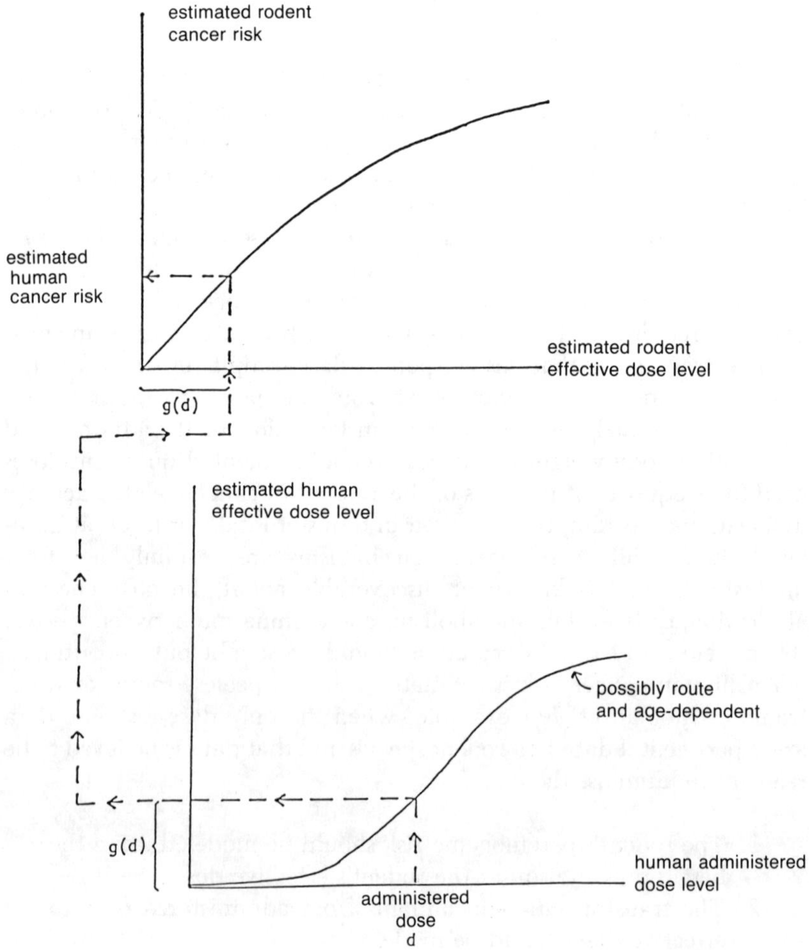

Many regulatory agencies, review boards, and researchers have realized that one of the best ways to make quantitative risk assessment more scientific is to incorporate more of what can be learned about what happens to an administered dose.[2–10]

MODELING WITH DISTRIBUTIONS OF INDIVIDUAL SUSCEPTIBILITIES AND BACKGROUND EXPOSURES

Although most carcinogenic processes and mechanisms are not completely understood, the quantitative dose-response models used in risk assessment should reflect the available experimental information and be the most scientifically creditable representations of the dose-response relationships possible. The multistage model and its time-to-response analogs are currently the most widely used models in the risk assessment process. Unfortunately, these models do not realistically reflect the biological, physiological, pharmacokinetic, etc., diversity among individuals, nor do they reflect the differences in the lifestyles and environments of individuals. However, these models (as well as most of the other existing models) can be extended to encompass some of the randomness of these individual characteristics. The remainder of this section describes such an extension.[11] Extensions of the new models emphasizing cell proliferation are described in the next section.[9,15] It should be emphasized that, while the extensions described in these sections make it possible to incorporate more of the current scientific understanding of carcinogenesis and more biological reality, these extensions still involve some simplifying assumptions. Hence, the extended models are envisioned only as improvements and not exact biological models. Even better biological models should emerge as research continues.

For numerous reasons, each individual's susceptibility to a particular carcinogen may be different. An individual's susceptibility frontier is the dose level below which the dose makes essentially no impact on that individual's normal body functions. Each individual has his own particular value f for the susceptibility frontier F. For a population of individuals, F is a random variable.

The concept of a susceptibility frontier is fundamentally different from that of a threshold. A threshold is the same for every member of the population, whereas the susceptibility frontiers for different individuals are different. (In fact, the concept can even be extended so that an individual's susceptibility frontier does not have to be constant over time). Also, the concept of a susceptibility frontier is not as rigid as the concept of a threshold in that, when the dose level drops below the threshold, the cancer probability drops immediately to the background

Figure 2. An example of a scenario where a susceptibility frontier would arise.

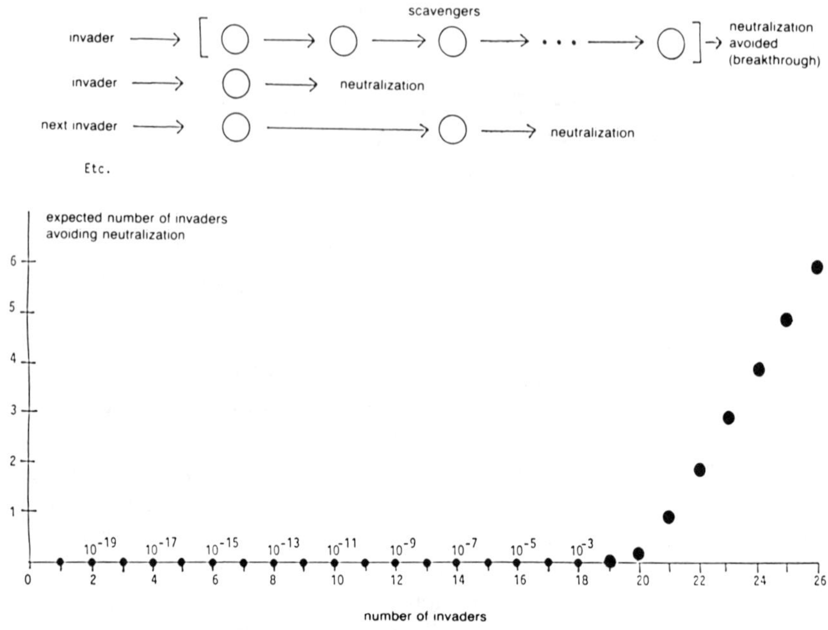

probability. On the other hand, when the dose level drops below the susceptibility frontier, the cancer probability only speeds up its approach to the background probability. In the numerical examples to follow, the rapid descent to the background probability below the susceptibility frontier has, for simplicity, been modeled as an immediate drop to the background probability. In many cases, it may be reasonable to somewhat conservatively model the decrease in the cancer probability below the susceptibility frontier as an exponential decline; in this case the decrease would become essentially a linear decrease for very small doses.

A susceptibility frontier is also different from a tolerance. If the dose level becomes greater than a tolerance, then the cancer probability jumps to 1.0; however, if the dose level becomes greater than the susceptibility frontier, then the cancer probability begins increasing more rapidly but does not immediately reach 1.0.

The concept of a susceptibility frontier can be illustrated in simple terms (Figure 2): Suppose that a chemical exposure causes a number of specific molecules to reach and invade a cell. The cells for a particular individual each contain, for example, 20 scavenger molecules with each scavenger having a 0.9 probability of intercepting and combining with an invader, thereby neutralizing both the invader and the scavenger. In such circumstances, the expected number of invaders which avoid the random neutralization process can be shown to be approximately

$$1 \times 10^{-(21-X)} \quad \text{if there are X invaders and } X \leq 20$$

and

$$X - 20 \quad \text{if there are X invaders and } X > 20$$

If the probability of a carcinogenic response is proportional to the number of invaders which avoid neutralization, then the probability of a carcinogenic response decreases steadily as the dose decreases until the dose of the chemical generates only approximately 20 invaders; at this point further decreases in the dose correspond to very rapid decreases in the probability of a carcinogenic response. In this example, the dose level corresponding to the generation of 20 invaders would be called the individual's susceptibility frontier f. In general, if an individual has a nonzero susceptibility frontier, then it occurs where, as the dose decreases, the effective dose shifts from a steadily decreasing function of the administered dose to a more rapidly decreasing function of the administered dose.

In addition to an individual's susceptibility frontier, each individual has a background exposure to the carcinogen. The background exposure is denoted herein by the dose D_0. Each individual has his own particular value d_0 for the background dose D_0. An individual's background exposure is dependent upon such factors as environment and diet for animals. A human's d_0 can also depend upon occupation, lifestyle, etc. For a population of individuals, D_0 is a random variable.

The individualized response model is the result of the merger of the random susceptibility frontier F and the random background dose D_0 with the concept of a multistage carcinogenic process and its extensions based on time and cell proliferation. In the individualized response model, the proportion of the entire population at risk which is expected to develop a specified carcinogenic response by time t for an administered dose d is

$$P(t;d) = \int_0^\infty \int_0^\infty P[t;g(d,d_0,f)] \times p_{D_0}(d_0) \times p_F(f)\, dd_0\, df$$

where

$P[t;g(d,d_0,f)]$ = the probability of the specified carcinogenic response occurring by time t at dose d when the individual's susceptibility frontier is f and his background dose is d_0

$p_{D_0}(d_0)$ = the probability density function of D_0 evaluated at d_0 (i.e., the likelihood of $D_0 = d_0$)

$p_F(f)$ = the probability density function of F evaluated at f (i.e., the likelihood of F = f)

$g(d,d_0,f)$ = the biologically effective dose.

Roughly speaking, P(t:d) is the weighted average of $P[t{:}g(d,d_0,f)]$ where the weights reflect the proportion of the population that have a susceptibility frontier F equal to f and a background dose D_0 equal to d_0. In the numerical examples to follow, the biologically effective dose has the form

$$g(d,d_0,f) = 0 \qquad \text{if } y(d,d_0) < f$$

and

$$g(d,d_0,f) = y(d,d_0) - f \quad \text{if } y(d,d_0) \geq f$$

where $y(d,d_0)$ is the physiologically based delivered dose (see the previous section) corresponding to an administered dose d and a background dose d_0.

For example, the individualized response model corresponding to a multistage model that is linear in the delivered dose but without extensions based on time and cell proliferation would have

$$P[t;g(d,d_0,f)] = 1 - \exp\{-\alpha_0\} \qquad \text{if } y(d,d_0) < f$$

and

$$P[t;g(d,d_0,f)] = 1 - \exp\{-\alpha_0 - \alpha_1 \times [y(d,d_0) - f]\} \quad \text{if } y(d,d_0) \geq f$$

Ideally, $y(d,d_0)$ would be based on physiology, pharmacokinetics, etc. However, if $y(d,d_0)$ were to be approximated directly from the epidemiological or animal chronic bioassay carcinogenicity data, then $y(d,d_0)$ might be approximated by

$$y(d,d_0) \doteq (d + d_0)^{\alpha_2}$$

A reasonable approximation for the distribution of the susceptibility frontier F may often be the lognormal distribution since it agrees with $F \geq 0$ and has been successfully applied in similar biological situations in the past. Similarly, the distribution of the background dose D_0 may be reasonably approximated by a lognormal distribution. Such a distribution can be very skewed to match chemical carcinogens for which D_0 is usually very near zero. On the other hand, the lognormal distribution can be nearly symmetrical and more like the normal distribution if D_0 is frequently moderately large.

Figures 3 through 6 illustrate the fit of the individualized response model to experimental data with

$$P[t;g(d,d_0,f)] = 1 - \exp\{-\alpha_0\} \qquad \text{if } (d + d_0)^{\alpha_2} < f$$

and

$$P[t;g(d,d_0,f)] = 1 - \exp\{-\alpha_0 - \alpha_1 \times [(d + d_0)^{\alpha_2} - f]\} \text{ if } (d + d_0)^{\alpha_2} \geq f$$

when D_0 is zero and F is assumed to have a lognormal distribution. This model is linear in the delivered dose when the delivered dose (estimated by $(d + d_0)^{\alpha_2}$) exceeds the individual's susceptibility frontier f. The individual values f for F are randomly distributed over the population with ln(F) being normally distributed with unknown mean μ_F and variance σ_F^2. Thus, there are five parameters α_0, α_1, α_2, μ_F, and σ_F^2 which are being estimated using maximum likelihood estimation.

Figure 3 refers to the Chemical Industry Institute of Toxicology chronic bioassay data on the incidence of nasal cavity squamous cell carcinoma in rats inhaling formaldehyde.[12] Here the delivered dose is estimated to be a nonlinear function, approximately $(d + d_0)^{2.1633}$, of the administered dose, and the susceptibility frontiers have ln(F) approximately normally distributed with estimated mean −0.41232 and estimated variance 0.81620. The distribution of susceptibility frontiers implies that 99.9% of individuals can be expected to have susceptibility frontiers greater than approximately 0.044 (since $\exp[-0.41232 - 3.0 \times (0.81620)^{0.5}] \doteq 0.044$), so that for these individuals the probability of the specified carcinogenic response is essentially zero below 2.36 ppm (since $(d + d_0)^{2.1633} \geq 0.044$ implies $d + d_o \geq 0.236$ (ppm/10) or $d + d_0 \geq 2.36$ ppm).

Figures 4 and 5 refer to the mice sacrificed at 24 months in the ED_{01} study.[13] Not only is the contrast between the two specified responses (the incidence of a bladder neoplasm and the incidence of a liver neoplasm) interesting, but also, from Figure 5, it is evident that the concepts of a delivered dose and a susceptibility frontier do not preclude the possibility of a nearly linear dose-response relationship.

Figure 6 illustrates the approximation of a very steep dose-response relationship using the combination of nonzero susceptibility frontiers and a nearly linear delivered dose.

Since the susceptibility frontier F, the background dose D_0, and the

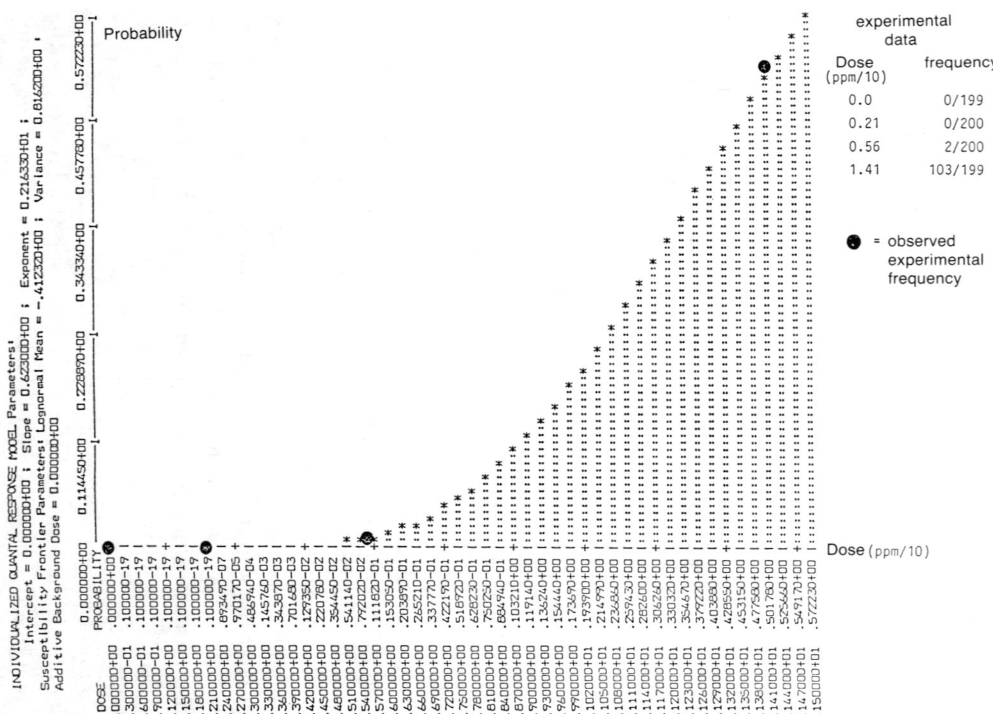

Figure 3. The individualized response model with lognormally distributed individual susceptibility frontiers and linearity in delivered dose: fit to the rat nasal cavity squamous cell carcinoma data in the CIIT Formaldehyde Inhalation Study.

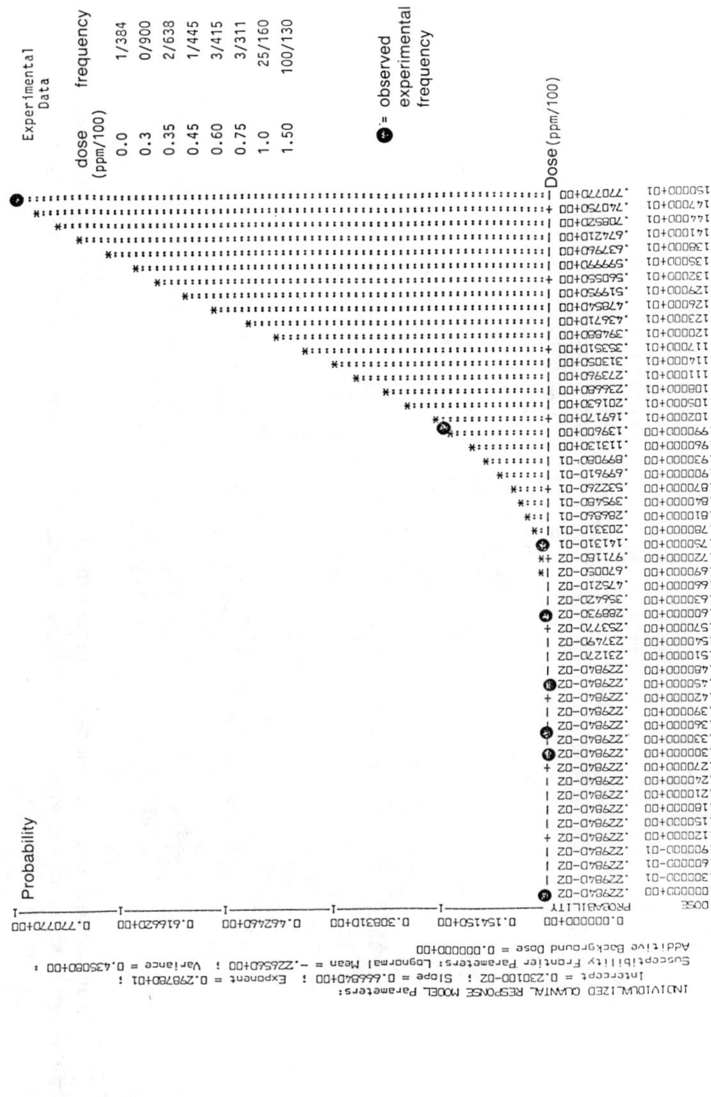

Figure 4. The individualized response model with lognormally distributed individual susceptibility frontiers and linearity in delivered dose: fit to the mouse bladder neoplasm data on mice sacrificed at 24 months in the ED_{01} study of 2-AAF.

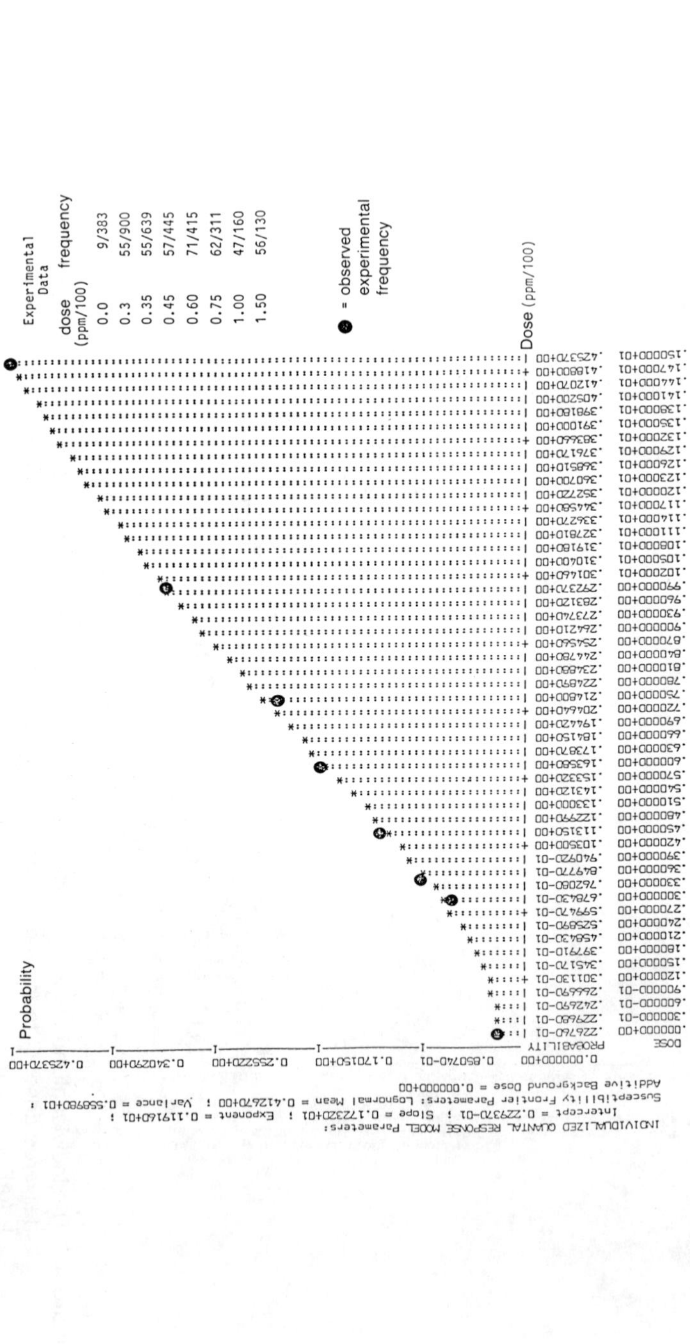

Figure 5. The individualized response model with lognormally distributed individual susceptibility frontiers and linearity in delivered dose: fit to the mouse liver neoplasm data on mice sacrificed at 24 months in the ED_{01} study of 2-AAF.

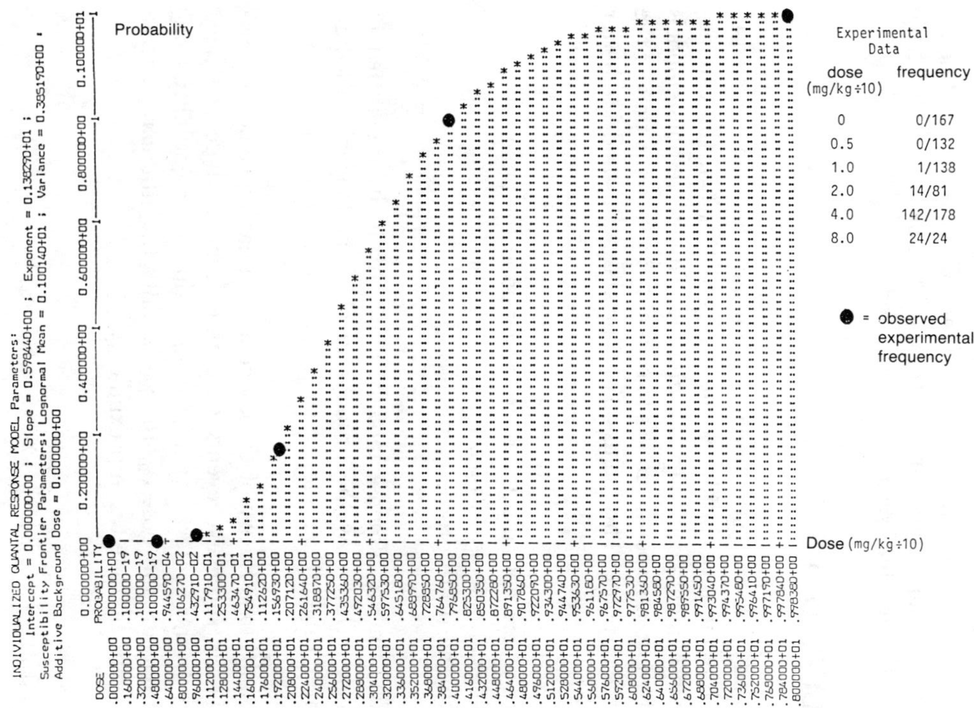

Figure 6. The individualized response model with lognormally distributed individual susceptibility frontiers and linearity in delivered dose: fit to the 1981 Khera study of ethylene thiourea.

delivered dose are all physical phenomena, the opportunity is there to determine their distribution or functional form through research and experimentation apart from the chronic bioassay for carcinogenicity. Before an individualized response model is fit to the observed dose-response frequency data, the distributions of the susceptibility frontiers and background doses and the functional form of the relationship between the biologically effective dose $g(d,d_0,f)$ and the delivered dose $y(d,d_0)$, the administered dose d, the individual's background dose d_0, and the individual's susceptibility frontier f should, ideally, have already been determined or at least estimated.

MODELING OVER TIME

In the past, dose-response models have usually been constructed to represent lifetime probabilities without regard to when the carcinogenic response might occur during a lifetime. Time-to-response models were the first dose-response models to explicitly include the time of the carcinogenic response. This inclusion of time allows the model to better reflect

- Data where not all subjects are observed at the same time
- Biological processes which differ over time
- Birth and death processes for cells.

In addition, time-to-response models allow risk to be quantified in terms which reflect how the probability of a carcinogenic response changes over time. (Examples of such risk quantifications are given later.)

In the multistage theory of carcinogenesis, a carcinogenic response occurs when k successive events stemming from a single cell have occurred.[14] In this context, for a given individual's susceptibility frontier f and background dose d_0, the probability that the carcinogenic response occurs by time t is approximately

$$P[t;g(d,d_0,f)] = 1-\exp\{-\int_0^{t-\tau} \lambda_k(t_k) \times N_{k-1}(t_k)dt_k\} \quad (1)$$

where

$$N_{k-1}(t_k) = \int_0^{t_k} \lambda_{k-1}(t_{k-1}) \times N_{k-2}(t_{k-1}) \times$$

$$\exp\{\int_{t_{k-1}}^{t_k} [\beta_{k-1}(v) - \delta_{k-1}(v)]\, dv\}\, dt_{k-1}$$

•

•

•

$$N_1(t_2) = \int_0^{t_2} \lambda_1(t_1) \times N_0(t_1) \times \exp\{\int_{t_1}^{t_2} [\beta_1(v) - \delta_1(v)]dv\}\, dt_1$$

Here, τ represents the time for a malignant cell to become the specified carcinogenic response (a detectable tumor, death from tumor, etc.), $N_0(t_1)$ represents the number of normal cells (stage 0) at age t_1, $N_i(t_{i+1})$ represents the number of cells which are in stage i at a particular time t_{i+1}, $\lambda_i(t_i)$ represents the transition rate from stage i to stage i + 1 at time t_i, and $\beta_i(v)$ and $\delta_i(v)$ represent the birth and death rates respectively of a cell in stage i at time v. Any or all of the parameters (λs, βs, and δs) can depend on the age of the individual as well as the biologically effective dose $g(d,d_0,f)$. The lag time τ is usually treated as a constant but can be more generally represented as a random variable.[15-16] The above form for $P[t;g(d,d_0,f)]$ assumes that each cell at a particular stage has the same probabilistic behavior and that all cells behave independently.

The above relationship for $P[t;g(d,d_0,f)]$ can be motivated in fairly rough terms as follows: Assume that cancer is a k-stage process. Then cancer occurs after at least one cell reaches the kth stage. The number of cells which reach the kth stage at time t_k depends on

1. The transition rate from the $(k-1)^{th}$ stage to the kth stage at time t_k (namely, $\lambda_k(t_k)$) and
2. The number of cells in the $(k-1)^{th}$ stage at time t_k (namely, $N_{k-1}(t_k)$)

The number of cells in the $(k-1)^{th}$ stage at time t_k depends on

1. The number of cells which make the transition from the $(k-2)^{th}$ stage to the $(k-1)^{th}$ stage at time t_{k-1} (namely, $\lambda_{k-1}(t_{k-1}) \times N_{k-2}(t_{k-1})$)
2. The growth in the number of $(k-1)^{th}$ stage cells from the time (t_{k-1}) when the parent cell makes the transition from the $(k-2)^{th}$ stage to the $(k-1)^{th}$ stage until the time t_k (namely, the multiplicative growth factor

$$\exp\{\int_{t_{k-1}}^{t_k} [\beta_{k-1}(v) - \delta_{k-1}(v)]dv\}$$

which incorporates the birth rate $\beta_{k-1}(v)$ and death rate $\delta_{k-1}(v)$ for $(k-1)^{th}$ stage cells for the intervening times v between t_{k-1} and t_k).

The interrelationships continue backward until the number of cells in the first stage which are available for possible transition to the second stage at time t_2 (namely, $N_1(t_2)$) depends on the number of normal cells which made the transition to a first stage cell at time t_1 (namely, $\lambda_1(t_1) \times N_0(t_1)$) multiplied by the growth rate

$$\exp\{\int_{t_1}^{t_2}[\beta_1(v) - \delta_1(v)]dv\}$$

for first stage cells between times t_1 and t_2. The number of normal cells $N_0(\cdot)$ can be different at different times if necessary to reflect changes during early childhood, puberty, menopause, etc.

If there are only two stages, then $k = 2$, and Equation 1 becomes

$$P[t;g(d,d_0,f)] = 1 - \exp\{-\int_0^{t-\tau} \lambda_2(t_2) \times \int_0^{t_2} \lambda_1(t_1) \times N_0(t_1) \tag{2}$$

$$\times \exp\{\int_{t_1}^{t_2} [\beta_1(v) - \delta_1(v)]dv\}\ dt_1\ dt_2\}$$

Here λ_1, λ_2, β_1, and δ_1 may depend on the biologically effective dose $g(d,d_0,f)$. Furthermore, the administered dose d may not be the same value for all times but rather may depend on time. In order to emphasize these possible dependencies on age, dose, and time, and to emphasize that dose may be a function of time, Equation 2 can be rewritten in expanded form as

$$P[t;g(d,d_0,f)] = 1 - \exp\{-\int_0^{t-\tau} \lambda_2\{t_2,g[d(t_2),d_0,f]\} \tag{3}$$

$$\times \int_0^{t_2} \lambda_1\{t_1,g[d(t_1),d_0,f]\} \times N_0(t_1)$$

$$\times \exp[\int_{t_1}^{t_2} \{\beta_1\{v,g[d(v),d_0,f]\} - \delta_1\{v,g,[d(v),d_0,f]\}\}\ dv]\ dt_1dt_2\}$$

This expanded form shows explicitly how to incorporate a dose which is not constant over time.

The models described by Moolgavkar and Venzon,[17] Moolgavkar and Knudson,[15] and Thorslund et al.[9] are special cases of the form of Equation 3 for $P[t;g(d,d_0,f)]$. These models assume that D_0 has the same value for every individual and that F is zero for every individual. Moolgavkar and Knudson[15] do not suggest parametric forms for the possible dependence of λ_1, λ_2, β_1, and δ_1 on the biologically effective dose g(d,d_0,f). Furthermore, in Moolgavkar and Knudson,[15] the λs, βs, and δs are assumed to be independent of time, so that

$$P[t;g(d,d_0,f)] \tag{4}$$

$$= 1 - \exp\{-\lambda_1\lambda_2 \int_0^{t-\tau} \int_0^{t_2} N_0(t_1) \times \exp[(\beta_1-\delta_1)(t_2-t_1)]\, dt_1\, dt_2\}$$

Allowing the administered dose d to have time-varying levels as opposed to assuming that d has the same value for all times, simply means that, if $k = 2$, then the modeler must utilize Equation 3 instead of Equation 4.

Both the susceptibility frontier and the background dose can be extended to allow for age-dependence. For example, Equation 3 can be generalized to

$$P[t;g(d,d_0,f)] \tag{5}$$

$$= 1 - \exp\{-\int_0^{t-\tau} \lambda_2\{t_2,g[d(t_2),d_0(t_2),f(t_2)]\}$$

$$\times \int_0^{t_2} \lambda_1\{t_1,g[d(t_1),d_0(t_1),f(t_1)]\} \times N_0(t_1) \times$$

$$\exp[\int_{t_1}^{t_2} \{\beta_1\{v,g[d(v),d_0(v),f(v)]\} - \delta_1\{v,g[d(v),d_0(v),f(v)]\}\}dv]\, dt_1 dt_2\}$$

In Equation 5, the random susceptibility frontier F taking on a particular value f for an individual means that, when that individual reaches age T, his susceptibility frontier (denoted by f(T)) can be calculated from the number f and the age T. The functional form of f(T) may or may not have to be estimated and may or may not contain parameters which have to be estimated. Two individuals with the same value f for the random variable F would be assumed to have the same values f(T) for any T. (The next generalization would be to allow F to be a stochastic process with the f(T)s for different Ts being different (possibly dependent) random variables.) The interpretation of an age-dependent background dose $d_0(T)$ is analogous to that for the age-dependent susceptibility frontier f(T).

Several of the earlier time-to-response models (e.g., Krewski et al.[18])

like the multistage Weibull model[19] and the Hartley-Sielken model[20-22] can be derived as special cases of the much more general form of Equation 1; however, these earlier models do not emphasize the potential roles of cell proliferation or time-dependent changes in the number of normal target cells.

The cell birth, death, and transition rates as well as the susceptibility frontier and the background dose are physical phenomena, and hence are potentially subject to scientific investigation separate from any chronic carcinogenicity bioassay or epidemiological study. The careful analysis of a chemical's effects in such supplementary research may well indicate which of these rates are dose-dependent and which of these phenomena are age-dependent and even indicate the functional form of any dependence. The results of such analyses should determine the particular form of $P[t;g(d,d_0,f)]$, and estimates for most of its unknowns should have been determined before the corresponding model is fit to the dose-response data to estimate the remaining parameter values.[9]

MODELING IN THE PRESENCE OF COMPETING RISKS

Not only should competing risks (heart attacks, accidents, etc.) be incorporated into the estimation of a time-to-response model, but also the stated probability of a carcinogenic response should refer to the physical world where competing risks exist and should be the probability that the carcinogenic response will actually occur. This latter probability requires that the carcinogenic response occurs before any of the competing risks occur. Without the inclusion of this requirement, the stated probability is larger than it should be (Figure 7).

Kalbfleisch et al.[23] developed the concept of a cause-specific model which includes the possibility of a competing risk occurring before the carcinogenic response. In a cause-specific model

$$P(t;d) = \int_0^t [\frac{d}{dT^*} H_1(T^*;d)]_{T^*=T} \exp[-H_1(T;d) - H_2(T;d)]\, dT$$

where $H_1(t;d)$ and $H_2(t;d)$ are interpreted as the cumulative cause-specific hazard functions for the carcinogenic response and the combination of all competing risks respectively. The dose d may or may not affect $H_2(t;d)$.

Kalbfleisch et al.[23] discuss the general maximum likelihood estimation of the parameters in $H_1(t;d)$ and $H_2(t;d)$ for both the case where the carcinogenic response is observable when its occurs and the case where the carcinogenic response cannot be observed at the time it occurs and

Figure 7. The difference between time-to-response models, which do not account for competing risk, and cause-specific models, which incorporate competing risks.

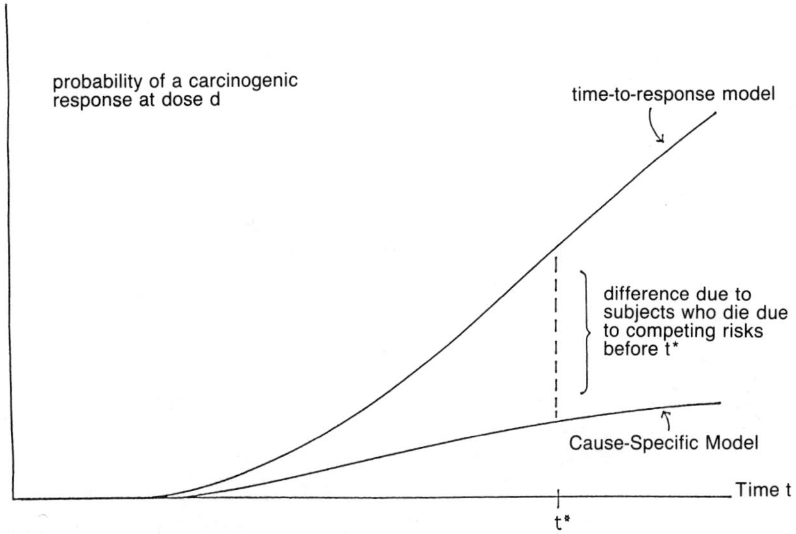

can only be detected after death. For the latter case, Berdine[24] proposes that, when the cause of death is uncertain, and "death from tumor" cannot be accurately determined, but the incidence of a tumor is believed to be potentially life-threatening, the risk should be characterized in terms of the increase in the cause-specific hazard rate for "death with tumor" relative to the cause-specific hazard rate for "death with no tumor" (Figure 8).

A reasonable choice for the parametric form of $H_1(t;d)$ would be the same form as that for a time-to-response model. Berdine[24] discusses the use of census data to estimate $H_2(t;d)$ and how this estimate could be combined with the estimate of $H_1(t;d)$ from a time-to-response model.

BEST ESTIMATES VS BOUNDS

It is very important to differentiate between *estimates* of risk and *bounds* on risk. The value of the fitted model at a particular dose is the estimate of risk at that dose which is most consistent with the presumed family of models. The best estimate of risk under the presumed family of models is the value of the fitted model.

In addition to a model family's best estimate of risk at a dose level, a 95% confidence limit on that risk can also be constructed. The purpose of 95% upper confidence limits is not to estimate the risk, but rather to

Figure 8. The cause-specific hazard rates for tumor onset and death in the presence of competing risks: a new definition of the specified carcinogenic response when the cause of death is uncertain.

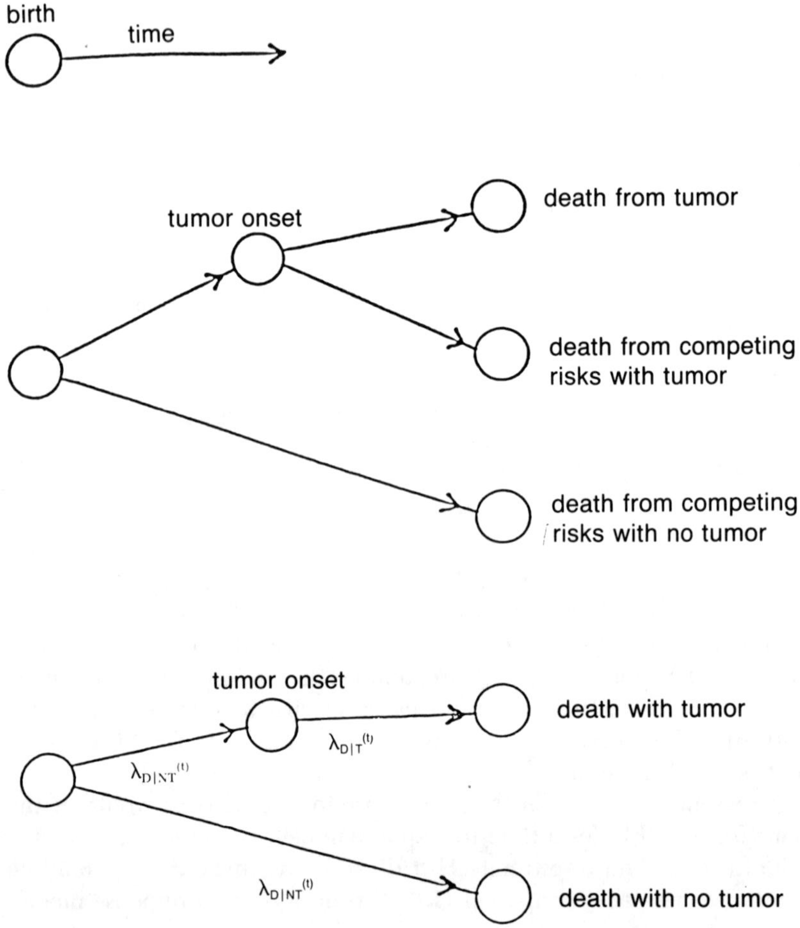

Carcinogenic Response ≡ Increase in the hazard rate for death $= \lambda_{D|T}^{(t)} - \lambda_{D|NT}^{(t)}$

be sufficiently large so that 95% of the time they exceed the true risk. Thus, 95% upper confidence limits are *upper bounds* on risk; they are *not estimates* of risk.

Figure 9. An example of the difference between the 95% upper confidence limits on risk and the fitted model values for risk.

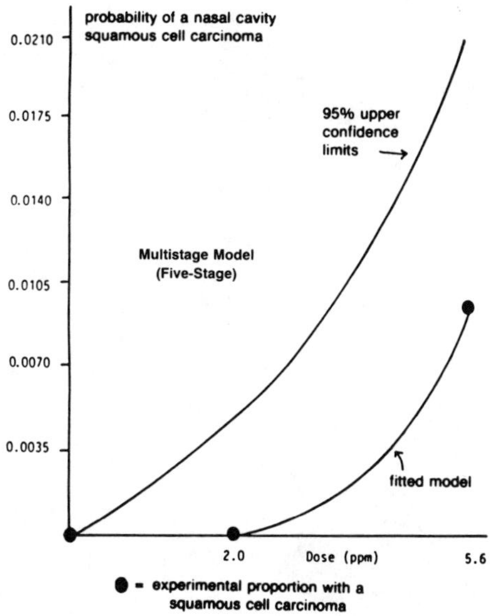

The Five-State Multistage Model

Dose (ppm)	95% upper confidence limit / maximum likelihood estimate
5.6	2
2.0	41
1.0	368
0.5	3,140
0.1	417,000
0.01	423,000,000

The difference between an estimate of risk and a bound on risk is more than a semantic difference. The 95% upper confidence limits on risk can be much larger than the fitted model's estimate of risk; for example, there can be a fortyfold difference even for a nonzero experimental dose level and a thousandfold or millionfold difference at lower dose levels (Figure 9).

The linearized multistage model is one collection of upper bounds on risk (95% upper confidence limits on the probability of a carcinogenic response) while the fitted values of the multistage model are the multistage model's best estimates of risk. Since the shapes of the linearized multistage model and the fitted multistage model can be very, very different, their implications about the dose-response relationship can be extremely different. In addition, the linearized multistage model is *not* very responsive to the experimental data (Figure 10). When large changes in the shape of the experimentally observed dose-response relationship make only very minor changes in the corresponding bounds, then the "stability" which is often ascribed to such bounds might be better described as "nonresponsiveness." In simulation experiments with the dose-response relationship known to be a particular quadratic multistage model, 99.6% of the time the 95% lower confidence limit on the virtually safe dose, corresponding to an increased probability of 1/1,000,000 and the linearized multistage model, was more than 100

Figure 10. The linearized multistage model's nonresponsiveness to even very different experimental results.

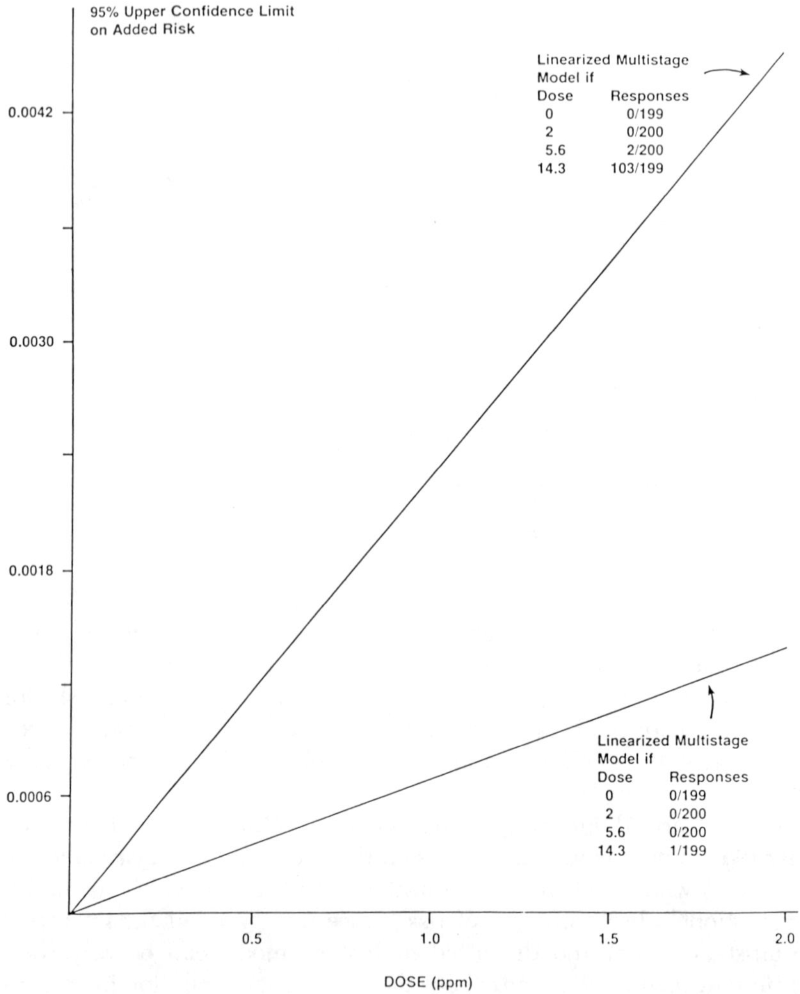

times smaller than the true virtually safe dose.[25] In the same study, 85% of the time the dose level corresponding to the linearized multistage model's 95% lower confidence limit on the virtually safe dose for an increased probability of one in a million had a true increased probability less than one in a hundred billion.

There is no standard way to compute 95% confidence limits in risk assessment.[26] Furthermore, different procedures can produce very different bounds (Figure 11). The earlier confidence limit procedures were

Figure 11. An example where two quantal response models (multistage and Weibull) have very similar fits to the experimental data but very dissimilar upper bounds in the medium to low dose region.

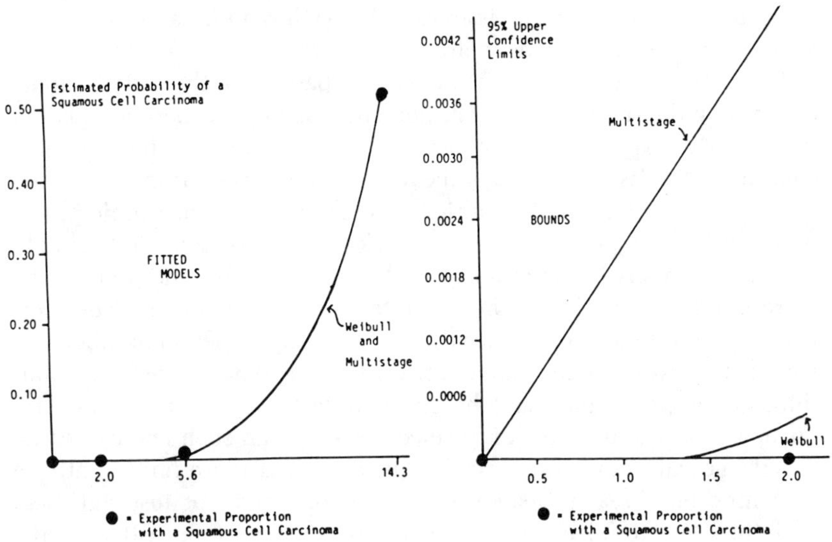

based on the asymptotic distributions of the model parameter estimates or the likelihood ratios which would theoretically apply if the sample sizes were sufficiently large and certain regularity conditions were satisfied.[27-29] Advances in bootstrap techniques offer some promising alternatives to these earlier confidence limit procedures.[18,20,30]

The structure of some models makes it almost impossible to develop a confidence limit procedure which will consistently generate bounds near the true value of the risk characteristic. For example, the multistage model for quantal response data represents the lifetime probability of a carcinogenic response at dose d as

$$P(d) = 1 - \exp[-(q_0 + q_1 d + q_2 d^2 + \ldots + q_k d^k)]$$

If the experimental data have a nonlinear shape, then the experimental data have a very strong influence on the estimates of the higher-order parameters q_k, q_{k-1}, etc., but very little influence on the estimate of q_1; yet, the estimate of the virtually safe dose corresponding to a small increase (e.g., 1/1,000,000) in the lifetime probability of a carcinogenic response over the corresponding background probability is almost entirely determined by the estimate of q_1.

Upper bounds should be carefully interpreted. When the upper bound on the risk is small, the risk is small. However, when the upper bound on the risk is not small, the risk may still be small. The values less than a 95% upper confidence limit are not all equally likely to be the true value of the quantity being bounded. The likelihood of being the true value diminishes (sometimes rapidly) as the candidate value moves away from the fitted model value.

The estimate of a risk characteristic based on the fitted model depends on the randomness in the epidemiological or experimental data (e.g., Portier and Hoel[31] and Smith and Sielken[25]). While statistical confidence limits give some indication of the uncertainty associated with the estimate, the difficulties with these confidence limit procedures and the model structure may make it more reasonable to simply vary the numbers in the experimental data a little bit and observe the corresponding variation in the estimate. Such an assessment, however, involves considerable care since the more the experimental numbers are varied, the less likely they are to reflect the true dose-response relationship. Great care should also be given to varying a zero response frequency to a positive frequency since not only can such changes cause very large variations in estimates, but also (and more importantly) a zero may be a real reflection of the biology and not just statistical variation. A careful evaluation of the sensitivity of the fitted model estimate for the particular data situation at hand is a far better strategy than always rejecting such an estimate in favor of a bound only because it is possible to construct a few hypothetical examples of experimental outcomes where the fitted model estimate is overly sensitive.[9]

If the uncertainty in a quantitative risk assessment is described in terms of bounds on risk, then both upper and lower bounds should be reported as well as their method of determination including any assumptions made and their impacts. The fitted model value is not a lower bound on the actual risk. However, the same procedures which give rise to upper bounds can also be used to generate lower bounds and vice versa. The difference between the upper and lower bounds, as well as the difference between these bounds and the fitted model values, can put any lack of specificity or lack of tightness in the bounds into perspective (Figure 12). Such a perspective is not gained by simply saying that the lower bound on risk might be zero.

RISK QUANTIFICATIONS WHICH REFLECT THE TIME TO RESPONSE

Risk can be quantified and communicated in several ways (e.g., The SOT ED_{01} Task Force[32]). Three-dimensional plots can indicate how the estimated probability of a carcinogenic response changes with both dose

Figure 12. The substantial differences between the upper and lower confidence limits generated by the linearized multistage model bounding procedure and the fitted five-stage multistage model for the rat nasal cavity squamous cell carcinomas in the CIIT Formaldehyde Inhalation Study.

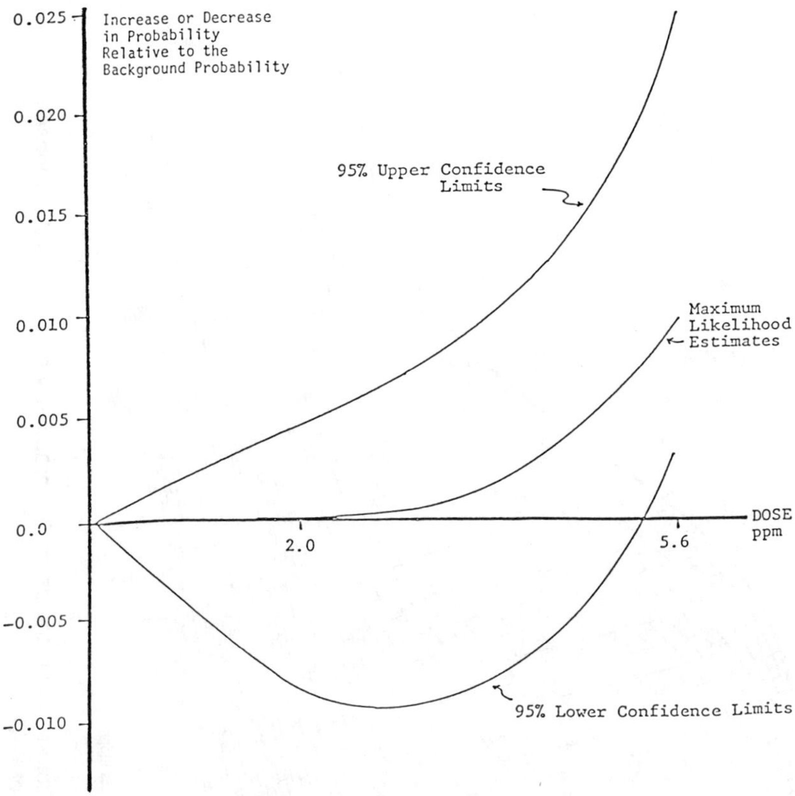

and time (Figure 13); however, many people find such plots difficult to describe and compare. Several simpler two-dimensional plots can be informative such as those which plot

- the mean amount of time without a specific carcinogenic response (the mean response free period) vs the dose level (Figure 14)
- the probability at a fixed dose vs time (Figure 15)
- the probability at a fixed time vs dose (Figure 15)
- the amount or time until the probability reaches a specified level vs the dose level (Figure 16)

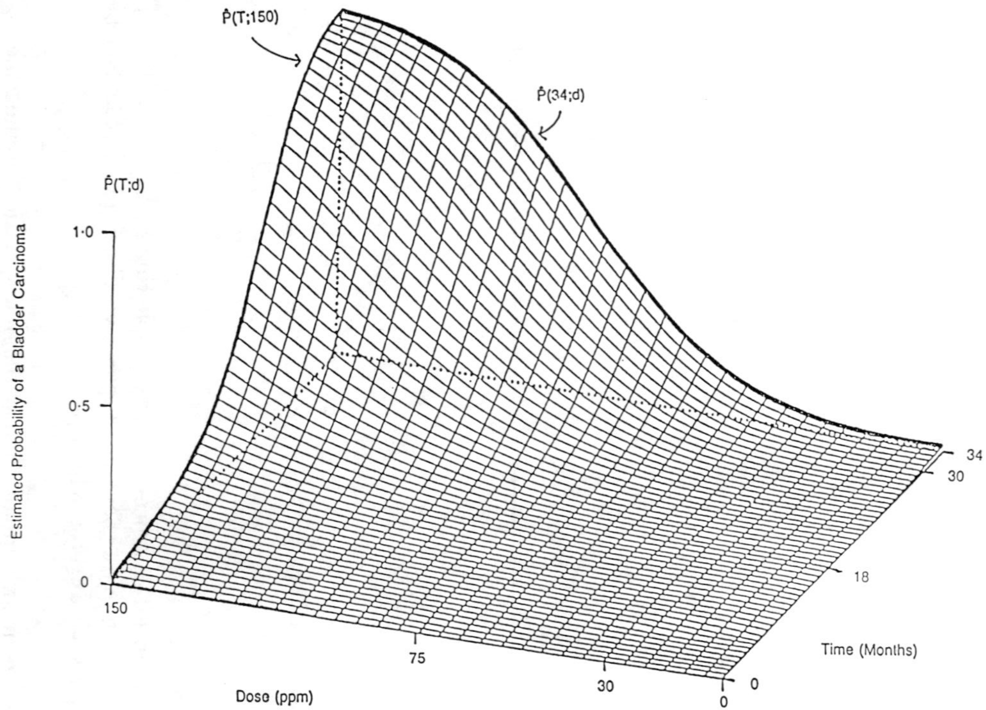

Figure 13. A three-dimensional plot illustrating how the estimated probability of a specified response changes with both dose and time (ED_{01} study of 2-AAF).

Figure 14. An example of how the estimated mean response free period within a 30 month period changes with the dose level and the specified response (ED_{01} study of 2-AAF).

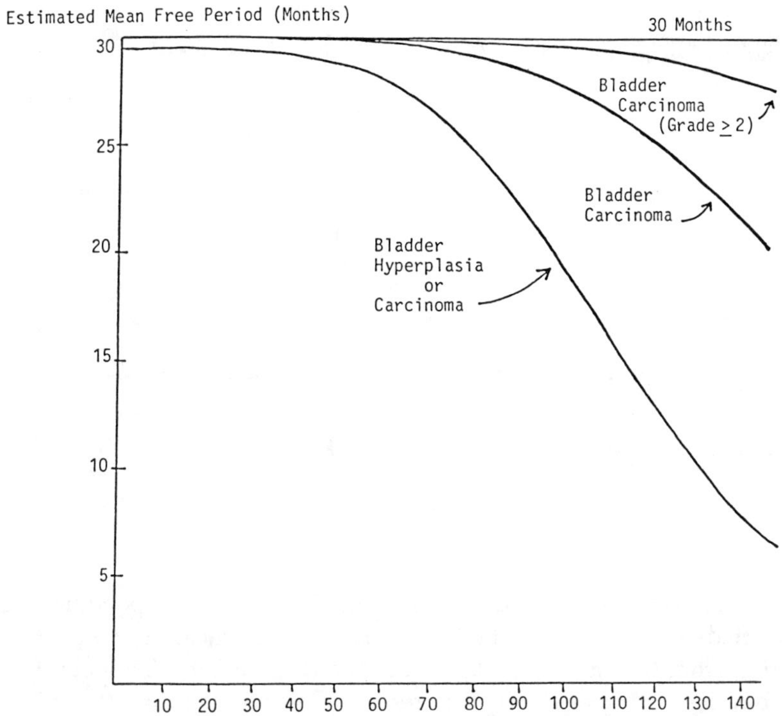

An interesting contour map of the probability of a carcinogenic response can be formed by combining several of the latter two-dimensional plots (Figure 17).

No single number fully describes how the estimated probability of a specified carcinogenic response changes with both dose and time; nevertheless, single number quantifications are often computed. One example of these oversimplified quantifications is the "virtually safe dose," or VSD, which is the dose level corresponding to a specified maximum acceptable increase in the probability of a carcinogenic response by one particular time over the rate at the background dose (Figure 18). Of course, two chemicals with very different health effects could have the same VSD if, for instance, one chemical's probability of a carcinogenic response increased very rapidly following exposure initiation, while the other chemical had very little effect until many years later (Figure 19). Another example of a single number quantification is

Figure 15. Two-dimensional plots illustrating how the probability of a specified response changes with dose for fixed times and changes with time for fixed doses (ED_{01} study of 2-AAF).

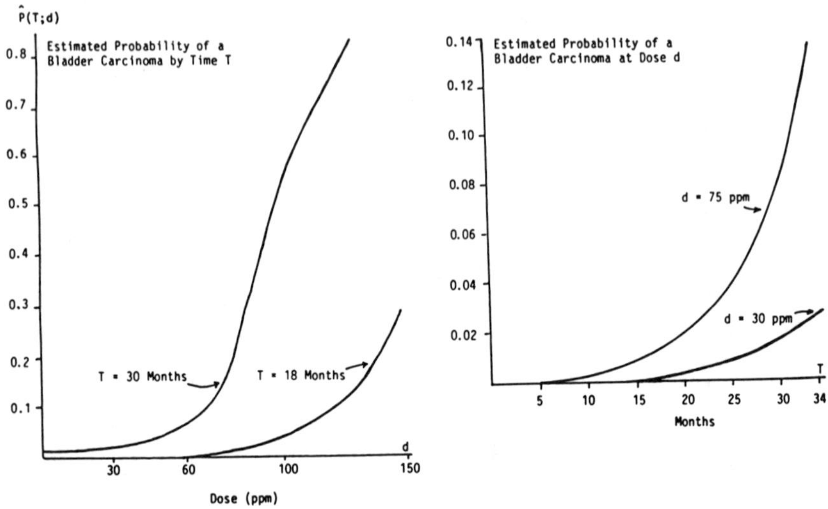

the "mean free dose" (MFD) which is the dose level corresponding to a specified maximum acceptable decrease in the mean response free period from that at the background dose (Figure 20). The mean free dose reflects more of the time-to-response information than the VSD.

Some risk quantifications can be more accurately estimated than others. For example, a VSD corresponding to a maximum acceptable increment of 0.001 can be estimated, relatively, more accurately than a VSD corresponding to 0.000001. Thus, it can be important to evaluate the relative empirical behaviors of estimators before deciding on the particular risk quantifications to be estimated and the weight to be attached to these estimates in the risk management process.

Many people do not readily grasp the meaning of small probabilities like 0.000001 or such numbers expressed in scientific notation. Also many people do not readily differentiate between different "small" numerical values for a probability but rather tend to group these probabilities together into simply "a chance." The risks may be more easily understood if, in addition to presenting the estimated probabilities, the risks are also presented in terms of comparative risks and the background risk, as well as in terms of particular populations, the expected number of carcinogenic responses, and any shortening of the mean response free period in those populations.

Figure 16. An example of how the estimated amount of time until the probability of a specified response reaches a specified level changes with the dose level and the severity of the specified response (ED_{01} study of 2-AAF).

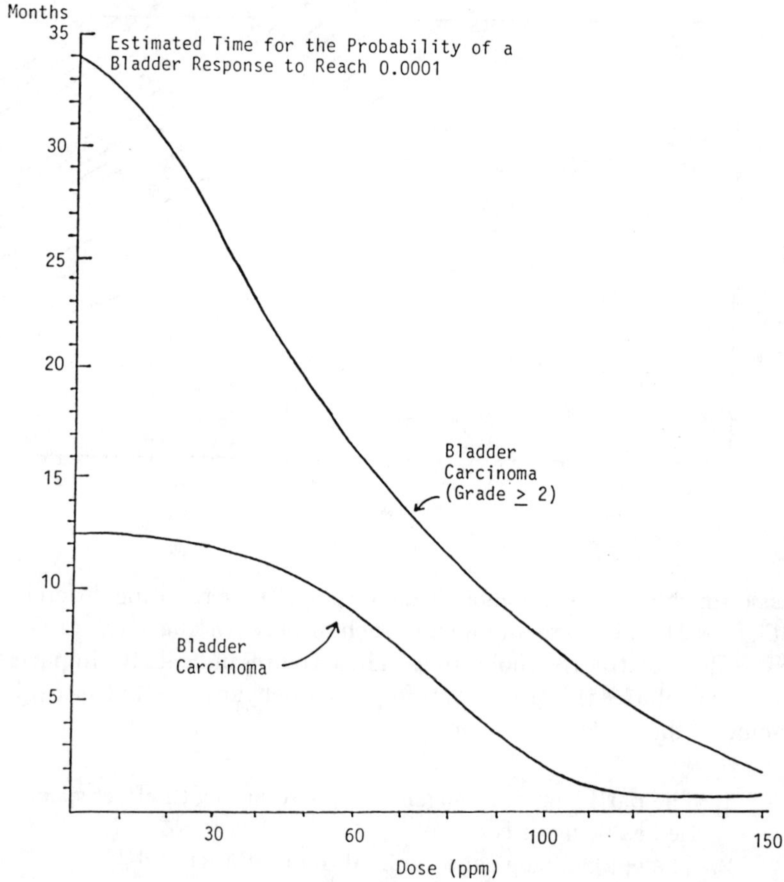

RISK QUANTIFICATIONS WHICH REFLECT THE CHOICES IN A RISK ASSESSMENT

The quantitative risk assessment can be very sensitive to the assumptions, policy decisions, and value judgements made during its determination.[33] Therefore, these choices should be clearly denoted, and their *quantitative* impact on the quantitative risk assessment evaluated and reported. Probably the clearest indication of the practical impacts of these choices can be obtained by carrying out the quantitative risk

Figure 17. A "contour map" showing the combinations of dose and time with equal values for the probability of the specified response (ED_{01} study of 2-AAF).

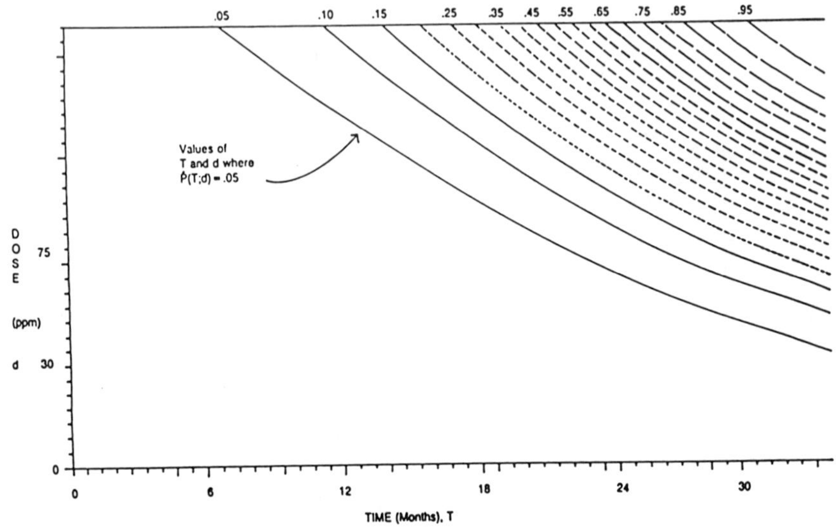

assessment for different choices and displaying the resulting differences. Figures 21 and 22 are examples of such displays and have been referred to as decision trees or choice trees. These figures indicate the impacts on the quantitative risk assessments for formaldehyde and ethylene oxide of some of the choices pertaining to

1. The particular carcinogenic response or health effect specified as being of concern
2. The epidemiological or animal data set analyzed
3. The inclusion or exclusion of the time-to-response data
4. Whether the risk quantification focuses on the increased probability at one particular time or the decreased mean response free period and which particular time or period is used
5. The dose-response model and whether it is considered in a general or restricted form
6. Whether the value of the risk quantification is an estimate, a lower bound, or an upper bound
7. The dose scale (administered, delivered, biologically effective) used in the dose-response modeling
8. The method used for any species-to-species extrapolation

Figure 18. The definition of the virtually safe dose: one definition of a maximum acceptable dose.

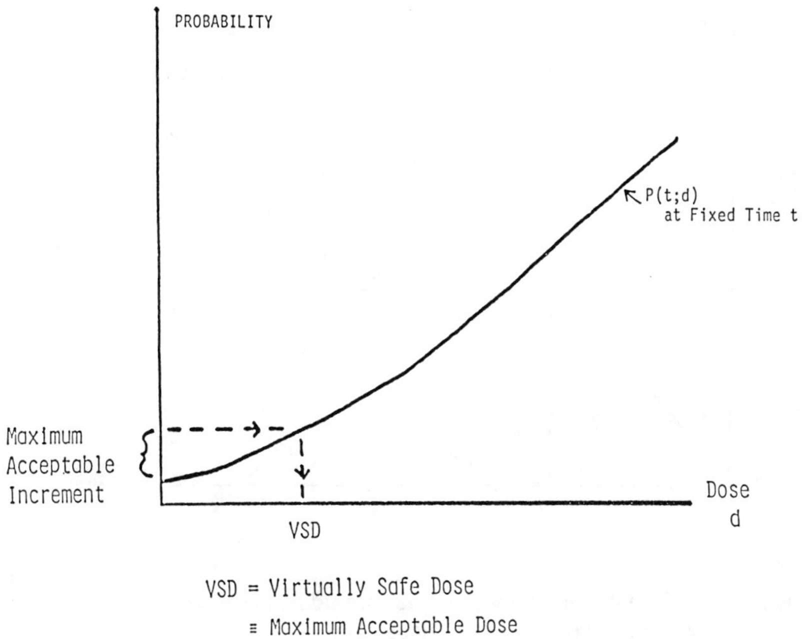

There are, of course, other areas requiring choices.[34] It should be noted that statistical variability in the epidemiological or experimental data may have only a small impact relative to the impact of the choices involved in the risk assessment procedure.

Although the different quantitative risk assessments indicated in a choice tree do reveal the sensitivity of a quantitative risk assessment to some of the assumptions, policy decisions, and value judgements, etc., being made, the different quantitative assessments are generally not all equally relevant to a human health assessment and should not be given equal weight in the final assessment. For example, if there are five independent choices with three possibilities for each choice (a highest risk possibility, a most likely possibility, and a lowest risk possibility with the probabilities of these cases being the real case of 0.1, 0.8, and 0.1, respectively), then the worst case (all five cases being the highest risk possibility) has only a probability of 1/100,000 of being the real case.

Figure 19. Two risk profiles with the same probability of a specified response late in a lifetime but very different probabilities earlier.

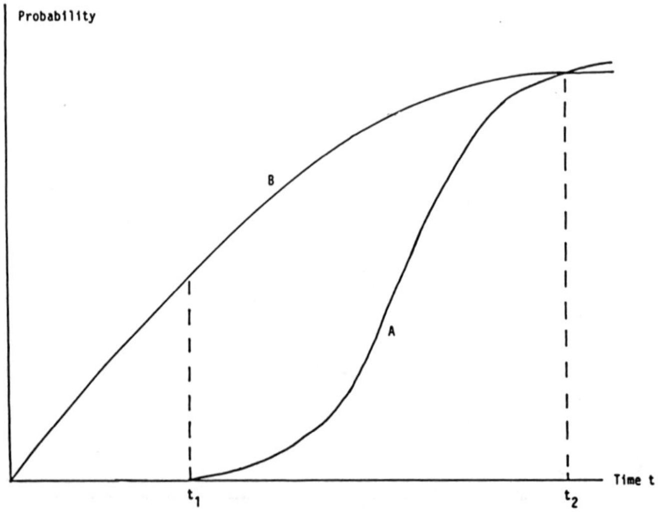

Figure 20. The definition for the mean free dose: one definition of a maximum acceptable dose.

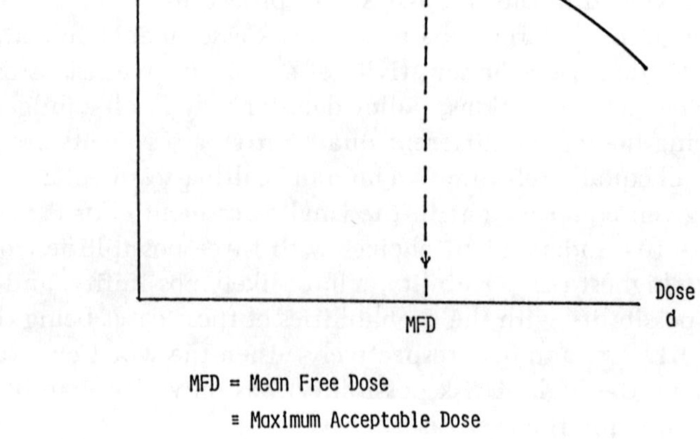

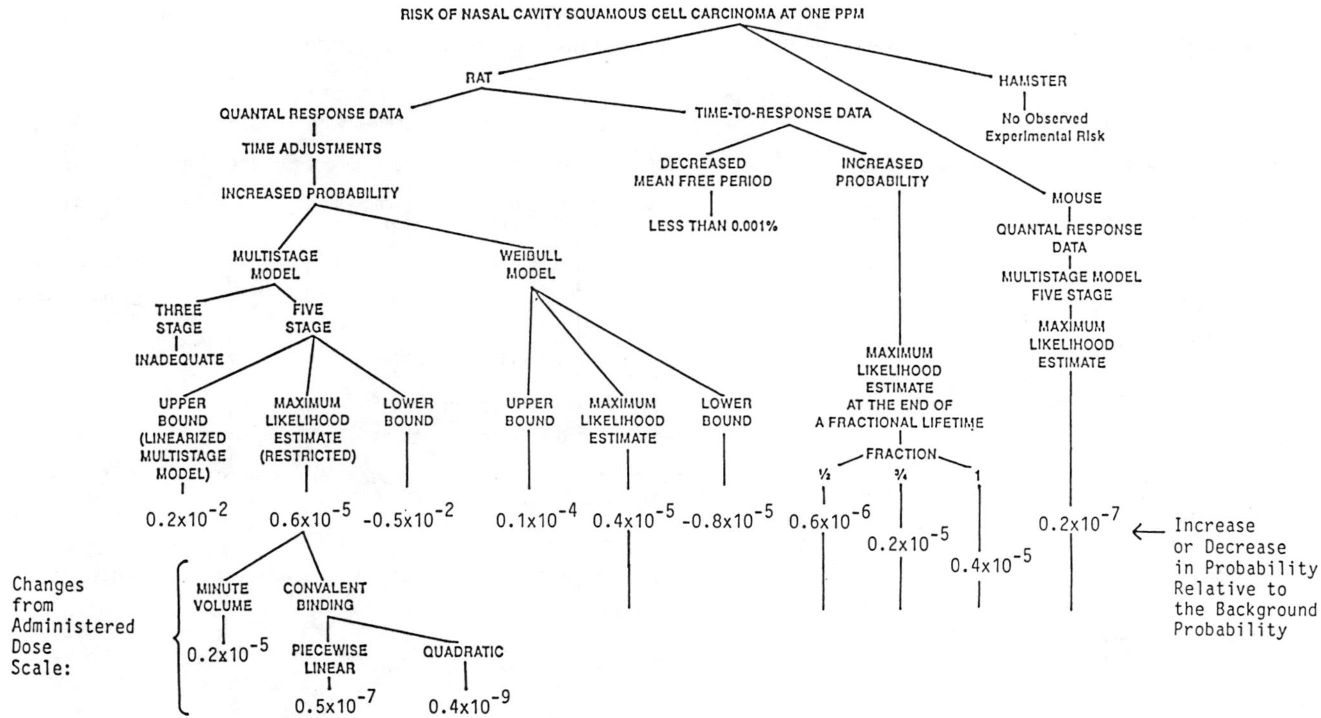

Figure 21. A choice tree illustrating the quantitative impacts of several of the choices in the quantitative risk assessment for formaldehyde inhalation.

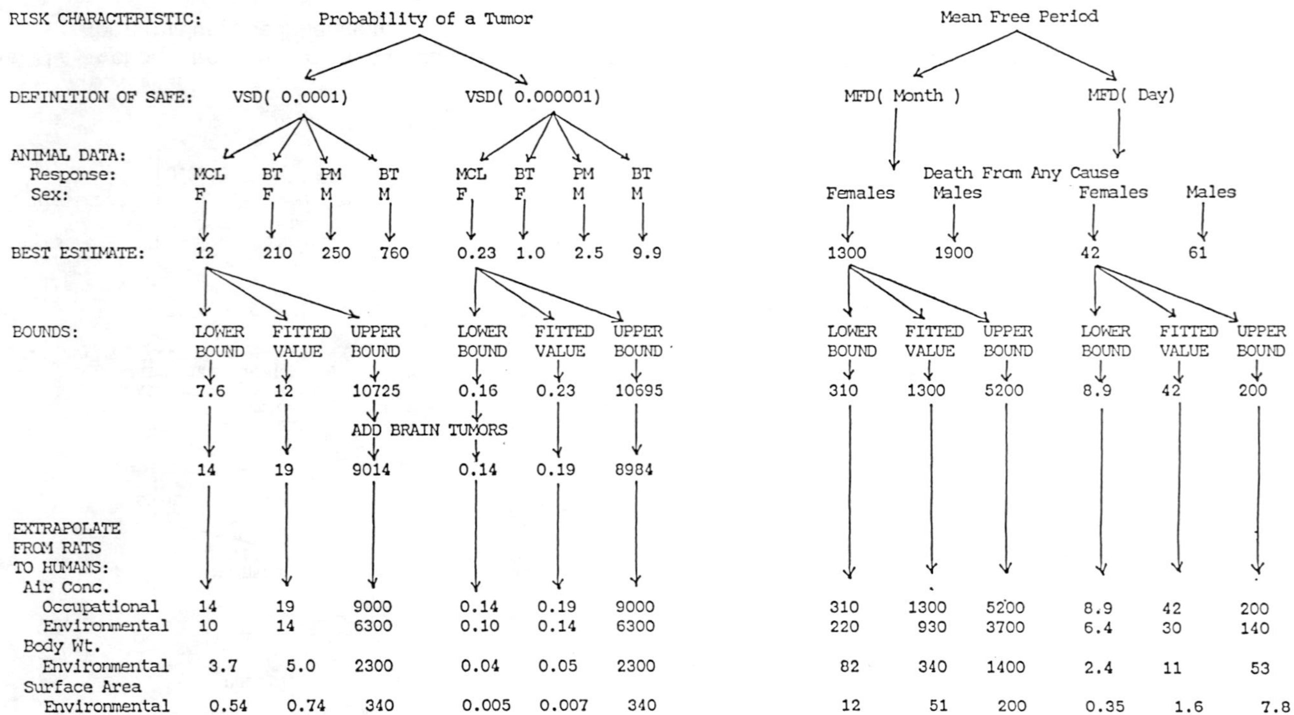

Figure 22. The quantitative impacts of several different choices in the quantitative risk assessment for ethylene oxide inhalation.

CONCLUSIONS

The quantitative dose-response modeling and risk quantification components of cancer risk assessment are both the subject of intensive research. It is not possible to make the risk quantifications more reflective of the time to the response, competing risks, and the choices inherent in risk assessment. New dose-response models can incorporate research on more biologically relevant dose scales, age-dependent and time-dependent changes, cell proliferation, and distributions of individual background exposure levels and individual susceptibilities to low levels of a chemical. Even though future research should provide further improvements, the new capabilities for risk quantification and dose-response modeling provide the potential for substantial improvements in quantitative cancer risk assessment now.

RECOMMENDATIONS

1. Cancer dose-response models should reflect the available information on biologically effective doses, age-dependent and time-dependent changes, cell proliferation, and distributions of individual susceptibility frontiers and individual background exposures.

2. Risk communications should be more reflective of the available information on time to response and competing risks, the differences between an estimate of risk and bounds on risk, the lack of mathematical certainty in risk quantifications, and the quantitative impacts of the choices in a risk assessment.

REFERENCES

1. Consensus Workshop on Formaldehyde. 1984. *Environ. Health Pers.* 58:323.
2. Andersen, M.E., Clewell, III, H.J., Gargas, M.L., Smith, F.A., and Reitz, R.H. 1986. Physiologically-based pharmacokinetics and the risk assessment process for methylene chloride. *Toxicol. Appl. Pharmacol.* In press.
3. Clewell, III, H.J. and Andersen, M.E. 1985. Risk assessment extrapolations and physiological modeling. *Toxicology and Industrial Health*, Vol. I(4):111.
4. Gehring, P.J. and Blau, G.E. 1977. Mechanisms of carcinogenesis: dose response. *J. Environ. Pathol. Toxicol.* 1:163.
5. Gehring, P.J., Watanabe, P.G., and Park, C.N. 1978. Resolution of dose-response toxicity data for chemicals requiring metabolic activation: Example-vinyl chloride. *Toxicol. Appl. Pharmacol.* 44:581.

6. Hoel, D.G., Kaplan, N.L., and Anderson, M.W. 1983. Implication of nonlinear kinetics on risk estimation in carcinogenesis. *Science* 219(4588):1032.
7. Starr, T.B. 1985. Improving quantitative risk estimates with mechanistic data: the case of formaldehyde, Ch. 8. In: Risk Analysis in the Chemical Industry. Chemical Manufacturers Association, Pub. by Government Institutes, Inc., Rockville, MD, p. 142.
8. Starr, T.B. and Buck, R.D. 1984. The importance of delivered dose in estimating low-dose cancer risk from inhalation exposure to formaldehyde. *Fund. Appl. Toxicol.* 4:740.
9. Thorslund, T.W., Brown, C.C., and Charnley, G. 1986. The use of biologically motivated mathematical models to predict the actual cancer risk associated with environmental exposure to a carcinogen. *J. Risk Anal.* In press.
10. Watanabe, P.G., Reitz, R.H., Schumann, A.M., McKenna, M.J., Quast J.F., and Gehring, P.J. 1980. Implications of the mechanisms of tumorigenicity for risk assessment. In: Witschi, H.R. (Ed.). Scientific Basis of Toxicity Assessment. Elsevier/North-Holland Biomedical Press, p. 273.
11. Sielken, Jr., R.L. 1986. An individualized response model for quantitative cancer risk assessment. Presented at the Joint Meeting of the Risk Assessment Subcommittee of the American Industrial Health Council and the Health and Safety Committee of the Chemical Manufacturers Association. Washington, DC, June 27, 1986.
12. Kerns, W.D., Pavkov, K.L., Donofrio, D.J., and Swenberg, J. 1983. Carcinogenicity of formaldehyde in rats and mice after long-term inhalation exposure. *Cancer Res.* 43:4382.
13. Staffa, J.A. and Mehlman, M.A. (Eds.). 1979. Innovations in Cancer Risk Assessment (ED_{01} Study). Pathotox Publishers, Inc., Park Forest South, IL.
14. Armitage, P. and Doll, R. 1961. Stochastic models for carcinogenesis. In: Neyman, J. (Ed.). Proceedings of the Fourth Berkeley Symposium on Mathematical Statistics and Probability. Vol. 4, p. 19. University of California Press, Berkeley and Los Angeles.
15. Moolgavkar, S.H. and Knudson, A.G. 1981. Mutation and cancer: a model for human carcinogenesis. *J. Nat. Cancer Inst.* 66:1037.
16. Whittemore, A. and Keller, J.B. 1978. Quantitative Theories of Carcinogenesis. *Siam Rev.* 20:1.
17. Moolgavkar, S.H. and Venzon, D.J. 1979. Two event models for carcinogenesis: incidence curves for childhood and adult tumors. *Math. Biosci.* 47:55.
18. Krewski, D., Crump, K.S., Farmer, J., Gaylor, D.W., Howe, R., Portier, C., Salsburg, D., Sielken, R.L., Van Ryzin, J. 1983. A comparison of statistical methdos for low dose extrapolation utilizing time-to-tumor data. *Fund. Appl. Toxicol.* 3:140.
19. Daffer, P.Z., Crump, K.S., and Masterman, M.D. 1980. Asymptotic the-

ory for analyzing dose response survival data with application to the low dose extrapolation problem. *Math. Biosci.* 50:207.

20. Hartley, H.O. and Sielken, R.L. 1977a. Estimation of "safe doses" in carcinogenic experiments. *Biometrics* 33:1.
21. Hartley, H.O., Tolley, H.D., and Sielken, R.L. 1981. The product form of the hazard rate model in carcinogenic testing. In: Csorgo, M., Dawson, D., Rao, J.N.K., and Saleh, E., (Eds.). Statistics and Related Topics. p. 185. North-Holland, Amsterdam.
22. Sielken, R.L., Jr. 1985a. Use of the Hartley-Sielken model in low-dose extrapolation, ch. 8. In: Clayson, D.B., Krewski, D., Munro, I. (Eds.). Toxicological Risk Assessment, Volume I, Biological and Statistical Criteria, p. 165. CRC Press, Inc., Boca Raton, FL.
23. Kalbfleisch, J.D., Krewski, D.R., and Van Ryzin, J. 1983. Dose-response models for time-to-response toxicity data. *Can. J. Stat.* 11:25.
24. Berdine, R.J. 1986. Improving risk characterizations based on time to response. Ph D dissertation. Dept. of Statistics, Texas A&M University.
25. Smith, L.A. and Sielken, R.L. 1985. Bounds for "safe" doses in quantitative risk assessment. Presented at the Biometric Society (ENAR) Joint Meeting with the American Statistical Association, Raleigh, NC, March 25. Submitted to Biometrics.
26. Crump, K.S. and Howe, R.B. 1985. A review of methods for calculating statistical confidence limits in low-dose extrapolation, ch. 9. In: Clayson, D.B., Krewski, D., and Munro, I. (Eds.). Toxicological Risk Assessment, Volume I, Biological and Statistical Criteria, p. 187. CRC Press, Inc., Boca Raton, FL.
27. Crump, K.S., Guess, H.A., and Deal, K.L. 1977. Confidence intervals and tests of hypothesis concerning dose response relations inferred from animal carcinogenicity data. *Biology* 33:437.
28. Crump, K.S. 1981. An improved procedure for low-dose carcinogenic risk assessment from animal data. *J. Environ. Pathol. Toxicol.* 5:675.
29. Krewski, D. and Van Ryzin, J. 1981. Dose response models for quantal response toxicity data. In: Csorgo, M., Dawson, D., Rao, J.N.K., and Saleh, E. (Eds.). Statistics and Related Topics, p. 201. North Holland Press, Amsterdam.
30. Smith, L.A. and Sielken, R.L. 1986. Bootstrap bounds for "safe doses" in quantitative risk assessment. Submitted to Communications in Statistics.
31. Portier, C. and Hoel, D. 1983. Low-dose-rate extrapolation using the multistage model. *Biometrics* 39:897.
32. The SOT ED_{01} Task Force. 1981. Re-examination of the ED_{01} study: risk assessment using time. *Fund. Appl. Toxicol.* 1:88.
33. Sielken, Jr., R.L. 1985b. Some capabilities, limitations, and pitfalls in the quantitative risk assessment of formaldehyde, ch. 6. In: Risk Analysis in the Chemical Industry. Chemical Manufacturers Association, p. 74. Published by Government Institutes, Inc. Rockville, MD.

34. Sielken, R.L., Jr. 1985c. Some issues in the quantitative modeling portion of cancer risk assessment. *Reg. Toxicol. Pharmacol.* 5:175.

26. Standards and Regulations Committee Conclusions and Recommendations

On a Model Unicode

The Conference for Food Protection recognizes the need for a model unicode for food protection and sanitation in retail food establishments (retail food stores, food service establishments, and food vending machines). It also emphasizes that the process used to develop a unicode should include the formation of a task force of representatives of key national organizations, both regulatory and industry, to work with the Food and Drug Administration on a cooperative and consensus basis.

RECOMMENDATIONS

1. The Conference endorses the development of a model retail food protection unicode as a priority project.

2. The Conference endorses the cooperative consensus approach to developing and maintaining a unicode that involves participation by FDA and several key national organizations.

3. Industry should be given an appropriate vote at the committee level in the development of the unicode.

On Quantitative Risk Assessment

The Conference recognizes the applicability of risk assessment in many food safety decisions and reaffirms the recommendations of the 1984 Conference with the following recommendations:

RECOMMENDATIONS

1. The Conference recognizes the potential benefits of an expanded quantitative risk assessment model, accepts Robert L. Sielken, Jr.'s evaluation of the topic (see Chapter 25), and recommends its consideration to the Toxicology Committee.

2. Methods reflecting biologically effective dose scales, age- and time-dependent changes, cell proliferation and distributions of individual susceptibilities, and background doses should be examined in order to determine if they can be usefully incorporated into current risk assessment practices.

EDUCATION AND TRAINING

27. Current Consumer Interests as Reflected by Consumer Complaints

Lilyan M. Goossens

U.S. Food and Drug Administration

INTRODUCTION

Consumers' interests are often reflected through their complaints and/or by information they are seeking. Their complaints may support the idea expressed in the Proceedings of the Second National Conference for Food Protection,[1] "that the protection of our food supply from contamination . . . is essentially a 'people problem.' " The people problem can occur at any phase of processing—distribution, service, and in commercial operations, as well as the private home. Often the consumer complaint file can serve to pinpoint the focus of an educational program. In reviewing the complaint file in the U.S. Food and Drug Administration's (FDA) Indianapolis office, the following examples stand out.

CONSUMER COMPLAINTS

A Cooperative Extension specialist from Purdue University called the FDA to inquire about flour turning a bright yellow when water or liquid was added. A homemaker had also called with this question, and the local county agent observed that the flour turned bright yellow when water was added. A sample of the flour was sent to a Purdue specialist who confirmed the unusual yellow color in the flour. When an FDA Consumer Safety Officer visited the home he, too, verified the appearance of the yellow flour when water was mixed with it. It turned out that the thrifty homemaker also had a problem with ants and sprayed her cupboards and counter tops. When she made biscuits any flour that spilled on the counter top was carefully brushed back into the flour canister. The yellow color came from the pesticide she had used to rid the kitchen counter of ants.

Misused pesticides were also the cause of an emergency plane landing in Indianapolis. Several passengers on a commercial flight became ill after the stewardess sprayed the plane with a pesticide prior to opening the plane doors upon returning from Europe. Unfortunately, the caps were not secured on the liquor bottles, and all the scotch drinkers on the flight became ill.

Currently the FDA is examining the practice of truck drivers using a variety of pesticides to keep ants and other insects off trucks carrying foods. The drivers are not always careful or knowledgeable about handling and using pesticides near food products.

Based on these three incidents, to target a group for a food safety education program including instruction in the safe use of pesticides, the FDA would have to develop different types of specially targeted information for the different groups of people involved.

In 1985, the Indianapolis Resident Post received 100 complaints about food products that ranged from broken glass, illness, and taste, to label tampering and allergic responses. The complaints that seem to be related to possible food poisoning cover a variety of products.

The presence of *Salmonella* was confirmed in one frozen pizza incident. It was not determined if the salmonellosis occurred because of mishandling of the product in the home, in the store, in transit, or at the plant. The family reported that they had cramps and diarrhea. The same symptoms were also reported for incidents involving yogurt-covered raisins, canned corn, and vegetable beef soup. Nausea and/or vomiting was reported from consuming a variety of products, including creamy peanut butter, soda pop (one incident involving soda pop resulted in a pregnant woman staying overnight in the hospital with IV therapy), canned oysters, Stresstabs, a chicken TV dinner, generic tea, instant coffee (with a roach), and suntan tablets. The true cause of these complaints were not determined.

About a fourth of the complaints reported by consumers involved symptoms that were both confirmed and unconfirmed allergies to Yellow Dye No. 5, sulfites, and aspartame. In many cases, the consumers were trying a new product but did not read the ingredients listed on the label. A large group of complaints indicated findings of foreign objects in food products. They ranged from a whole mouse found in a sealed bag of potato chips to red worms in canned oysters.

CONSUMER INFORMATION

Consumers also contact the FDA for information on the types of problems they perceive in the marketplace. In 1985, the FDA's Washington staff answered 31,518 inquiries. The most frequently asked questions dealt with sodium (4,115), fiber (1,128), sulfites (796), weight control products and diets (724), and food color safety (670), followed by food additives, aspartame, food irradiation safety, and caffeine.

From January 1 to April 30, 1986 the inquiries were on sodium, bee pollen, Gerber baby food, Brie cheese, methylene chloride in decaffeinated coffee, food additives, weight loss products, and irradiated foods.

Consumers seek food safety information from the FDA and the Cooperative Extension Service after storms, floods, tornadoes, and whenever electric service has been interrupted. Thanksgiving has always been a time for food safety questions. A few questions that come from "the Gleaners" and other food for the poor operations are usually about dented or swollen cans or the age of a product.

FOODBORNE DISEASE OUTBREAKS

The Centers for Disease Control (CDC) in Atlanta, Georgia, provide information to assess the need for food safety and determine which etiologic agents should be selected as topics for an educational program. The CDC data for 1982 showed that the greatest number of foodborne disease outbreaks were caused by *Salmonella* (25%), *Staphylococcus aureus* (12.7%), *Clostridium perfringens* (10%), *Clostridium botulinum* (9.5%), *Bacillus cereus* (3.6%), and *Yersinia enterocolitica* (0.9%). Chemical foodborne outbreaks were due to Scombrotoxin (8.2%), Ciguatoxin (3.6%), and heavy metals (2.3%).[2]

To determine where to target education programs, the FDA must look at where foodborne outbreaks occur. Home canners were responsible for all but one *Clostridium botulinum* outbreak in 1982 (the last year that figures are available). Restaurant catering was listed as the site where food was mishandled or consumed in 41% of the outbreaks caused by the eight etiologic or chemical agents most commonly associated with foodborne disease. Thirty-six percent of the illnesses took place in private homes. This group included a firehouse and a Brownie troop. Nursing homes and hospitals were listed as sites for 7% of the foodborne illnesses; schools were next with 5.5%, followed by churches with 4.5%. The last two sites listed were reform schools and prisons (2%) and "statewide" (1%).

EDUCATION OF THE PUBLIC

To select targets for educational topics, all cases of foodborne illness may be included, rather than the eight highest causes of illness as illustrated in the previous example. The example chosen here yields three major targets for education: restaurants, caterers, and homemakers. The materials developed for the restaurant employees may be applicable to hospital and school personnel as well. The private home, however, where 36% of foodborne illness occurred, is more difficult to approach with an educational program.

In 1984, it was noted that 26.7% of the population 25 years and older have not finished high school. The U.S. census data noted that 100% of school children took English, and in 1982, 94.3% of the students took health and physical education. These two subjects may provide the best mechanism for the delivery of food safety in the schools for the majority of students. It was interesting to note that science courses might not be a good vehicle because not all students elect to take science, and only 24% take home economics, which may *not* include subjects pertaining to food safety.

The "people problem" posed by the population over 25 years old is a little more difficult. Sixty-four percent of the women 18 to 44 years of age are working outside the home. This frequently results in both men and women sharing food responsibilities for the family. It also means that fast foods or instant foods are desirable, and frequently children fix their own snacks. Superimposed on the "time crunch" is the fact that 75% of Americans have changed their eating habits in some way as a result of health concerns. For example, they eat more fresh fruits, vegetables, and other foods that contain less fat, sugar, and salt. Anything labeled "fresh and natural" seems to sell. All of these factors impact on the excellent recommendations made at the 1984 Food Protection Conference. The Proceedings of the Second National Conference For Food Protection[3] contains an excellent guide to educating the consumer in food protection. It may be that the social and behavioral changes of contemporary Americans will necessitate modifying the recommendations just a little.

RECOMMENDATIONS

1. It should be mandatory to incorporate food safety principles as part of the required curriculum at primary and secondary levels of education. Data from the *National Data Book and Guide to Sources, Statistical Abstract of the United States* 1986,[4] indicate that English and health and physical education were the subjects taught to 94% or more of American students. Textbooks currently in use contain very little, if any, information on food safety. Supplying information to teachers in the form of ready-to-use materials is as effective as having a state mandate for these educational programs. A redesign of familiar brochures may also help capture interest.

2. A federal agency should act as a focus for timely food safety press releases. USDA and the FDA have issued food safety rules as a public

awareness tool. But, we must remember that in 1985, 60 million Americans were functionally illiterate. One million functionally illiterate citizens come from the public school system each year. Kathryn Stechert[5] predicted that, by the year 2000, approximately 80 to 100 million Americans will be classified as functionally illiterate. They will be the most difficult group to reach.

Busy people don't always take time to read an entire newspaper either. As a former employee of the Detroit Health Department who issued weekly press releases on many health topics, I know many papers will only use these releases as fillers and most will never use them at all. The Education and Training Committee should consider working with the local electronic and print media to plan news features around each holiday. Radio is a good medium for talk shows since people listen throughout the day.

3. Another recommendation of the 1984 conference called for the establishment of a multidisciplinary task force to develop a plan for consumer food safety guidance; to endorse the development of materials aimed at high-risk groups; to urge federally funded food programs to devote more resources to education and training; and to endorse a national "Food Protection and Safety Week." These are all meritorius activities that deserve further consideration and active efforts for implementation.

REFERENCES

1. U.S. Dept. of Health and Human Services. 1984. Proceedings of the Second National Conference for Food Protection. U.S. Dept. of Health and Human Services—U.S. Food and Drug Administration, Washington, DC.
2. Centers for Disease Control. September 1985. Foodborne Disease Surveillance. U.S. Dept. of Health and Human Services, PHS, Washington, DC.
3. U.S. Dept. of Health and Human Services. 1984. Proceedings of the Second National Conference for Food Protection, May 9–11, 1984. U.S. Dept. of Health and Human Services—U.S. Food and Drug Administration, Washington, DC, pp. 321–324.
4. Centers for Disease Control. 1986. National Data Book and Guide to Sources, Statistical Abstract of the United States. U.S. Dept. of Health and Human Services, Washington, DC.
5. Stechert, K. 1985. Illiteracy in America. *Better Homes and Gardens*, November 1985, pp. 27–28.

BIBLIOGRAPHY

Natfield, D. 1986. Marketing Nutrition is Good for Business—And You. *Forecast for Home Economics* March 1986, pp. 18–21.

The Survey of Buying Power, Data Service 1985. *S & MM*, a Bill Publication, p. 1–3.

New Era for Employers/Employees? *Food Processing* March 1986, p. 23.

Maturing "Boomers" Boost Food Volume. *Food Processing* May 1986, p. 130.

28. Competency-Based Adult Learning in Food Safety Programs

Linda B. Rhodes

Dairy Council of Michigan

STATEMENT OF THE PROBLEM

Advances in food technology increase our daily food choices, while foods available in the marketplace change from week to week. The way in which we plan our meals, shop, prepare, and even consume food has changed markedly in the past twenty years. With the home introduction of highly processed convenience foods, microwave and convection ovens, slow-cooking pots, and sterile packaging, some new and old challenges face the consumer. The safety of our food supply continues to be a concern amidst the technological gains in the food industry.

A 1981 report by the U.S. Centers for Disease Control in Atlanta emphasized the extent of the food safety challenge. The report indicated that more than 14,000 people were affected by foodborne illnesses in 568 outbreaks reported in 1981. It is interesting to note that 251 of the outbreaks (44%) resulted from food consumed in restaurants or cafeterias. The improper heating or cooling of food, mishandling of foods, use of contaminated equipment, and consumption of foods from unsafe sources account for a large percentage of the outbreaks.

Consumers' food safety concerns extend beyond contamination. Consumers list their concerns as: chemical residues and pesticides on food; the safety of salt, colors, and additives; the safety of caffeine, sugar substitutes, fat, and cholesterol; microbial hazards; the use of antibiotics in animal products; nonchemical contamination (foreign objects in food); and the use of sterilization and/or irradiation procedures. While some alterations in this listing are noted, the list remains fairly stable since the mid 1960s.

Significant changes in consumer attitude toward food protection regulation are noted, however, during this same time period. From the 1960s to 1970s, the public tended to oppose government initiated consumer protection regulations. From 1980 to 1984, a trend toward increased consumer desire for strict industry regulation by governmental agencies emerged. A 1984 Roper poll found that 56% of those surveyed felt that food safety regulations were not adequately enforced. A majority of those surveyed favored more regulations. Today, consumers rely more on themselves and the government to monitor the food supply, and less on the manufacturers of food products.

A challenge exists to provide consumers and food handlers with basic competencies and skills necessary to monitor their personal food safety and safety of others as it relates to them. Specific education programs complete with appropriate strategies geared to the needs of the audience may be outlined. Effective means of evaluating the degree of learning can be built into these food safety education models. Through effective education programs designed for those who handle food, levels of competency in this area may be elevated and key food safety skills may be developed.

ASSESSING THE AUDIENCE

According to the First National Conference on Food Protection, the protection of our food supply from contamination is a "people problem." Either as food handlers on the job or consumers at home, almost every American will handle food. The growing number of meals consumed in restaurants, the large size of the food industry, and food industry personnel produce a potential "classroom" of millions of learners. Add to this the number of consumers preparing food at home for themselves and others, and the learner group must be defined as anyone handling food for consumption. Essentially all adults and older children fit this broadest definition.

An effective educational model for development of skills or competencies in food safety must be designed for the adult consumer. Strategies based on adult learning styles will theoretically exert the most impact on food handling behavior of adult consumers, production workers, and food handlers.

TRAINING PROGRAMS

A wider network of training and testing programs for food handlers is available today than ever before. Industry and trade groups such as the National Restaurant Association, and state and local health and agricultural departments sponsor a variety of existing programs. Many food chains and large companies provide in-house training for their personnel. Additional food sanitation courses are offered at universities, community colleges, and vocational schools. Over 300 training programs exist today.

The general consumer receives food safety information from the media, from course offerings in adult learning sites, such as school districts and extension programs, or through public health agencies. These programs vary.

More formalized testing and certification procedures for food service managers and handlers have been recommended to ensure appropriate levels of food protection knowledge. The Food Protection Certification Program, organized and implemented by the Food and Drug Administration in July 1985, represents an example of a formalized testing procedure. The FDA program includes a testing procedure for all managers to be provided and scored by the Educational Testing Service (ETS) of Princeton, New Jersey. Candidates receive information regarding their test outcome and the specific questions answered incorrectly. The certification program is voluntary and the FDA has no plans to make participation mandatory. No training component exists.

ASPECTS OF ADULT LEARNING

The concept that adults have particular learning traits, styles, and preferences is well-supported. Many educational theorists have extended or projected cognitive, social, and affective developmental theory into adulthood. Adults learning about food safety through formal courses, in-service training, or programmed self-instruction have key characteristics in common. Some factors which seem to influence adult learning include:

1. Taking into account the array of personal experiences the adult relates to school and learning
2. Providing constant and continued class and training time for successful experiences
3. Recognizing the need for involvement
4. Recognizing the need for practical application and practice
5. Understanding the immediacy of adult goals
6. Identifying the course participants as a prime learning and teaching resource
7. Considering the learning speed of adults; physical and mental fatigue are determining factors in adult learning
8. Identifying key factors which create a climate for motivation of the adult learner

Adults learn any subject best when principles of androgogy are translated into a process for planning and implementing education or training programs.

According to Knowles, key phases must be considered when training adults. In terms of food safety, these phases may be easily applied.

- Create a climate for learning: establishment of rapport with the adult food service managers, food handlers, or consumers is essential.
- Structure participative program planning: allowing handlers to be involved in the development of their programs ensures ownership.
- Assess needs: prior diagnosis of skills and competencies necessary to handle food safety is important.
- Formulate and implement objectives and activities: recognition of adult characteristics as cited above provides insight into program design.
- Evaluate: testing of skills development generates feedback for retraining programs.

Food safety training programs for adults generally have not addressed these specific phases of learning nor the characteristics of adult learners. Perhaps the size of the audience or their diversity as learners has stymied the consumer educator.

Many competency-based, adult-focused education programs exist today. Teachers, physicians, dietitians, and other professionals are trained through these methods. The Cooperative Extension Service has developed specific models for developing competencies for successful farming. However, programs for the adult consumer are seldom competency-based.

A variety of teaching methods may be employed in adult education and training programs including new and traditional presentation techniques (lecture, computer-based learning, video, panels, slides, films, audio cassette, exhibits, programmed instruction), discussion techniques (case discussion, problem solving discussion), simulation techniques (role playing, critical-incident process), and skill practice exercises. The type of method used to educate adults for skills development is important. The method must meet the objectives and allow participation. The importance of the learning activity lies in its relation to the adult learners' characteristics and not in the medium itself.

IDENTIFIED COMPETENCIES

Based on the Proceedings of the Second National Conference for Food Protection, four major food safety objectives to be met by consumer food handlers would produce a safer food supply. These are translated

Table 1. Food Safety—Skill/Competency Assessment Scale

	Current Level of Skill	Desired Level of Skill
	0 1 2 3 4 5	0 1 2 3 4 5
1. Ability to apply personal hygiene techniques related to food safety:		
a. wash hands properly		
b. dispose of personal refuse properly		
2. Ability to apply appropriate food storage techniques:		
a. hot foods hot, cold foods cold		
b. foods covered		
c. foods off floor		
3. Ability to identify and select foods which are safe to consume		
4. Ability to prepare foods which are safe to consume		

into adult competencies in Table 1. They represent the well-documented needs as assessed by consumer trainers and food scientists.

Food safety programs for adults may be designed around these competencies. Objectives and strategies may be planned which focus on developing the level of skills to be reached. Finally, evaluation of programs based on development of skills, attitudes, and behaviors associated with these competencies may be put into place.

A MODEL FOR COMPETENCY-BASED FOOD SAFETY PROGRAMS

The challenge for consumer educators is to develop food safety programs based on adult learning theory which create the climate for skills development. The proposed model (see Figure 1) is flexible, allowing for the introduction of varied teaching strategies appropriate to the audience, the locale, the instructor, and the resources available. Critical areas of concern are the participative assessment of competencies before and after training and the adherence to adult learning principles.

The use of strategies appropriate to the learning setting deserves some comment. Proceedings from the Second National Conference for Food Protection discussed the use of video as a tool for effective education and training. This is an important area. National and local video teleconferencing and use of video tapes both offer new and effective strategies when used as part of a total plan.

"Consumer Week" activities, exhibits, programmed learning, television, audio cassettes, and behavior modification workshop techniques

Figure 1. Model for development of ongoing food safety training program.

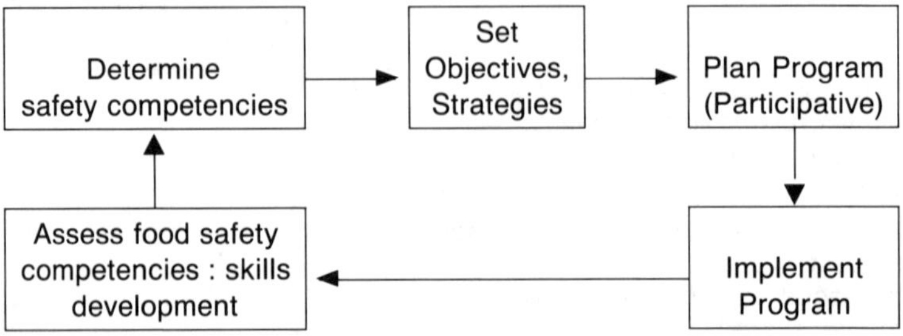

are also documented as appropriate for adult learning. The proposal is to "find the fit" or choose the strategy which best suits the learning opportunity.

Broadcasting media is an excellent partial implementation strategy. A well-marketed television program about food safety, designed for a general audience and given a peak hour evening placement, is likely to reach about 10% of the adult population depending on available alternatives. A radio talk on a highly specialized topic for a selective audience may have an audience of only 0.1% of the population. This still may represent a cost-effective way of reaching adults.

But is this effective adult education? Jessup suggests that effectiveness of education through broadcast media depends on various factors, such as the address of the program to basic adult interests, the number of times the adult hears the message, and the setting where the message reaches the adult. Use of broadcast media in partnership with other educational tools is suggested.

In summary, a model for effective adult food safety education will address the skills and competencies needed to achieve goals. Strategies used to develop skills must be based on the characteristics of adult learners. Appropriate activities for learning focussing on these characteristics can be developed. The process of skill development is gradual and continued, just as learning is a life long project. The model may change, but the principles for adult education to affect behavioral change remain the same.

BIBLIOGRAPHY*

1. "Putting Retail Food Safety to the Test," FDA Consumer, May, 1985, V. 19 #4.
2. "Current Issues Report: Consumer Food Safety Concerns," National Restaurant Association Report, 1985.
3. "Considering the Needs of Educators as Learners," *Education*, Winter, 1985, V. 106 #2.
4. "The ABC's of Food Storage," FDA Consumer, March, 1986, V. 20 #2.
5. Proceedings of the Second National Conference for Food Protection, 1984.
6. Proceedings of the 1971 National Conference for Food Protection.
7. Wilsey, Cathy."Making Staff Development Programs Work," *Educational Leadership*, V. 40 #1.
8. Krupp, Judy-Arin. *The Adult Learner: A Unique Entity*. Adult Development and Learning, Connecticut, 1982.
9. Brookfield, Stephen. *Understanding and Facilitating—Adult Learning: Analysis of Principles and Practices*, Columbia Press, 1986.
10. Brookfield, Stephen. *Self-Directed Learning*. Columbia Press, 1986.
11. Scupham, John. "The Media of Mass Communication," *Life Long Learning: A Symposium of Continued Education*, Pergamon Press, Ltd., London, 1969.
12. Moore, Donald E., Jr. "Assessing the Needs of Adults for Continuing Education," *Assessing Educational Needs of Adults*, Jossey-Bass, Inc., 1980.
13. Nored, Arnold. "Competency Based Adult Education," *Materials in Adult and Continuing Education*, Klevens Publications, Inc., 1982.
14. Knowles, Malcolm. *The Modern Practice of Adult Education*, Follett Publishing Co., Chicago, 1980.

*References to these sources were not provided within the text of this chapter.

29. Producing Consumer Education Material: A Practical Guide

John Knapp

National Coalition for Consumer Education
Eastern Michigan University

One of the puzzlements that arise in writing about consumer education materials in the food safety area is the diversity of the audience. Besides growers, sellers, packagers, and users, there are government officials, academics, environmentalists, and advocates for various causes. Making a fairly specific topic relevant to such a diverse audience—and the list is not exhaustive—can be difficult.

Acknowledging that this problem confronts anyone writing about food safety, I would like to list several of the assumptions that brought the topic into clearer focus for me.

First, I assume that the need for ultimate consumers of all ages to be educated about food safety is not in question. Indeed, in the prologue to the recommendations made by the Education and Training Committee at the Second National Conference for Food Protection, the point is emphatically made: "Either in the home or on the job, virtually every American handles or will handle food to some degree. . . . Therefore, when viewed in the broadest sense, every person is a potential food handler and needs to have some knowledge of the importance of personal hygiene and sanitary food handling."[1]

Another assumption is that the act of setting down certain guidelines, tips, and advice concerning the production of consumer education materials is not a mere exercise in composition—that, in fact, there is a modest need for such information, especially among business persons or government officials whose consumer education experience may be slight. I'm not implying that such persons don't understand the needs of their customers or constituents, but only that a little assistance might be useful. Accordingly, this chapter summarizes the points that should be considered by someone either producing consumer education material or responsible for its production. Food safety and food will be the subject of some if not all of these resources.

Before getting into the do's and don'ts, it might be useful to repeat the distinction which SOCAP, the Society of Consumer Affairs Professionals in Business, draws between three related types of material. These are found in the brochure entitled, "Guidelines For Business Sponsored Consumer Education Materials,"[2] a 1982 pamphlet offering excellent advice to business people. *Promotion materials*, according to the Guidelines, are designed primarily to sell a company's products or services or to enhance its image or justify its viewpoint. Characteristically, promotional materials "have an emotional, enthusiastic tone, present only favorable information, and prominently display the sponsor's corporate identification or product brand names."[2] *Information materials* are more straightforward and factual, more neutral in tone. "The facts are presented without bias, but they usually pertain exclusively to the sponsor's business."[2] "*Education materials*, ideally, are statement of fact presentations."[2] They are noncommercial, dispassionate in tone, and balanced, that is, representing all relevant facts and points of view. Sponsor identification is included but product brand names are not.

These distinctions help to define what is meant by consumer education material although the difference between information and education is perhaps somewhat artificial.

RECOMMENDATIONS

There are three important considerations for someone planning a consumer education work.

1. The first consideration is to determine the need. (One of the best indicators of need is the phone calls or letters an organization gets from consumers who have questions, complaints, or compliments.) If a business or unit of government has some indication of a gap in the education of the consumers it deals with, that could signal a go ahead. If not, there is little point in filling a hole that does not exist. But even bare need is not enough. The would-be producer should undertake a materials search to determine what has been done in the past and whether it needs to be redone. A list of organizations helpful in such a search is appended to the chapter.
2. The second consideration is to determine what your objectives are in producing the work. Since education of consumers is the long-term goal, the educational purpose and value of the resource should be scrutinized carefully. A handy collection of such concerns appears in the checklist

entitled "Educator's Guide for Using Business Sponsored Resources in Consumer Education Classrooms."[3] Although it approaches the subject from the teacher/student side of the coin, the issues it raises are relevant to producers as well. What values are implied by the consumer education resource? Who is the intended audience? What socioeconomic level is being targeted? Are learning activities supplied or suggested? Does the material motivate students to further inquiry or creative thinking? These questions all help to clarify the producer's aims.

In creating these materials, if an organization tries to mix the educational motive with the promotional impulse, "almost inevitably," in SOCAP's words, " that will undermine the educational integrity of the project"[2] — a warning that does not apply to business alone.

3. The third consideration deals with the qualities of content and medium, which are closely related. There is likely agreement on the proposition that the printed word will be the medium of choice for those producing consumer education material. Of course, one should not overlook the impact that video can have, as Dr. Robert Gravani pointed out at the 1984 conference.[4] Indeed, ultimate consumers whose reading skills may not be up to par can often be reached by video, whereas material in print will literally remain a closed book. Yet, works on paper will dominate. The reasons for this — feasibility, reasonable cost, permanence, historical precedent — are obvious enough and need not be expanded upon here. Teachers using resource materials supplied by business or government especially appreciate the qualities of sturdiness, convenience, easily replaceable parts, and ease of storage.

Regarding content, both the SOCAP Guidelines and the Michigan Guide use many of the same words to describe high-quality consumer resources. Here are six of those adjectives:

Accurate. The work should be factual, easily verifiable, and sufficiently documented. It should take into account current laws and regulatory decisions. It should be up-to-date not only when first issued, but for as long as the sponsor expects to provide it. A date of publication should be included.

Objective. Although the sponsor's viewpoint can be included in the material, it should be identified as such. When controversial issues are touched on, the pros and cons should both be given or at least alluded to.

Balanced. Omissions of relevant information that might discomfit the sponsor should be avoided, as should exaggerations or distortions.

Appropriate. The level and style of writing will be appropriate to the expected audience. Echoing the advice given at the 1984 conference by Edwin Traisman, the consumer education work should avoid technical terms and jargon. What the consumer needs, says Traisman, is a "limited number of sensible, easily understood concepts" about food safety.[5]

Nondiscriminatory. The consumer resource should not exhibit derogatory attitudes toward particular groups in its language, subject matter, and illustrations. Care should be taken to avoid inadvertent discrimination based on race, sex, creed, age, or ethnic group.

Noncommercial. A great deal has been written on this topic, especially following the 1979 book by Sheila Harty entitled *Hucksters in the Classroom: A Review of Industry Propaganda in Schools.*[6] Because that volume evoked such vehement reactions upon publication, it might be instructive to touch on a few of its points.

Harty, who can be fairly described as a "Nader raider," to use the pop journalism term, sets forth her thesis in the book's title. She examines business-sponsored educational materials in the areas of food and nutrition, energy (especially nuclear power), environment, and economics. They contain, she finds, propaganda, bias, one-sidedness, omissions, and commercialism. The fact that budget-conscious school systems and hard-pressed classroom teachers can receive attractive materials at little or no cost compounds the problem. She concludes that: "Corporations are inappropriate sponsors of educational materials precisely because they have something to sell."[6]

Her prohibition against business in education has been emphatically rejected by many companies and trades who believe their viewpoint of a subject close to them can be validly expressed in educational material.

Has Harty's book caused business to draw back from its consumer education efforts? Without studying the matter, it would seem that the quantity of business-sponsored material in 1986 is about the same as that which existed prior to Harty's book. However, there is much greater awareness by business of the potential for criticism if, in dealing with controversial issues, care is not taken so that overt promotion is not mingled with the educational purposes of the material. Harty's book did some consciousness-raising in corporate public relations and consumer affairs departments.

One way for business to avoid commercialism, in her view, is to refrain from mentioning brand name products or picturing corporate trademarks and advertising logos. She says that company sponsorship should definitely be indicated but only by a typewritten statement.

CONCLUSION

In the last several years, the practice of using qualified educational consultants to assist business in preparing consumer resources has increased. Although large corporations have educators on staff, the use of outside helpers can result in greater diversity of opinion. But again, business needs to assign meaningful roles to its consultants, not just to purchase their seal of approval.

Similarly, teacher review boards to examine materials prior to publication, and field consultants to pretest materials in the classroom are becoming more common. Besides involving educators, business has also given thought to ways of evaluating its material. One suggestion that has been picked up is the development of a feedback mechanism to be included along with every piece of consumer education material distributed. The National Association for Industry-Education Cooperation has produced a handy one-page evaluation form which is sometimes used.[7]

Although business has shown an interest in getting help from educators, the actual finding of qualified consultants sometimes presents a problem. But several nonprofit consumer education organizations are willing to provide advice and assistance, much of the time without charge. The previously mentioned list of consumer education organizations in the appendix to this chapter may be helpful. The Michigan Consumer Education Center has, in the past, conducted reviews of research and materials, identified consultants, assembled review boards, and located teachers able to field-test materials. Other organizations provide similar services.

There is one final caveat on dissemination of materials. Some corporations have been so enthusiastic about a newly developed consumer education work that they have attempted a saturation mailing to every school district in the country. Though well-intentioned, this tactic can backfire. It is better to advertise in educational publications or display the resource at consumer education conferences and wait for orders than to distribute it broadcast across the land.

APPENDIX

Consumer Education Material—Where to Go for Help

American Council on Consumer Interests
238 Stanley Hall
University of Missouri
Columbia, MO 65211

Center for Consumer Services
Oklahoma State University
Home Economics West
Stillwater, OK 74074

Consumer Information Center
General Services Administration
Pueblo, CO 81009

Consumers Union Educational Services
256 Washington Street
Mount Vernon, NY 10550

Joint Council on Economic Education
School Services Division
2 Park Avenue
New York, NY 10016

Michigan Consumer Education Center
207 Rackham Building
Eastern Michigan University
Ypsilanti, MI 48197

U.S. Office of Consumer Affairs
1009 Premier Building
1725 I Street NW
Washington, DC 20201

REFERENCES

1. Education and Training Committee. 1984. Proceedings of the Second National Conference For Food Protection. U.S. Department of Health and Human Services—Food and Drug Administration, Washington, DC, p. 321.
2. Anonymous. 1982. Guidelines For Business Sponsored Consumer Education Materials. Published by Society of Consumer Affairs Professionals in Business, 4900 Leesburg Pike, Suite 311, Alexandria, VA, 22302, pp. 3–4.
3. Michigan Consumer Education Center. 1978. Educator's Guide for Using Business Sponsored Resources in Consumer Education Classrooms. Michigan Consumer Education Center, Eastern Michigan University, Ypsilanti, MI.
4. Gravani, R. 1984. Can Video Be Used as an Effective Training Tool in

Food Protection. In: Proceedings of the Second National Conference For Food Protection, Washington, DC.

5. Traisman, E. 1984. Educating the Consumer in Food Protection. In: Proceedings of the Second National Conference For Food Protection, Washington, DC, p. 276.
6. Harty, S. 1979. Hucksters in the Classroom: A Review of Industry Propaganda in Schools. Center For Study of Responsive Law, Washington, DC, p. 139.
7. National Association for Industry-Education Cooperation. 1976. A Guide For Evaluating Industry-Sponsored Educational Materials. National Association for Industry-Education Cooperation, 235 Hendricks Boulevard, Buffalo, New York, 14226.

30. Uniform Facilities Planning and Plan Review for Food Service Establishments

James L. Brown

National Restaurant Association

INTRODUCTION

The food service industry in the United States is regulated by more than 3000 different governmental agencies. Federal, state, and local jurisdictions often overlap in their control of the safety of the American food supply.

Model codes, such as FDA's 1976 Foodservice Sanitation Code have been prepared and recommended for state and local adoption and use. These model codes received considerable input from well-informed individuals representing the regulatory community, the food service industry, and the consumer. The inclusion of a general requirement for plan review prior to construction was obviously intended to anticipate sanitary defects in the establishment and to design for prevention of those anticipated defects. We might conclude that the model codes represented the "state of the art" at the time they were written, and provided for the maximum attainable level of consumer protection.

However, the model codes were not always perceived at the local regulatory level as "state of the art." Because the states are sovereign units, deriving their authority to regulate food service establishments from their general police powers to protect the public health, safety, and welfare, there were a number of state and local programs already operating before the "model codes" were issued. Indeed, the 1971 Conference on Food Protection noted that state and local agencies historically regulated foods years before the federal government adopted the responsibility, and that the state and local units have a better grounded authority for such regulation, with the proper federal role being supportive and advisory.

Thus, not all agencies adopted the model code verbatim. Locally perceived requirements crept into state codes, which were again modified by local jurisdictions, generally to provide more stringent standards than the parent state documents. Finally, agencies and individuals made widely varying interpretations of the language in the model codes, including the requirement for review of "properly prepared plans." Such plan review may vary from formal engineering blueprints to literal thumbnail sketches, to no review at all.

The result is a potpourri of codes, which, although developed with the intention of protecting the consumer, are at best confusing and lead to cost and time delay to the food service industry and regulators.

Regulations can only be justified in terms of their capacity to promote the safety and well-being of the individual and the group of which the individual is a part. Regulators should document tangibly the degree of imperfection which exists in the condition to be regulated, and the cost-benefit outcome that can be guaranteed by the regulation. Codes governing food service operations often fail this test, in that they may be couched in vague language subject to interpretation, or based on arbitrary or technologically obsolete standards. More to the point of this chapter, food service codes against which plans are reviewed are simply not uniform, either between or within their jurisdictional applications. The majority of codes are in need of review, update, and revision to achieve the uniformity which would benefit the regulatory community, the affected industry, and the consumer. Uniform procedures for plan review would have similarly widespread benefits, but rest on concurrent code revision.

BACKGROUND AND HISTORY

In 1897, the forerunner of what is now the Association of Food and Drug Officials (AFDO) met to promote unity among chaotic state food and drug laws. FDA developed the Pure Food and Drug Act in 1906 and began to write model codes and ordinances in the 1930s. The model food service code was updated in 1943, 1962, and most recently in 1976. There remains today a need to again review this model code in the light of current demands and technology.

As noted earlier, the only basis on which food service codes may be justified is protection for the consuming public. To provide that protection, FDA is promoting the Hazard Analysis Critical Control Point (HACCP) concept of inspection, to reestablish emphasis on the factors which are known to cause foodborne disease. Concurrently, the food service industry is promoting a similar self-inspection tool, the Sanitary Assessment of Food Environment (SAFE). Both the HACCP and SAFE programs have the same goal of consumer protection, and both use similar approaches to achieve that goal; stressing safe food handling practices to prevent bacterial contamination of foods and to block survival and growth of pathogenic microorganisms in foods. However, both concepts are dependent on the "physical plant," that is, the building design and construction and equipment layout and installation, to achieve that goal. Without uniform plan review, the proper emphasis on food protection becomes mired in a conflicting maze of widely varying physical plant and construction requirements, some of which may appear to serve little purpose.

Consider the plight of the food service chain which operates in several different regulatory jurisdictions, facing different construction requirements in different areas, all of which are applied to the same basic food service operation, all in the name of public health protection. A simple item such as lighting in work areas may vary from 20 to 50 foot-candles in different states or counties. Although the FDA has issued interpretations on conversion of hot water dishwashing machines to employ chemical sanitizers, some areas of the country will not permit such conversion, and others prohibit chemical sanitizing machines outright. Some codes require floor drains in refrigerated units; others prohibit these drains. In some regions, nontoxic, cleanable beverage tubing, cold plates, and carbonators are allowed to contact potable ice; other areas prohibit this practice. Most areas call for 3-compartment sinks for utensil washing; others require 4 compartments. The FDA's 1976 recommended Food Service Sanitation Ordinance accepts 45°F or less as an acceptable temperature to hold potentially hazardous foods, but in at least one jurisdiction, the mandated temperature is 40°F or less. (A difference of five degrees may seem trivial, but to achieve this temperature drop requires a geometric increase in refrigeration capacity and energy demands, as well as causing inconsistencies in chain operation and training procedures.)

Such disparity extends to the plan review process, and the cost to the industry is extreme, with little or no demonstrable improvement in the level of public health. Presumably, these and other various requirements were originally justified and based on good faith information and an intent to protect the public. But the result of such jurisdictional variance is to produce confusion and increased cost for both industry and consumers and to reduce the credibility of the regulator.

Both the 1971 and 1984 National Conferences on Food Protection called for better uniformity and reciprocity of regulatory codes and plan review processes. It would appear that the problem originally attacked by AFDO nearly 100 years ago is far from solved.

Now is the time to review current codes, and bring them into perspective and agreement with current technical data. With this in mind, we should determine to what extent we can unify requirements for the physical plant and the plan review process. The physical plant, the building, facilities within the building, and processing are all critical to the safe and sanitary function of the food service operation. HACCP and SAFE programs, although well-intentioned, are to a considerable degree at the mercy of facility and equipment performance and thus plan review.

WHAT CAN BE DONE?

- We can recommend the development of a standardized food service facility plan review process.
- We can recommend that food service facility plan reviewers be trained and certified by a central authority or credentialing group.
- We can recommend reciprocity between jurisdictions for standardized facility plans which have been previously approved by a competent plan review process.

This appears to be a tremendous undertaking. However, with concurrent action by the FDA, the regulatory community, the food service industry, professional organizations, and third party standards development groups, these goals can be achieved.

Implementation of this approach would ensure that the plan review process meets its intended aims: to guarantee physical compliance with code requirements, to prevent procedural misunderstandings and delays, and to anticipate and prevent later errors which may result in additional cost or hazard to the operation.

RECOMMENDATIONS

1. Specific guidelines for plan preparation, format, review, approval, and subsequent action should be described. The National Sanitation Foundation (NSF) reference guide for sanitation aspects of food service facility plan preparation and review would be an appropriate beginning for this centralized guideline. This guide addresses many of the administrative factors which should be included in the development of a uniform plan review process.

2. Competent review of the plans and specifications should be provided by trained personnel, certified in these functions by independent standards or credentialing organizations.

3. The process should provide for prompt and professional review against nationally recognized codes and standards applicable to the facility.

4. A central clearinghouse should be established through which an operator or agent could be guided and assisted in review of plans and specifications. This clearinghouse could be national, but regional operations might be more logistically feasible. The clearinghouse could be governmental or an acceptable third party certification group.

5. Regulatory agencies should accept standardized multiple-unit plans and specifications which have been accepted by the central agency or other recognized jurisdictions. Initially, such acceptance could be accomplished by memoranda of agreement between state and local agencies, different states, or state and federal agencies.

The Conference of State Health and Environmental Administrators has hosted conferences which identified some of the problems that exist in the plan review process. The findings of those meetings further demonstrate the need for a centralized and uniform plan review process.

CONCLUSION

Standardized plan review processes, with training and credentialing of plan reviewers, will provide the basic elements necessary to promote improved uniformity of facility and equipment requirements, based not upon the whims of local programs, but upon the true operational needs of the food service establishment and the achievable levels of public health protection consistent with current technology.

Survey data have demonstrated the need for better uniformity in codes and the plan review process. The complexity of plan and specification review calls for establishment of training and credentialing programs for plan reviewers. The advent of multiple-unit companies with construction and operations based on standardized plans and specifications calls for reciprocity between jurisdictions, recognizing previously accepted plans.

Centralized, reciprocal, and uniform plan review processes will benefit all parties. Industry and consumer cost will be reduced; regulatory agency costs will be reduced; and protective programs such as HACCP and SAFE will have a sound basis in uniform facilities and equipment to provide the consumer protection for which we are all striving.

31. Education and Training Committee Recommendations

1. The Conference for Food Protection endorses the creation of a standardized food service plan review process that includes guidelines for plan preparation, format review, approval, and subsequent action.

2. The Conference will send a letter to the National Sanitation Foundation (NSF) requesting that NSF create a task force to establish uniform plan review processes. NSF will be asked to keep the Conference abreast of developments as they occur.

NEW FOODS, PROCESSING, AND PACKAGING

32. Genetically Engineered Foods and Ingredients—Technical Basis and Industrial Applications

Nanette Newell

Biotechnology Consultant

NATURE OF THE PROBLEM

The advent of new genetic technologies over the last decade allows for novel modifications of living organisms, including those that we use for food. Because the technologies involved in these genetic modifications are substantially different than those traditionally used, concerns have been raised as to the safety of the resulting foods. This discussion centers on the issues over food safety that directly result from genetically engineering new foods.

Because the techniques of genetic engineering are so powerful, the potential for new foods and ingredients are nearly limitless. In the simple understanding we have now of molecular genetics, we are able to conceive, for instance, of more nutritious foods due to changed amino acid content and increased amounts of vitamins. As we gain more understanding, it is possible that we will be able to modify crop plants so they store better after harvesting, resist spoilage, and can be processed more efficiently. With all of these potential improvements to our food supply, though, care must be taken to assure that detrimental effects are not produced concurrently with the improvements.

Any discussion of genetic engineering must first begin with an agreement on a working definition. For the purposes of this discussion, genetic engineering of food organisms will include any transfer of genetic information into a food organism that cannot occur in nature because of evolutionarily determined species barriers.

HISTORICAL PERSPECTIVE

The genetic manipulation of food organisms is probably the oldest form of technology known to humankind. The earliest farmers saved seed from their most productive plants for planting the following year and selectively mated their most productive animals. In this way, the agricultural gene pool was manipulated, and genes for such complex traits as yield and tolerance to environmental conditions were selected. This selection process has continued up to present time, albeit in more sophisticated ways.

Until the last decade, agriculturalists have only had the tools associated with conventional breeding within species to transfer genetic traits to give us improved food commodities. Now, with the advent of genetic engineering, new tools have been added to the agriculturalists' technical toolbox for the introduction of additional genetic characteristics. These new technologies allow genetic material to be inserted into agriculturally important organisms from virtually any existing organism. In other words, the breeding barriers imposed by species can now be crossed.

Two technologies fall within our definition of genetic engineering: recombinant DNA and cell fusion. Recombinant DNA technology involves the joining, in vitro, of two DNAs from different species. In most cases, these experiments involve taking a gene of interest from any source—bacterial, viral, or higher organism—and joining it to a vector DNA so that it can become part of a new microorganism or can be stably transferred into the cells of food organisms. Cell fusion technology refers to the joining of two cells from different species such that their genetic material is combined, very much like the result of traditional breeding but without the need for sexual compatibility. Except in the case of a new microorganism, the engineered cells need to be grown into whole organisms before they are useful to the food producer. In the case of plants, the genetically engineered cells are regenerated into whole plants, and the resulting plants are placed in a breeding program. In the case of animals, only recombinant DNA experiments are applicable currently, and the new genes are inserted into very young embryos. These genetically engineered embryos are transplanted into surrogate mothers to produce the genetically engineered whole organism. Then, the new animal can be used in a breeding program. It should be understood that genetic engineering is not a concept that is completely new to agriculturalists. It is only an extension of the traditional technologies for moving genetic traits among different organisms that have been practiced for centuries.

POTENTIAL PROBLEMS AND SOLUTIONS

The newly acquired ability to cross species barriers has caused concerns on two fronts. The first is a purely religious one—a belief that species barriers exist for some existential reason, and thus, people have no right to interfere with the "ways of nature." The relevance of this concern within our own beliefs must be decided by each of us, but in no way does it have any relevance to the issue of food safety. The second concern is that we know so little of life's molecular basis that we cannot predict the results of genetically recombining material from different species. Among many practitioners, this concern is considered unfounded, and it is believed that we possess enough knowledge to, at least, ask the right questions and do the right experiments.

A discussion of safety considerations for genetically engineered foods and ingredients can be divided into three areas because of three distinct kinds of products that arise from the use of genetic engineering:

- *Case 1*: Foods where the newly introduced DNA *does not* cause a substance novel to food to be expressed in the part of the organism that is eaten, e.g., crops developed for different amino acid content, vitamin content, or higher levels of fructose.
- *Case 2*: Foods where the newly introduced DNA *does* cause a substance novel to food to be expressed in the part of the organism that is eaten, e.g., crops developed for herbicides tolerances or where a new sweetener, such as thaumatin, is expressed.
- *Case 3*: Food ingredients from genetically engineered organisms or in vitro bioprocesses derived from genetic engineering, e.g., flavors, fragrances, dyes, and modifiers.

Case 3 can be considered of only minor concern, because the food ingredients that result are no longer associated with genetically engineered organisms and should be treated like any other chemical being considered as a food ingredient. In fact, this strategy was proposed by the FDA several years ago for any product of genetic engineering that falls within its purview and is accepted as the working system for the FDA's analysis of genetically engineered products. Whether or not the FDA's regulatory requirements are appropriate with respect to food safety may be a matter for discussion, but *the requirements are, and should be, independent of the source of the food ingredient*. Purity of the ingredient, its functionality, and standard of identity are examples of such requirements.

Cases 1 and 2 present both unique and classical problems to food safety. Two unique problems can be recognized in these two cases. The first, present in both Cases 1 and 2, is that people will actually be eating recombined genetic material from different species. Among many practitioners, this situation is not considered a problem. Since the entire basis of genetic engineering is predicated on the fact that the building blocks determining the structure and function of the genetic material are the same for all living organisms, the source of that material should be of no consequence to the digestive tract of animals or humans. Traditional genetic recombination produced through breeding has certainly never been an issue in food safety, and, scientifically, it makes no sense that novel genetic recombination should be a food safety problem.

The second problem unique to genetically engineered foods is that cellular metabolites—either protein or secondary metabolites—not already present in our diet will be expressed in some genetically engineered food organisms. This situation is exemplified by Case 2 and is only relevant in the case of recombinant DNA technology. It is essential that developers of these new foods assure the public that the new substances are not toxic. Two scenarios are discussed to illustrate this situation.

One likely scenario is that the first genetically engineered food organisms to enter the marketplace will be those that have had one gene, thus one protein, added to them; an herbicide-tolerant plant is a good example. This plant would be producing a novel protein that causes a particular herbicide to be nontoxic to the plant, and it is possible that this protein would be expressed in the edible portions. (In fact, the presence of an herbicide that remains on the plant is probably more of a concern.) Since virtually all proteins are degraded into their constituent amino acids by either enzymes or stomach acid, it seems unlikely that any new protein would be toxic to animals or humans upon ingestion. At this point in time, and for the next several years, protein addition is the only kind of recombinant DNA experiment scientists have the knowledge to perform.

Another scenario is that, as more is understood about the mechanisms of gene regulation, it becomes possible to manipulate entire metabolic pathways. One could conceive of the introduction of a metabolic pathway into a food organism that produces a secondary metabolite, such as the pathway to produce a novel sweetener. In cases such as this, it will be essential to test these and concomitantly produced new compounds for toxicity. It should be understood, though, that this complex genetic manipulation is many years away.

A classical problem that is not unique to genetically engineered foods, but could be encountered for any newly developed food, is not knowing whether a particular genetic change presumed to be beneficial has unknown, but linked, deleterious effects. When using recombinant DNA technology to insert a new gene into an organism, however, we can more accurately predict potential problems than can be predicted using traditional breeding. With the transfer of only one or a few genes, the amount of DNA that is being introduced is less than one millionth of the total DNA in the organism. Since the new gene is being introduced into a characterized genetic background, the outcome of the transfer is much more predictable than that of classical breeding where two complete sets of genetic material are intermixed relatively randomly. The cell fusion case is basically equivalent to a traditional breeding situation and, thus, inherently does not present any additional issues.

Preventing hazardous foods, either traditionally generated or genetically engineered, from entering the marketplace is the main food safety issue. The agricultural community, using a nonpredictive system of self-monitoring, has been extraordinarily successful over the past decades in this respect. Considering the predictive capability associated with genetic engineering, we can assume that the assessment of food safety can only improve with the advent of this new technology.

BIBLIOGRAPHY

Harlander, S.K. and Labuza, T.P., Eds. 1986. *Biotechnology in Food Processing*. Park Ridge, NJ: Noyes Publications.

Hauptli, H. et al. May 1985. Genetically Engineered Plants: Environmental Issues. *Bio/Technology* Vol. 3, pp. 437–442.

Office of Technology Assessment (OTA), U.S. Congress. 1984. *Commercial Biotechnology: An International Analysis*. OTA-BA-218.

Wood, F.A. et al. 1984. Genetically Engineered Food and Food Ingredients, In *Proceedings of the Second National Conference for Food Protection*. U.S. Department of Health and Human Services.

33. Genetically Engineered Foods and Ingredients – Legal Considerations

Gerad McCowin

U.S. Food and Drug Administration

NATURE OF THE PROBLEM

One of the most difficult tests of a statute and its accompanying regulatory process is whether they can adjust to new situations. Genetic engineering presents just such a challenge to the Federal Food, Drug, and Cosmetic Act ("the Act") and to the Food Additives Amendment.

A basic premise of a statute is that it provides broad guidance of congressional intent as to how an agency shall react to particular situations. Although there are exceptions, a statute generally leaves to the agency the responsibility of issuing regulations that describe the specific procedures to be followed to implement that congressional intent. The Federal Food, Drug, and Cosmetic Act as enacted in 1938 and as amended in 1958 creates a number of categories under which the U.S. Food and Drug Administration (FDA) is to assess the safety of food.

The types of processes and resulting products created today as part of genetic engineering were virtually unknown and not specifically considered during either the development of the statute or the issuance by FDA of implementing regulations. This discussion will cover the variety of the resulting legal issues raised by genetically engineered foods or food ingredients. In particular, it will cover those points that need to be evaluated by the FDA in order to determine the safety of such products.

HISTORICAL PERSPECTIVE

Section 402 of the Act contains those parts that pertain to whether a food is adulterated and possibly unsafe. Paragraph (a)(1) of this section states, in reference to an *added* substance, that a food shall be deemed to be adulterated "if it bears or contains any poisonous or deleterious substance which may render it injurious to health." In case the substance is *not added*, the food "shall not be considered adulterated under this clause if the quality of such substance does not ordinarily render it injurious to health."

The next paragraph of the Act identifies specific categories of added substances that may serve to adulterate food. There are two that are of specific relevance for genetic engineering:

> (a) If it bears or contains any added poisonous or added deleterious substance . . . which is unsafe within the meaning of section 406 (Section 402 (a)(2)(A)).
> (b) If it is, or it bears or contains, any food additive which is unsafe within the meaning of section 409 (Section 402 (a)(2)(C)).

The significance of these sections revolves around two questions. First, what is the standard to be used in establishing safety? Does the substance *ordinarily* render a food injurious or *may* it render a food injurious? Second, who has the burden of demonstrating safety—the government, or the industry, or some other interested person?

Whether the more stringent "may render" safety standard applies to a situation depends upon whether the substance is added or occurs naturally in a food. If a substance has been added, it falls under the "may render" standard. As one might expect, there have been a number of court cases involving this issue. The FDA has held, and been supported by the courts, that any substance that is not an inherent constituent of food may be regulated as an "added substance." An example of such a situation would be the presence of methyl mercury in swordfish as a result of its bioaccumulation. Such environmental or unavoidable contaminants may be regulated under Section 406 or, more informally, through action levels. In such a case, however, the government has the burden of producing the data necessary to establish safe levels of these substances. The provisions of Section 406 relate to substances, such as methyl mercury in swordfish, that are unavoidably and unintentionally added.

If an added substance is intended, the determination of the safety of such use is subject to Section 409, the food additive provisions. The definition of food additive is stated in Section 201 (s):

> The term "food additive" means any substance the intended use of which results or may reasonably be expected to result, directly or indirectly, in its becoming a component or otherwise affecting the characteristics of any food (including any substance intended for use in producing, manufacturing, packing, processing, preparing, treating, packaging, transporting, or holding food; and including any source of radiation intended for any such use), if such substance is not generally recognized, among experts qualified by scientific training and experience to evaluate its safety, as . . . safe.

In the case of a food additive, the proponent of the food additive has the burden of producing the data necessary to support safety. Until this support has been produced to the satisfaction of the government, and a regulation has been issued prescribing safe conditions of use, the substance is considered an unsafe food additive, and any food containing it is considered adulterated under Section 402 (a)(2)(C).

As noted in the definition, there is one exception to the definition of food additive that may pertain to genetically engineered substances. If the substance is "generally recognized as safe" (GRAS), the substance is not a food additive and is not subject to the premarket clearance provisions of Section 409. There are two bases upon which the determination of GRAS can be made: (1) through scientific procedures or (2) based upon a safe history or common use in food prior to January 1, 1958.

POTENTIAL PROBLEMS AND SOLUTIONS

Since genetic engineering may be used to produce various new food products or food ingredients, it is reasonable to ask: how will the statute apply to them and how will the FDA regulate them? So far, most of the interest in this area has involved the manipulation of bacteria to produce new sources of enzymes. In 1986, in fact, the FDA filed a GRAS affirmation petition for approval of an enzyme produced by recombinant DNA technology for use in food processing.

The FDA has concluded that genetically engineered foods fall under the provisions of the act pertaining to added substances. As indicated above, it has interpreted broadly the term "added" to include substances that may be present in food as a result of human activity as well as those that are directly added. Clearly, an enzyme from a genetically engineered microorganism may fall under the definition of a food additive. Not so clear are the situations where these techniques are used to produce a new food, for example, a new type of tomato. The FDA has generally held that the "may render injurious" standard is applicable to a food that has been genetically modified and, consequently, might contain high levels of a toxicant that has not been "added" in the traditional sense. The GRAS regulations state that the FDA will review the GRAS status of ingredients of natural biological origin that have been widely used for food and that have been significantly altered by breeding and selection. Investigators are encouraged to consult with the FDA early in their research so that it may evaluate the significance of any changes in the composition of the food.

Since it would be impossible to anticipate each new application, the FDA believes that the review of genetically engineered products should be done on a case-by-case basis. Much of this review would be based on

FDA's considerable experience in evaluating the safety of the more traditional manipulations of microorganisms and other food-related products.

The microbiological review plays a leading role in initially assessing this potential for toxic products to be present as a result of genetic manipulations. The FDA believes that it is important to adequately characterize, through techniques such as restriction mapping, the genetic alterations introduced in order to establish the absence of harmful sequences. This information, in conjunction with information on the donor, vectors, and recipient organisms, is needed to establish that the recombinant microorganism is not pathogenic or toxigenic.

Concurrent with the microbiological review, other technical divisions also review relevant information. The disciplines covered include chemistry, toxicology, nutrition, environment, and others as necessary. The chemistry review considers the identity of the ingredient, impurities that might be present, necessary specifications, and dietary exposure estimates for the ingredient and any impurities. Toxicology focuses on the potential for substances present in food to cause adverse reactions.

Generally, the need for toxicological data is based on available information pertaining to the identity of the substance, potential impurities, and the potential dietary exposure. For recombinant microorganisms, however, the microbiological assessment of the nature of the organism and the gene products is especially important in establishing the level of concern and the extent of toxicological testing necessary to address this concern. More stringent toxicological studies may be required in cases (1) where new microorganisms are used as sources of enzymes or other food ingredients, (2) of microorganisms that are related to pathogenic organisms and for which current information does not clearly establish nonpathogenicity, (3) of high dietary exposure, or (4) where adequate information on the genetic manipulations employed is not available for safety review. Conversely, minimal testing may be required for ingredients produced by microorganisms that are clearly nonpathogenic and nontoxigenic and for which information on the cloning procedures establishes the absence of harmful sequences.

The question often asked is whether a GRAS food ingredient can lose its GRAS status in cases in which it is manufactured through genetic engineering; this is conceivable. The ability to transfer genetic information, including regulatory sequences that control gene expression among diverse organisms, raises the potential for significant alterations to the final food product. Because of this potential, the FDA believes that the regulatory status of new source microorganisms, new enzymes, and new uses of traditional ingredients must be evaluated on an individual basis.

The FDA believes that new laws or regulations are not necessary to assess the safety of food products produced by genetic engineering either through recombinant DNA technology or other applications of biotechnology. It will continue to review products developed through this technology on an individual basis.

BIBLIOGRAPHY

Office of Science and Technology Policy. December 31, 1984. "Proposal for a Coordinated Framework for Regulation of Biotechnology: Notice. 49 FR 50856.

Office of Science and Technology Policy. June 26, 1986. Coordinated Framework for Regulation of Biotechnology: Announcement of Policy and Notice for Public Comment. 51 FR 23302.

34. Food-Packaging Interaction—Technical Aspects of Polymeric Materials

Robert W. Keown

University of Delaware

NATURE OF THE PROBLEM

In the past several years, a number of new polymers and polymer systems have been introduced and suggested for potential use in food packaging. This move to polymeric packaging materials from the traditional metal and glass containers continues unabated. Few of the new food products or processes currently being introduced incorporate either cans or glass containers. The majority of these new packages are based on laminates of polymer films. Packaging structures with as many as seven layers of polymer, paper, and thin metal films are in use today. The general concerns stated in the chapters on flexible packaging[1] and the interaction of packaging materials[2] in the Proceedings of the Second National Conference for Food Protection are still operative.

From a chemistry standpoint, this move to polymeric materials has brought the chemistry of the packaging much closer to the chemistry of the food contents. The many polymers that are finding their way to food applications have a great variety of chemical functionality. No one polymer or combination has been found that is suitable for all applications. With this congruency of chemistry comes a major problem of a more facile interaction of the food contents with the package and the potential compromise of package integrity. This then brings the concerns of flexible packaging and interaction of packaging materials with regard to total package performance into sharp focus. These concerns are:

- Seal damage
- Pinholing
- Loss of barrier property
- Reduction of shelf life
- Microbial recontamination
- Contamination from polymer adjuncts

Since the outside of the package is also polymeric, external chemical stresses are also possible. Some of the more typical chemical stresses related to the package performance are:

Atmospheric	Environmental	Graphic	Processing
Oxygen	Pesticides	Ink	Heat
Carbon dioxide	Microorganisms	Adjuncts	Chemicals
Pollutants		Solvents	Irradiation

Other prepackage chemical stresses are also possible. For example, potential degradation is possible by the processing and converting operations performed during the preparation of the packaging structure or web. These include thermal oxidation, treatment for adhesion promotion, and the presence of adjuncts used during the polymerization to manufacture the base polymer.

The chemical stresses on the inner layer are the flavorants (terpenes, esters, aldehydes), nutrients, preservatives, and special modifying gas flushes.

From the above lists, it becomes obvious that the chemical stresses are numerous, and the variety of potential reactions is great. Most of the reactions that could be predicted may be very subtle. However, the results with regard to the polymeric structures and their physical and chemical behavior can be quite large, and the resultant effect on packaging performance may be catastrophic.

HISTORICAL PERSPECTIVE

The dramatic changes in the gas (i.e., oxygen) barrier properties of ethylene-vinyl alcohol copolymers due to their known sensitivity to water is a good example of what can occur when small amounts of plasticizing chemicals interact with polymers. The concentrations of some critical ingredients in foodstuffs may be in the parts per billion (ppb) to parts per million (ppm) range (i.e., some flavors and vitamins). This range has not been of much interest to polymer chemists; consequently, there is a very small amount of information on the behavior of polymers and their interaction with small molecules in this concentration region. The information on the behavior of laminar composites of interest to the food processor and packager is almost nonexistent. Most of the experimentation reported on organic or "aroma" compounds has thus far been with single materials and in concentration ranges far removed from those experienced in food. Work done by Gaicin and reported by Hagenbaugh[3] and Zobel[4] is beginning to show some of the dramatic effects that small amounts of organics can have on the barrier properties of polymeric materials. What is considered satisfactory under one set of conditions may be inadequate with different foods.

This chemistry has many potential ramifications. The change in physical properties as induced by small concentrations of food ingredients or external chemicals has implications beyond normal shelf life testing, namely, the safety of the food contained in the package. For example, variable outside influences could affect the seal character and compromise microbial integrity. These influences may have different kinetics and may never be noted during normal shelf life testing.

POTENTIAL SOLUTIONS

The crucial issue of this change in packaging technology is the dramatic effect that small amounts of organic materials can have on the mechanical and chemical properties of polymeric materials. It is a universal problem that may affect food protection. The kinetics of the various reactions of food constituents with polymeric packaging materials, which will be concentration dependent, needs to be addressed. Some polymer suppliers have recognized that specifications for food packaging polymers will be difficult to develop without information on the concentration dependence of the chemicals of interest.

The solution to this lack of information requires that methods be established to discern these effects and that the data be readily accessible to packaging technologists. In particular, the following steps should be taken:

- Establish methodology to determine the chemical and physical effects of small concentrations of food ingredients on polymeric packaging materials
- Establish techniques that may allow for the prediction of potential reactions
- Begin to collect information for a database on the interaction of organic food ingredients with food packaging materials

This database should include both barrier and mechanical effects. Determination of the systems and methodology should be a joint effort of government, industry, and universities. Close cooperation will be required to assure uniform methods and techniques.

REFERENCES

1. Montgomery, E., Oyler, H., Winslow, R.L., and Brody, A.L. 1984. Food Processing and Preservation. Proceedings of the Second National Conference for Food Protection, p. 197.

2. Emery, D.F., Farkas, D.F., Schwartz, P.S., and Siedler, A.J. 1984. Interaction of Packaging Materials with Foods and Food Components, Proceedings of the Second National Conference for Food Protection, p. 353.
3. Hagenbaugh, J.R. 1986. "Measurement of Organic Vapor Permeation Rate in Oriented Polypropylene Films," Presented at the April 1986 Food Packaging Seminar, University of Delaware, Newark, DE.
4. Zobel, M.G.R. 1985. *Polymer Testing* 5:153–165.

35. Food-Packaging Interaction—Regulatory Guidance: Migration Testing of Plastic Packaging

Gregory M. Cramer

U.S. Food and Drug Administration

NATURE OF THE PROBLEM

Innovations in plastics packaging dominate the food packaging news. They also raise many questions about food-package interactions. The Food, Drug, and Cosmetic Act places the responsibility of premarket safety assessment of food packaging materials on the U.S. Food and Drug Administration (FDA). Although companies trying to convert their food packaging from glass or metal to plastics are concerned about product safety, they are also concerned about flavor scalping by plastics that may affect a food's flavor profile as well as its expected shelf life. Although changes in food composition can result from the diffusion of food components into the packaging, the primary purpose of FDA's premarket safety assessment of food packaging materials is to evaluate probable changes in food composition that may result from migration in the opposite direction—from the packaging into the food.

Substances shown to become components of food, or which may reasonably be expected to become components of food as a result of their use in food packaging, are considered as indirect food additives and legally require premarket safety approval by the FDA. As part of the petition process to amend food additive regulations to permit the use of new materials in food packaging, the FDA requires industry to perform extraction tests to determine the nature and amount of packaging materials that are likely to migrate to food and to supply appropriate toxicological data on the migrants. In order for the FDA to reach appropriate conclusions about the safety of food packaging materials, the extraction test procedures must be able to reliably predict migration levels in likely food packaging applications. With all of the innovations taking place in plastics packaging, the FDA is critically evaluating recent work in migration testing to decide whether existing test protocols can be improved to provide more appropriate migration estimates for foods packaged in these new structures.

HISTORICAL PERSPECTIVE

Due to the complexity of food matrices as well as the minimal levels of migration of package components that normally result from food-package interactions, the FDA's premarket safety assessment of packaging materials seldom relies on analyses of foods. Over the years, the FDA has depended instead on extraction tests with solvents to simulate the leaching action of aqueous, acidic, alcoholic, and fatty foods.[1] While pure solvents allow an assessment of the total level of migrants as well as specific migrants, considerable work has been performed to find solvents that can adequately simulate the leaching action of food. Although the use of radiolabeled migrants may allow estimates of specific migration into specific foods and may eliminate some of the uncertainty associated with solvents and their ability to simulate food, such an approach may not always be feasible. In the case of laminates and coextrusions, manufacturing procedures make it impractical to incorporate radiolabeled components into such composite structures.

While the emphasis is on food-simulating solvents and extraction tests to simulate the leaching action of food, most FDA test protocols do not in fact involve true "food simulants." Solvents used in accelerated (short-time, elevated-temperature) tests designed to indicate the amount of leaching by food at lower temperatures over extended periods should really be referred to as "food correlants," since such solvent results are expected to correlate with long-time food leaching. If solvents that contact packaging materials under the same conditions as food are found to leach packaging components at the same rate as food, then the solvents can be thought of as simulating the leaching action of food. When high-temperature food contact is anticipated, the FDA recommends that test solvents contact the food packaging material for periods of time and at temperatures corresponding to the most severe contact envisioned with food. For such test procedures, the test solvents are expected to function as food simulants rather than as food correlants. For the sake of convenience, these test solvents will be referred to as food simulants; the distinction between food simulants and food correlants will be covered later.

The use of laminated and coextruded packaging to hold food during hot fill or thermal processing operations raises a number of questions about the ability of a food contact layer to serve as a functional barrier to nonfood contact layers or materials that are not permitted to be used intentionally in direct contact with food. If the food contact layer does not function as a barrier during high-temperature contact with food or food simulants and if migration of unapproved components of nonfood contact layers is detected, then it is necessary to amend 21 CFR Section 177.1390 (High-Temperature Laminates) to provide clearance of the unapproved components. For example, a thin layer that provides heat seal properties to a laminate and functions as a barrier during contact with food under ambient conditions may not necessarily function as a barrier when stressed during high-temperature contact with food. While the FDA bases its assessment of safety either on the level of migration or on the limit of measurement, migration experiments with laminated and coextruded structures need to be performed with food simulants in order to attain the necessary level of sensitivity.

As an example, laminates utilizing 2,4-toluenediisocyanate-based urethane adhesive were found to yield measurable levels of the animal carcinogen 2,4-toluenediamine in aqueous extracts when the laminates were subjected to simulated boil-in-bag and retort pouch applications. Until recently, the FDA has held the position that packaging materials containing carcinogenic impurities that migrate to food would be subject to the Delaney Clause. The FDA now takes the position that as long as the packaging material or any of its functional additives is not carcinogenic, unavoidable carcinogenic impurities may become components of food provided that, when probable exposure and carcinogenic potency are considered, the risk does not exceed acceptable limits.[2] In order for the potential exposure to 2,4-toluenediamine to approach dietary levels that might pass risk assessment, migration levels to food simulants or the limit of measurement for 2,4-toluenediamine would need to be no greater than around 1 ppb. Even with pure food simulants, it is obvious that the most powerful analytical techniques are required to make such measurements.

POTENTIAL SOLUTIONS

As many know, Arthur D. Little, Inc. completed, in July, 1983, a multiyear contract with the FDA to assess the reliability of existing migration test protocols.[3] Using radiolabeled migrants and eight polymer migrant systems, they compared the level of migration into test foods stored under typical storage conditions with the level of migration found in food simulants following accelerated migration testing using existing FDA extraction protocols. Their evaluation, however, did not address the reliability of test protocols to simulate food-package interactions at elevated temperatures.

These studies showed that migration of the additives generally involved polymer-controlled diffusion and that extractions to equilibrium, according to earlier FDA guidelines, would occur only if migration was limited by the solubility of the migrant in food or food simulants or if the migrant was completely depleted from the test article. A. D. Little found that test solvents were reasonably effective in exaggerating migration to food. There were instances, however, where they felt that better estimates of migration could be derived by using improved food simulants. Migration to water, for example, was found to underestimate migration to aqueous food in instances when the low solubility of the migrant in water resulted in a partition limit to the polymer-controlled diffusion. As a consequence, the FDA is implementing A. D. Little's recommended replacement of water by 8% ethanol as a simulant for aqueous food. The FDA believes that 8% ethanol will better approximate the solubilizing capacity of aqueous foods and will be less likely to result in partition-limited migration.

Another major change in FDA extraction protocols brought about by the work of both A. D. Little and the National Bureau of Standards[4] is the replacement of heptane as the recommended fatty food simulant. Due to the aggressive nature of this solvent with polyolefins as well as polystyrene, estimates of additive migration to fatty food were found to be of little value. Since such estimates greatly exaggerated migration to food, the FDA recognized that, in order to obtain more realistic estimates of contamination, the use of heptane as a fatty food simulant would need to be abandoned in favor of solvents that more closely approximate the leaching action of fatty food. The FDA now recommends that a pure liquid fat such as corn oil or the synthetic triglyceride HB307 be used to simulate worst case leaching by fatty food. Since a liquid fat presents some of the same analytical difficulties as food, simple solvents simulating the worst case are also necessary. It seems reasonable to conclude that acceptable alternatives to corn oil as a worst case fatty food will be found with solvents that exhibit comparable solvent-polymer interactions and similar migrant solubilities. While a number of solvents have been investigated as an all-purpose fatty food simulant, it seems unlikely that any one solvent will accurately mimic the leaching by fatty food of the many food contact polymers.

Work done by the National Bureau of Standards and studies conducted by the FDA laboratories have shown that aqueous ethanol mixtures of varying proportion may serve as a good alternative to corn oil for a number of polymer-migrant systems. The National Bureau of Standards found that, with respect to the extraction of olefin polymers, 95% ethanol and corn oil extracted comparable amounts of oligomers with a similar molecular weight distribution. Additionally, they found that the diffusion coefficients for BHT, octadecane, and dotriacontane migrating from high density polyethylene and low density polyethylene into corn oil were nearly identical in absolute ethanol, despite the fact that measurements were made over a 30C range and spanned over four orders of magnitude. Migration studies were conducted by a company using 100% ethanol and the synthetic triglyceride HB307 as fatty food simulants in support of a food additive petition for use of a hindered phenol stabilizer in low-density polyethylene, high density polyethylene, and polypropylene. The studies showed that for each resin, both fat simulants yielded essentially identical migration levels after 10-day contact. Since 95% ethanol is more practical to work with than absolute ethanol, the FDA now recommends the use of 95% ethanol instead of heptane when corn oil cannot be used as a fatty food simulant for olefin food contact polymers.

When investigating the leaching of styrene from polystyrene, the FDA laboratories found that the logarithm of the diffusion coefficient for styrene increased linearly with increasing aqueous concentrations of ethanol. Although 20% ethanol yielded migration levels that slightly exceeded those found with corn oil, it is recognized that use of a fatty food simulant having such a low organic content could result in partition-limited migration for migrants of higher molecular weights. Therefore, it is now being recommended that 50% ethanol be used to evaluate fatty food leaching from polystyrene food contact articles. Although this solvent may slightly accelerate leaching rates relative to corn oil, the FDA believes that it will serve effectively as an alternative to corn oil.

While these changes as well as others being instituted in the test protocols will certainly lead to better correlations of the results of accelerated leaching experiments with likely migration levels in room temperature-stored food, the FDA does not as yet have the benefit of such carefully designed studies to evaluate the adequacy of test protocols for packaging in contact with food at elevated temperatures. Although many of the new uses for plastics involve such contact where diffusion rates are considerably higher, the FDA expects that migration in these applications can be evaluated by assuming that this migration will primarily follow polymer-controlled diffusion processes. In view of the fact that polymer phase transitions (such as glassy phase to rubbery phase) can have a significant effect on diffusion rates through plastics, test protocols for high-temperature applications will need to address migration from plastics in the same phase that is expected for food contact. In instances where fatty foods are able to penetrate the polymer and accelerate migration, considerable differences are expected between the levels of migration for aqueous and fatty foods.

As in the experiments for evaluating leaching with food stored at ambient conditions, the FDA is now also recommending that 8% ethanol and a liquid fat such as corn oil be used as simulants to extract plastics for periods of time and at temperatures likely to be encountered during the thermal treatment of aqueous and fatty food in a container. Earlier test protocols for evaluating possible leaching by fatty food during retorting at 250F were carried out with heptane at only 150F. The FDA felt that, because of the aggressive nature of heptane, the results of the accelerated leaching tests at lower temperatures would correlate with the leaching of fatty food during retorting. Since the degree of heptane penetration depends on the nature of the polymer being extracted, both the degree of acceleration of migration and the extent of correlation with fatty food leaching would also depend on the nature of the polymer tested.

The FDA now recommends that extraction tests with corn oil be carried out at 250F, because the expected action by corn oil is one of simulation of fatty food leaching rather than acceleration of leaching. While it might be argued that a pure liquid fat at 250F will exaggerate the leaching of food which is only partially fatty, the FDA does not have any high-temperature data comparing the leaching rate of a liquid fat with those of foods of varying fat content, which might show that a less conservative approach could be taken.

While corn oil may be useful for evaluating migration of some package components during high-temperature contact with fatty food, it is obvious that alternative liquid simulants are also necessary. In addition to the inclusion of aqueous ethanol solutions into the room temperature extraction protocols, the FDA expects that ethanol solutions will also be used in protocols for evaluating high-temperature fatty food leaching. The extent of solvent penetration of plastics by corn oil and aqueous ethanol solutions during experiments designed to simulate high-temperature food contact is expected to vary for each resin and to depend on the extraction temperature. Since solvent penetration will accelerate diffusion rates in the polymer, test protocols will need to accommodate any differences that exist between the abilities of corn oil and ethanol solutions to penetrate plastics. Due to the varied pattern of solvent-plastic interaction, it is expected that with some resins ethanol solutions may be used as food simulants, while with other resins they may be used as food correlants. The FDA plans to focus future work with ethanol solutions on developing high-temperature extraction protocols for olefin and polyester resins, since they are used most extensively in high-temperature food applications.

In addition to the changes in the test protocols described here, there are a number of other modifications that have been made. A complete set of extraction protocols, including recent changes, will be found in the FDA revised Chemistry Guidelines. These guidelines were expected to be available from the FDA before the end of 1986. As the FDA has endeavored to improve the extraction protocols to arrive at more realistic estimates of food leaching, it has tried to remain true to some of the guidance offered in A. D. Little's Final Report: "Only by understanding the basic mechanism of migration (and food component penetration) is there hope to develop fair, yet conservative test protocols." Although the FDA recognizes that additional changes and improvements will be made in methods to evaluate food leaching while research is carried out in areas such as high temperature and fatty food leaching, it believes that the revisions in the Guidelines represent enough of a departure from previous recommendations that they will be immediately useful to an industry undergoing such rapid change.

REFERENCES

1. U.S. Food and Drug Administration. 1976. FDA Guidelines for Chemistry and Technology: Requirements of Indirect Food Additive Petitions. Bureau of Foods, U.S. FDA, Dept. of Health, Education, and Welfare, Washington, DC 20204.
2. Vinyl Chloride Polymers; Withdrawal of Proposal. 1986. *Fed. Reg.* 51(22):4173–4188, Feb. 3, 1986.
3. Little, Arthur D. July 1983. Final Summary Report. A Study of Indirect Food Additive Migration.
4. U.S. Dept. of Commerce. March 1982. Migration of Low Molecular Weight Additives in Polyolefins and Copolymers. U.S. Dept. of Commerce, National Bureau of Standards, Centers for Materials Science, Polymer Science and Standards Division, Washington, DC 20234, Final Project Report.

AFTERWORD

The rapidly expanding use of plastics as substitutes for glass and metal food packaging, discussed in Chapters 34 and 35, is resulting in exposure of plastic materials to a greatly expanded range of temperatures, pressures, and chemical interactions. Since the magnitude of potential extraction of packaging components by food is greatly increased with plastic packaging relative to glass and metal containers, reliable procedures for evaluating probable migration levels to food are clearly needed by the FDA and the packaging industry in order to reach appropriate conclusions about the safety of new plastic packaging. As a result of FDA-sponsored studies that compared extraction levels in food and food simulants for a number of plastics and additives, the FDA has revised some of its recommended room-temperature extraction protocols in order to approximate expected food-plastic interactions more reliably. Existing test protocols for plastics contacting food at elevated temperatures have not been subjected to such careful scrutiny. It is clear that broad-based, fundamental research is needed in order to address the myriad of possible food-plastic and solvent-plastic interactions that are likely at high temperatures. Such research may lead to the development of predictive databases and methodologies that should benefit the FDA and the packaging industry as well.

36. Task Force on Irradiation Processing—Wholesomeness Studies

Peter S. Elias

Federal Research Center for Nutrition

INTRODUCTION

Food irradiation is a feasible technology for the preservation of vital food supplies and for extending the shelf life of food. Its ultimate goal is the prevention of adverse changes by undesirable microbial or biochemical action, thus preserving quality.

As early as 1896, Mink discovered the bactericidal action of X-rays, but research on the biological effects of ionizing radiation was hampered largely by the low radiation energies then available from existing installations. In 1943, the first report appeared on the successful preservation of hamburgers by irradiation.[1] Since then, there has been extensive development of this technology, using not only X-rays but also gamma rays from radioactive isotopes and accelerated electrons from machine sources, that addressed the following objectives:

- Reduction or elimination of food spoilage and pathogenic microorganisms
- Elimination or inactivation of parasites and the control of insect infestation at all stages of their lifecycle
- Prevention of sprouting of root vegetables, onions, garlic, and other members of this plant family; and delaying of ripening, particularly of tropical fruits in transit
- Preservation of the freshness of food beyond its normal shelf life at ambient temperature
- Reduction in the use of, or the avoidance of, chemical preservatives, pesticides, or fumigants
- Reduction in energy expenditure on food preservation

NATURE OF THE PROBLEM

Consideration of the objectives of this new technology reveals four areas of potential impact of the process on the wholesomeness of the treated food:

- Possible chemical and physical changes in the food resulting from irradiation
- Microbiological and hygienic consequences
- Nutritional consequences
- Health implications of the formation of radiolysis products in irradiated food

Toxicological and Health Aspects of Wholesomeness

Since the introduction of food irradiation, reassurance has been sought by national and international authorities concerned with the safety of the food supply for the consumer that any changes in food composition caused by irradiation did not render the food unwholesome. Although the chemical changes in food components resulting from irradiation are quantitatively very small, they could conceivably result in the possible appearance of harmful substances in foods so treated. Such events have been known to occur occasionally in more conventional food processing technology.[2–3]

Other concerns were related to the possible loss of nutrients and the introduction of microbiological and toxicological hazards. (Recently, similar concerns have been raised about food treated by heat or microwaves, since the presence of mutagenically active compounds has been demonstrated in traditionally heat-processed foods.)[4,5] Much research has been carried out worldwide to evaluate any conceivable risk, and Barna[6] compiled a list of over 1200 such studies; many more have since been reported.[7]

In the past, the approach toward assessing the wholesomeness of irradiated foods has been based on the concept, originally developed in the U.S. legislation, that irradiation "adds" something to the treated food.[8] Food irradiation, therefore, was considered as a food additive and not as a process, although there is no question of any transfer of specific substances to the food. This view was adopted by the U.S. Food and Drug Administration (FDA) and also by some other national authorities. However, the Ministry of Agriculture, Fisheries and Food (MAFF) in the United Kingdom, among others, regulated food irradiation by controlling it as a process, generally prohibiting this form of processing except in specifically permitted instances.[9]

The "food additive" approach toward evaluating food irradiation implied that the evaluation of the toxicological aspects of wholesomeness had to be based on the concepts used for chemical food additives or pesticide residues and developed by the FAO/WHO Joint Expert Committee on Food Additives in 1956, i.e., the establishment of an acceptable daily intake (ADI) and of safety factors based on expensive, lengthy, and often repetitive feeding studies with laboratory animals.[10–11] Similar methodological complications did not arise with the nutritional and microbiological aspects of wholesomeness, which required equally thorough consideration in such a new process.

The general fear of anything related to radiation led to progressively more complex test protocols, more numerous and detailed examinations of tissues and organs of test animals, requests for tests in more animal species, and for tests for mutagenicity, embryotoxicity, teratogenicity, and others. This situation eventually demanded a complete reconsideration of the approach to the establishment of the toxicological acceptability of irradiated foods. In addition, many methodological difficulties were encountered in designing traditional laboratory animal feeding studies with irradiated foods so that other avenues for providing safety data had to be explored in order to replace the feeding studies hitherto employed.

Nutritional Aspects of Wholesomeness

Irradiation shares with other food processing techniques the ability to alter the nutrient composition of food by producing chemical changes. The nature and extent of these changes depends on many factors such as the composition of the food, the radiation dose, the processing temperature, the presence or absence of air, storage conditions, and others. A large amount of data shows that most components of food remain unchanged upon irradiation. Vitamins show differential sensitivity to irradiation. Riboflavin, niacin, and vitamin D are quite stable, while thiamin (B_1), vitamins A, C, E, K, and the polyunsaturated fatty acids are partially destroyed. Irradiation at high doses may cause considerable losses of radiation-sensitive vitamins, especially under aerobic conditions. The same occurs on heating, and such losses should cause no greater concern just because they are the result of irradiation. There is no evidence that technologically feasible irradiation with doses up to 1 kGy (1 kGy = 100 krad = 1000 Joule absorbed per kg mass) will cause major nutrient losses in any food, but higher doses may cause significant losses of some essential nutrients in some foods.

As most irradiated foods are also heated before consumption, the combined effects of radiation and heating on nutrients have been studied.[12] Indeed, the 1980 Joint Expert Committee on Food Irradiation (JECFI) concurred with the view that, up to an average dose of 10 kGy, irradiation introduced no special nutritional problems, but emphasized that the overall importance of nutrient losses in an irradiated food also depends on the importance of the specific food to the total diet.[13]

Microbiological Aspects of Wholesomeness

The major benefits of sterilization by irradiation and of reducing the microbial load arise from the possibility of controlling many rather common health hazards due to foodborne diseases such as salmonellosis; intestinal infections with Campylobacter or Yersinia; trichinosis, taeniasis, and toxoplasmosis from meat and poultry; or infections with shigella, vibrio parahaemolyticus and enteropathic *E. coli* in deep-frozen seafood.

High-dose irradiation for the purpose of achieving sterility of food raises no public health problems, but this technology is not yet generally applicable. However, at substerilizing doses, there is the possibility of public health problems arising from those microorganisms surviving the treatment. The problems are virtually the same as arise with those other methods of preservation that also do not lead to complete sterility of the preserved food. The usual hygienic precautions are, therefore, essential in handling irradiated foods to prevent growth of the pathogens that have survived irradiation. They include subsequent refrigerated storage or preservation at low pH or at low water activity. Furthermore, as irradiation does not affect enzyme activities to any great extent, a combination with chilled storage is necessary for fresh products.

The destructive effect of radiation on microorganisms has its basis in alteration of their genetic apparatus. There is a correlation between the number of surviving microorganisms and the dose applied. The particular assumption that growth of and toxin production by *Cl. botulinum* may be favored in irradiation-preserved food as a result of the destruction of the competing bacterial flora has remained unconfirmed. The problems due to suppression of spoilage organisms by low doses of irradiation are not likely to be greater than those encountered with pasteurization, curing, and vacuum packing.

Because of the differences in radiation resistance of microorganisms, even doses of 10 kGy may not be sufficient to eliminate microbiological concerns as well as to preserve the keeping quality of foods. In some cases, combination treatments may be required. Irradiation, however, creates a barrier to the transmission of pathogens through the food chain and renders the survivors more sensitive to heat, drying, and other technological treatments. No evidence could be found that radiation-induced genetic variations might lead to enhanced pathogenicity or enhanced radiation resistance.[14] Indeed, the 1980 JECFI concluded that the microbiological safety of irradiated food was comparable to that achieved by other currently accepted food treatments.[13]

HISTORICAL PERSPECTIVE

The initial research into the scientific and technological aspects of food irradiation was carried out in the United States in the late 1940s. Concomitantly, concern arose over the wholesomeness of irradiated food so that many individual irradiated foods were investigated as if they were food additives. Not surprisingly, this led to a large number of expensive animal studies being carried out in a number of countries.[15]

Early expectations that irradiated foods may become of commercial interest led the Food and Agriculture Organization of the United Nations (FAO), the International Atomic Energy Agency (IAEA), and the World Health Organization (WHO) to sponsor a committee of experts. Its task was to make recommendations aimed at influencing legislation on the control of the production and use of irradiated food in various countries. Such a common approach to legislation might facilitate international acceptance of the process.

The Joint FAO/IAEA/WHO Expert Committee on the Technical Basis for Legislation on Irradiated Food convened in 1964. Its terms of reference were to consider the available evidence of the effects of treatment with ionizing radiation on food and to make recommendations on the general principles governing the permission to produce and use irradiated food and on the procedures for evaluating the wholesomeness of irradiated food. Its report[16] set out technical procedures and tests required to permit an evaluation of the safety for consumption of irradiated food. The same report also reviewed the microbiological aspects of irradiated food.

In 1969, a Joint FAO/IAEA/WHO Expert Committee on Wholesomeness of Irradiated Food was convened for the first time. In its report[17] the methodologies and procedures for evaluating irradiated food were reviewed. It extended the safety considerations to nutritional studies and investigations of the mutagenicity of irradiated food. It assessed the biological evidence for irradiated wheat, potatoes, and onions and evaluated their safety for consumption.

To rationalize and coordinate more productively the efforts to provide wholesomeness data on irradiated foods, an International Project in the Field of Food Irradiation (IFIP) was set up in 1971, sponsored by IAEA, FAO, and OECD. The Federal Republic of Germany provided host center facilities at Karlsruhe. Membership comprised 24 countries when the Project terminated at the end of 1981. The objectives of IFIP included research into methodology and the coordination of wholesomeness testing in laboratory animals being carried out at various national contract laboratories. IFIP also organized the investigation of certain toxicological aspects related to irradiated foods, requested by the 1969[17] and 1976[18] sessions of JECFI, to help with the clearance of the process from the point of view of safety to health. After 1976, the project developed a sensitive methodology[19-21] based on short-term mutagenicity tests on digests of irradiated foods to supplement an extensive coordinated program of research into the radiation chemistry of food and food components.

As a result of these efforts, the 1980 JECFI[13] accepted the safety of the process of food irradiation for preservation up to an overall average dose of 10 kGy. Subsequently, the Codex Alimentarius Commission developed a General Standard for Irradiated Foods and a Recommended International Code of Practice for Operating Irradiation Facilities Used for Treating Food[22] to facilitate international trade in irradiated food.

Key Approaches towards the Wholesomeness Evaluation

It was previously pointed out that the conservative attitude of regarding the irradiation of food as an additive and not as a technological process implied the need to investigate each individual irradiated food by feeding studies with laboratory animals. There are, however, serious objections to the applicability of food additive testing methodology to the safety testing of complex foods. These objections arose mainly from the known logistical limitations inherent in crude biological toxicity tests when used for detecting radiolysis products formed only in minute amounts in irradiated food. Other objections related to the nutritional and toxicological effects of the distortion of the diet of test animals when large percentages of irradiated foods were incorporated. Apart from interfering seriously with the interpretation of the results, it made it impossible to study any dose-response relationship.[23]

Once the notion was accepted that the irradiation of food was a process, it became possible to examine its safety from this perspective. The advantages of this approach can be easily recognized. Evidence could be obtained to determine whether the process produced any toxic substances in physiologically active amounts. Furthermore, it was no longer necessary to consider the safety of each individual irradiated food, and extrapolation of results from one treated food to other treated foods should become feasible.

Elucidation of the detailed mechanisms of radiation chemical reactions in major food components and analyses of the radiolysis products in model systems led to the recognition that the actual chemical changes in irradiated foods would be quantitatively low and that most of the compounds produced are also present in foods treated by other accepted processes. (Quantitatively, the changes caused by cooking are much more pronounced than those caused by irradiation, even at high doses.[24])

Radiation Chemical Considerations

The absorption by matter of energy quanta from the photons of the incident radiation causes changes in its component molecular structure, manifested as ionizations and excitations. This primary process leads to the formation of ions and free radicals having short lifetimes. Free radicals undergo secondary reactions leading eventually to the formation of stable radiolysis compounds.

The extent of the chemical effects of irradiation is expressed in G-values, i.e., the number of molecules changed for every 100 electron volts (eV) of absorbed energy. For appropriately irradiated foods, a G-value of 1 is generally applicable. Assuming an average molecular weight of 100, a dose of 10 kGy would produce about 300 mg of radiolysis products per kg of food, most of which have been identified as occurring also in unirradiated, but otherwise processed, food. It has been assumed that a small fraction might be "unique" radiolysis products, not usually found in food;[25] but none has been so identified.

The formation and decay of free radicals in food components are the basic mechanisms leading to chemical changes detectable in food. Irrespective of the energy source generating these radicals, their reactions will be determined by the nature of the medium, the temperature, the concentrations of radicals and other reactive molecules, and the rate constant for each competing reaction. The concept of "chemiclearance" was therefore proposed based on the demonstration in model systems that similar radiolytic reactions occur with the same constituents of different foods, i.e., the proteins, fats, carbohydrates, water, vitamins, and metalloproteins. These reactions form common radiolysis products in roughly predictable yields in foods of similar chemical composition irradiated under similar conditions. Although the product yields cannot be predicted with precision, the upper limits of these yields can be estimated.[26-27] This commonality of behavior provides an important parameter for evaluating the wholesomeness of irradiated food.

The nature of most of these radiolysis products has been determined. For example, in radiation-sterilized beef, over 100 compounds have been identified by gas chromatography/mass spectrometry at concentrations of 1 to 700 mg/kg. Ninety percent of the total are aliphatic hydrocarbons; the remainder are aldehydes, ketones, alcohols, and traces of sulphur- and nitrogen-containing compounds. Other studies have shown that radiolysis compounds are of a similar nature in meats from different mammalian species. The yields of aliphatic hydrocarbons in irradiated beef, pork, chicken, and ham were shown to depend only on radiation dose and fat content. Radiation chemical studies, therefore, provide a basis for extrapolations from one food to another and from one set of irradiation conditions to another.

Toxicological Considerations

Feeding studies, particularly long-term studies, have been the mainstay of the toxicological assessment of irradiated foods until very recently, despite the problems and deficiencies of this methodology as mentioned earlier. Many substances naturally occurring in foods, i.e., phytates, oxalates, and solanine can be toxic in large amounts and, therefore, limit the amount of many foodstuffs that can be incorporated into animal diets. The nutritional imbalance or specific deficiencies resulting from the incorporation of large quantities of a complex foodstuff may affect the health of test animals and alter the incidence of background disease or tumor appearance without being a toxic effect of irradiation. Control groups receiving untreated food may suffer from the effects of spoilage absent in the irradiated sample. Hence, comparisons can be made in most cases only with foods preserved by some other technology and not with untreated material. Absence of precise knowledge of the nature of the potential toxic substances in irradiated foods prevents the prediction of the expected effects or of the target organs and obviates studies of the distribution and metabolism of the putative toxic constituents.

In studies on food, it is not possible to demonstrate a dose-response relationship and thus to administer a dose level high enough to produce a significant toxic effect attributable to irradiation. Hence, it is not possible to use the traditional hundredfold safety factor by limiting the intake of irradiated foods to 1% of the maximum dose at which no unfavorable effect is observed. Classical risk estimations, therefore, are impossible with irradiated foods. This difficulty cannot be circumvented by either increasing the irradiation dose or by feeding nutritionally improper doses. The problem is not unique to irradiated foods but applies generally to the testing of foods or food components.

Nevertheless, the JECFI was presented at its 1980 meeting[13] with the results of feeding studies on some 60 different irradiated foods and with comprehensive toxicological tests including long-term and multigeneration studies on some 20 food items.

Genetic Toxicity Considerations

One of the key approaches currently taken to overcome the inadequacy for feeding tests with irradiated food was developed in the laboratory of the IFIP. Genetic effects of certain chemicals, e.g., the induction of heritable mutations in germ cells and somatic mutations in body cells are receiving increasingly more attention. Somatic mutations are probably responsible for many developmental and other disorders as well as for cancer. Germline mutations are irreversible and normally have serious consequences. They may remain masked for many generations and become spread undetected throughout the population. Their induction, even at low frequencies, can be a serious problem in a large population.

Life span studies in animals are rather insensitive to genetic effects because of the small number of individuals involved and the rather high background frequency of mutations observed. For this reason, and also because of the expense in resources, time and money involved in life span studies, there has been considerable interest in the application of cheaper, quicker, and more sensitive short-term mutagenicity studies for the testing of irradiated food. Mutagenic principles can be formed in food and isolated food components upon irradiation under certain circumstances. The presence of mutagenic substances, however, was demonstrated only after irradiation at high doses, and then only when the tests were performed shortly after irradiation or on isolated components of food. By 1976, some 35 mammalian in vivo mutagenicity studies, e.g., dominant lethal assays, host-mediated assays, micronucleus tests, cytogenetic analyses of bone marrow or meiotic germ cells, had been done on various irradiated foods and animal feeds.[28]

The development in the IFIP laboratory of a novel modification for investigating irradiated foods by in vitro mutagenicity screening,[19-21] involving an in vitro enzymatic digestion step to allow the direct incorporation of food into genetic toxicity test systems made it possible to have results from another 70 mutagenicity tests for assessing the safety of irradiated food.

A critical evaluation of the few alarming published reports of mutagenicity tests on irradiated wheat showed these studies to be inadequately designed and not actually capable of demonstrating any mutagenic potential of irradiated wheat.[29]

No mutagenic effects in other foods have been demonstrated if the foods have been irradiated at technologically relevant doses and then stored or heat treated. Presumably, either active substances are not formed in biologically significant quantities at the irradiation doses used, or they are rapidly inactivated by reactions with other food components.

POTENTIAL SOLUTIONS

The impressive body of evidence on the wholesomeness of irradiated food, which includes data not only from feeding studies on some 80 irradiated food items and from over 105 mutagenicity studies,[28] but also from animal colonies maintained on irradiation-sterilized diets[30] and from commercial livestock reared on irradiated feedingstuffs, persuaded the JECFI in 1980 to recommend as safe for human consumption food irradiated up to an overall average dose of 10 kGy. The Committee concluded that irradiation of food presented no toxicological hazard and that further toxicological testing of foods so treated was no longer required.[13] Since many of the feeding studies were carried out with foodstuffs irradiated at much higher doses, e.g., up to 50 and 60 kGy, an ample margin of safety exists.

In a separate but relevant action, the FDA published in the Federal Register of April 1986, its Final Rule on Irradiation in the Production, Processing, and Handling of Food. Irradiation is now permitted for the control of *Trichinella spiralis* in pork carcasses or in fresh, nonheat-processed cuts of pork carcasses, for inhibition of growth and maturation of fresh foods, for disinfestation of arthropod pests in food, for microbial disinfection of dry and dehydrated enzyme preparations (including immobilized enzymes), and for microbial disinfection of dry or dehydrated aromatic vegetable substances, herbs, seeds, spices, teas, vegetable seasonings, and blends of aromatic vegetables.

Despite the extensive investigations of all aspects of food preservation and sterilization by ionizing radiation, the process still tends to be looked upon with some uncertainty. This attitude is heightened by both the public and national authorities becoming increasingly sensitive about potential health hazards involved in the use of chemicals for food preservation, not only from the chemicals themselves, but also from their interaction products and metabolic by-products. Consequently, governmental authorities charged with ensuring a safe and wholesome food supply for the population and protecting the consumer need to be reassured by objective evidence of the safety of the food irradiation process. Only on the basis of such objective and scientific evidence will it be possible to make an informed judgement about the acceptability of an irradiated food for human consumption.

The attitude of the consumer toward the irradiation of food is difficult to assess. The doubts and hesitations of various consumer associations are known, but there is no good knowledge of the actual attitude of the consumer to irradiated foods, especially in those countries where the marketing of irradiated foods is forbidden. Sooner or later, it may be possible to convince the consumer of the advantages of foods that are free of chemical preservatives or guaranteed to be free of *Salmonella* and other disease-producing enteric pathogens. Consumer acceptance tests conducted in different countries have revealed that the average homemaker does not object, in principle, to irradiated foods, but almost completely lacks any knowledge about this new technology. Public education and information, therefore, is an essential future goal. A requirement for the labeling of irradiated foods and, possibly, of irradiated ingredients in foods may be a price to pay in order to enable the consumer to decide whether he wishes to purchase irradiated products of better hygienic value, though possibly more expensive. Food irradiation would be a particularly important technique for tropical and subtropical Third World countries as a realistic means of reducing the enormous postharvest losses in agricultural products that constantly threaten the economy of these regions.

SUGGESTED RECOMMENDATIONS

1. Adopt the General Codex Standard for Irradiated Foods and the Recommended International Code of Practice for Operating Radiation Facilities for the Treatment of Food promulgated by the Codex Alimentarius Commission following the decision by JECFI in 1980.[6]

2. Communicate and follow the detailed model legislation for the control of trade in, and the marketing of, irradiated foods and for the supervision of irradiation facilities prepared on behalf of the IAEA by E.T. Bigwood of the Research Center for Food Law at the University of Brussels and published in 1979. If this legislation were to be adopted in national food laws worldwide, it would ease and simplify international trade in irradiated foods and provide assurances that high standards of safety are being maintained.

3. Expand the available database on radiation sterilization (involving doses above 10 kGy) of deep-frozen food to preserve its organoleptic properties, quality, and appearance. It is not yet sufficient to evaluate unequivocally all foods irradiated with sterilization doses, and such an expansion would help in carrying over the results from low-dose irradiation to higher dose applications.

4. Replace toxicologically unacceptable chemical sterilants by irradiation, which leaves no harmful residues in the treated foodstuffs.

5. Provide proper information and education to the public about the applications and limitations of food irradiation, about the form and meanings of labels on irradiated food, and about the basis for approving the process as safe and wholesome. Such education would minimize potential political exploitation of the understandable fears and disquiet of the public related to anything connected with nuclear energy and radiation.

REFERENCES

1. Goldblith, S.A. 1966. Historical Development of Food Irradiation. In: Food Irradiation. Proceedings of the International Symposium on Food Irradiation, STI/PUB 127, IAEA, Vienna, pp. 3–21.
2. Lang, K. 1973. Die physiologischen Wirkungen erhitzter Fette, insbesondere der Fritierfette. *Fette, Seifen, Anstrichm.* 75:73–76.
3. Magee, P.N. 1971. Toxicity of nitrosamines: Their possible human health hazards. *Food Cosmet. Toxicol.* 9:207.
4. Commoner, B., Vithayathil, A.J., Dolara, P., Nair, S., Madyastha, P. and Cuca, G.C. 1978. Formation of Mutagens in Beef and Beef Extract During Cooking. *Science NY* 201:913–916.
5. Nagao, M., Honda, M., Seino, Y., Yahagi, T., and Sugimura, T. 1977. Mutagenicities of Smoke Condensates and the Charred Surface of Fish and Meat. *Cancer Lett.* 2:221–226.
6. Barna, J. 1979. Compilation of bioassay data on the wholesomeness of irradiated food items. *Acta Aliment.* 8:205.
7. Elias, P.S. and Cohen, A.J., Eds. 1983. Recent Advances in Food Irradiation. Elsevier, Amsterdam.
8. U.S. Congress. 1958. Amended Federal Food, Drug and Cosmetic Act, Section 201(5) and Section 402(a)(7).
9. United Kingdom. 1972. The Food (Control of Irradiation) (Amendment) Regulations. S.I. 1972/205, HMSO, London.
10. Food and Agriculture Organization/World Health Organization (FAO/WHO). 1958. Second Report of the Joint FAO/WHO Expert Committee on Food Additives. Wld. Hlth. Org. Techn. Rep. Ser., No. 144, Geneva.
11. Food and Agriculture Organization/World Health Organization (FAO/WHO). 1961. Fifth Report of the joint FAO/WHO Expert Committee on Food Additives. Wld. Hlth. Org. Techn. Rep. Ser., No. 220, Geneva.
12. Diehl, J.F. 1981a. Effects of combination processes on the nutritive value of food. In: Combination processes in Food Irradiation. Proceedings of the International Symposium, Columbo 1980, IAEA, Vienna.
13. Food and Agriculture Organization/International Atomic Energy Agency/World Health Organization (FAO/IAEA/WHO). 1981. Wholesomeness of irradiated food. Wld. Hlth. Org. Techn. Rep. Ser., No. 659, Geneva.
14. Ingram, M. and Farkas, J. 1977. Microbiology of foods pasteurized by ionizing radiation. *Acta Aliment.* 6:123.
15. U.S. Atomic Energy Commission. 1954. Protocol for Wholesomeness Investigations of Radiation-Pasteurized Foods in the United States Atomic Energy Commission's Food Irradiation Program. Biology Branch, Division of Biology and Medicine, U.S. AEC.
16. Food and Agriculture Organization/International Atomic Energy Agency/World Health Organization (FAO/IAEA/WHO). 1966. The Technical

Basis for Legislation on Irradiated Food. Wld. Hlth. Org. Techn. Rep. Ser., No. 316, Geneva.

17. Food and Agriculture Organization/International Atomic Energy Agency/ World Health Organization (FAO/IAEA/WHO). 1970. Wholesomeness of Irradiated Food with Special Reference to Wheat, Potatoes and Onions. Wld. Hlth. Org. Techn. Rep. Ser., No. 451, Geneva.
18. Food and Agriculture Association/International Atomic Energy Agency/ World Health Organization (FAO/IAEA/WHO). 1977. Wholesomeness of irradiated food. Wld. Hlth. Org. Techn. Rep. Ser., No. 604, Geneva.
19. Phillips, B.J. and Elias, P.S. 1978. A new approach to investigating the genetic toxicology of processed foods. *Food Cosmet. Toxicol.* 16:509–510.
20. Phillips, B.J., Kranz, E., Elias, P.S., and Munzner, R. 1980. An Investigation of the Genetic Toxicology of Irradiated Foodstuffs Using Short Term Test Systems. Part I. *Food. Cosmet. Toxicol.* 18:371–375.
21. Phillips, B.J., Kranz, E., and Elias, P.S. 1980. An Investigation of the Genetic Toxicology of Irradiated Foodstuffs Using Short-term Test Systems. Part II. *Food Cosmet. Toxicol.* 18:471–475.
22. Codex Alimentarius Commission. 1984. Codex General Standard for Irradiated Foods. Recommended International Code of Practice for the Operation of Radiation Facilities Used for the Treatment of Food. *Codex Alimentarius*, XV, FAO, Rome, Italy.
23. Golberg, L. 1970. Chemical and biochemical implications of human and animal exposure to toxic substances in food. *Pure Appl. Chem.* 21:309.
24. Schubert, J. 1978. Toxicological studies on irradiated food and food constituents. In: Food Preservation by Irradiation. Proceedings of the International Symposium, Wageningen 1977, 2:3, IAEA, Vienna.
25. Takeguchi, C.A. 1983. Regulatory and Toxicological Aspects of Food Irradiation. *J. Food Safety* 5:213–217.
26. Merritt, C., Jr. and Taub, I.A. 1983. Commonality and predictability of radiolytic products in irradiated meats. In: Elias, P.S. and Cohen, A.J., Eds. Recent Advances in Food Irradiation. Elsevier, Amsterdam, 27–57.
27. Taub, I.A. 1984. Free Radical Reactions in Food. *J. Chem. Ed.* 61:313–324.
28. Diehl, J.F. 1981b. Wholesomeness of Irradiated Foods. Seminar on Food Irradiation for Developing Countries in Asia and the Pacific, Tokyo 1981, IAEA-TECDOC-271, IAEA, Vienna.
29. Kesavan, P.C. 1978. Indirect effects of radiation in relation to food preservation: Facts and fallacies. *J. Nucl. Agric. Biol.* 7:93–97.
30. British Industrial Biological Research Association (BIBRA). 1980. an Evaluation of the Irradiation of Animal Feedstuffs. Unpublished report dated 16/10/1980 to IFIP, Karlsruhe.

37. Irradiation Processing—Industrial Applications

Darrell F. Wood

Food Research Institute

NATURE OF THE PROBLEM

The food irradiation process has been extensively studied over the past 30 years. On a worldwide basis, more than 30 countries have approved over 40 different food items for irradiation treatment. In addition, an expert committee set up under the auspices of the World Health Organization/Food and Agricultural Organization and International Atomic Energy Agency (WHO/FAO/IAEA) declared, in November of 1980, that foods irradiated up to a dose of 1 million rads, or 10 kiloGray (kGy), presented no toxicological hazard for human consumption.

The irradiation of food may be achieved by three different methods: gamma irradiation using radioactive isotopes (cobalt-60 and cesium-137), electrons produced by electron accelerators, or X-rays produced by X-ray machines. Irrespective of the method, irradiation is considered a cold process and, at the appropriate dose levels, does not alter the physical appearance of the food. The irradiation process has several potential uses, which range from stimulation of plant growth and yield characteristics at extremely low doses to the production of shelf-stable sterile foods at high doses. The potential uses of food irradiation are summarized in Table 1 along with the dose level required and some examples of foods which may be treated.

In view of the many potential benefits that could result from the utilization of the food irradiation process, several concerns need to be addressed relating to its commercialization. These can be grouped generally into the regulatory, public health, nutritional, and educational areas.

Lack of Standardized Regulatory Approaches Among Various Countries

Despite the WHO/FAO/IAEA Expert Committee declaration as to

Table 1. Potential Uses of the Food Irradiation Process

Desired Effect	Dose (kGy)[a]	Food(s)
Low Doses (1 kGy)		
Stimulation of growth	0.001–0.01	Plant seeds such as grains, potatoes, vegetables
Sprout inhibition	0.05–0.15	Potatoes, onions, garlic
Destruction of insects and/or parasites	0.2–1.0	Cereal grains, pulses, fresh pork, dried fish, fresh or dried fruit
Delay of ripening	0.2–1.0	Bananas, mangoes, mushrooms
Intermediate Doses (1–10 kGy)		
Shelflife extension	1.0–3.0	Strawberries, fish
Elimination of pathogens	2.5–6.0	Poultry, red meats, shellfish, frog legs
Reduction of bacterial load	10.0	Spices
High Doses (10–50 kGy)		
Shelf-stable products	25–50	Special hospital diets, prepared foods, meats
Sterile ingredients	10–50	Spices, gum arabic, enzymes

[a]1 kGy = 1,000 Gray (1 Gy = 1 Gray = 1 Joule of energy absorbed per 1 kg material).

the safety and wholesomeness of foods irradiated up to a dose of 10 kGy, most countries still promulgate their own regulations and clearances on a food-by-food basis. While standard procedures exist for measuring dose received, there are no standardized regulatory procedures for monitoring irradiation facilities to ensure that appropriate irradiations have been carried out. Resolution of these concerns is important in that appropriate dose application is a must for optimum product quality. Excessive or repeat applications could result in a loss of product quality.

Lack of Information on Public Health Benefits

Irradiation has several potential applications in the food system, which range from reduction of overall microbial load, through complete elimination of pathogens, and finally to complete sterilization of foods. Although pathogen control or elimination has been widely studied, very little consideration has been given to a public health benefit that may result from irradiation of certain foods to eliminate foodborne pathogens. A recent review of the technical, public health, and economic considerations of food irradiation vs. alternative technologies[1] is an excellent example of the type of information that industry, government, and consumers need in order to properly assess the relative potential and effectiveness of the food irradiation technology.

Lack of Comparative Information on Nutritional Aspects of Irradiated Foods Vs. Other Food Processing Methods

A great deal of information has been published on the effects of irradiation on various nutrients such as vitamins, amino acids, etc., but much of this has been "model system" studies involving irradiation of single, pure nutrients. Such studies tend to show much more destruction than that which occurs when the same nutrients are irradiated at far lower levels in the complex foods actually consumed. Moreover, all forms of food handling and processing result in some nutrient loss. In many instances, in-home preparation causes the most significant damage to nutrient levels. Conclusions about the nutritional consequences of irradiation processing should be based on "real-life" studies involving foods themselves under customary conditions of purchase and use. The effects of irradiation should be compared with the consequences of other more traditional processes applied in the same circumstances.

Lack of Consumer Knowledge About the Technology

The subject of irradiated foods is an emotional issue with many consumers. Considerable misinformation has been published that directly impacts on the emotional fears of anything to do with radiation. Many consumers still have difficulty in separating the differences between Chernobyl and Three Mile Island nuclear power technology and food irradiation technology. They are not aware that medical products irradiation technology, which is the same as food irradiation technology, has achieved fairly wide acceptance. Concerns relative to handling and disposal of the radioactive isotopes such as cobalt-60 and cesium-137 must also be addressed. The consumer must be provided accurate, factual information to be communicated on a level which can be easily understood.

HISTORICAL PERSPECTIVE

The first attempt to commercialize food irradiation was initiated in Canada in 1963 when construction of a combination potato storage warehouse/cobalt-60 pallet irradiator was begun. The facility was opened in time for the 1965 potato harvest; however, within less than a year it was bankrupt, due to an almost complete failure of the potato crop and a lack of understanding of the technical feasibility of the process on a commercial scale.[2]

The first successful commercial food irradiation facility was built in Japan in 1973 for the irradiation of potatoes. Since that time, a yearly average of about 15,000 tons of potatoes have been treated for the table and processing markets.[3]

An up-to-date list of current world food irradiation facilities with the type and amounts of food being irradiated has recently been published.[4] Significant quantities of food are currently being irradiated in seven different countries, with two more countries scheduled to commence irradiating food in 1986. The amounts vary from 400,000 tons of grain/year at two linear accelerator units in the USSR to about 12,000 tons/year in The Netherlands and 7,000 tons/year in South Africa. The types of foods treated in The Netherlands are mainly spices, frozen frog legs, and shrimp. In South Africa, mainly fruit and vegetables such as mangoes, strawberries, onions, and potatoes are treated for shelf life extension.[4] At least nine other countries are currently constructing or considering constructing facilities for irradiating food.

Despite the number of countries irradiating food and the amounts being irradiated, most of the irradiated food goes to institutional, commercial, or export markets, except in South Africa where direct retail sale of irradiated foods is carried out.

POTENTIAL SOLUTIONS

There is a considerable wealth of scientific data available on the safety and benefits of food irradiation. The full review of this data led to the 1980 WHO/FAO/IAEA Expert Committee decision to support irradiation up to 10 kGy as a safe process. Continued monitoring of the scientific and technical literature since then has not led these international organizations to question their decision. What is needed now is to establish a basis for communicating to industry and to the public information that would allow informed judgment.

In this regard, a prioritization of the different food groups that could benefit from food irradiation is required. At the lower doses, i.e., less than 1 kGy, the disinfestation of fruit and vegetables is the most immediate candidate. For doses between 1 and 10 kGy, the elimination of pathogens from meats and seafood is of prime importance. A secondary side benefit from this treatment is an extension in the refrigerated shelf life for these products. For doses of 10 kGy, treatment of spices for microbial reduction is a major use.

Irradiation has been studied more than any food process in history. If other processes had to undergo such stringent scrutiny, they probably would fail to achieve the level of approval irradiation has already attained. For example, if the same test requirements were applied to canning as have been used for irradiation, canning probably would not be an approved food process under today's government regulations.

Another way to assist in the appropriate judgement of food irradiation is to have pilot-scale irradiation facilities available in several locations. The International Facility for Food Irradiation Technology (IFFIT) at Wageningen in The Netherlands has demonstrated the value of such pilot establishments. Such facilities would provide testing sites for industry to establish the technical and economic feasibility and, in addition, serve as training facilities.

Moreover, the consumer and regulatory officials must feel confident that standard regulatory procedures are being followed. Adoption of the Codex Alimentarius standard for irradiated foods and code of practice for operation of irradiation facilities[5] by all countries would help eliminate some of the confusion and provide a basis for international trade.

Full-scale international involvement and collaboration by committees such as the recently formed "International Consultative Group on Food Irradiation" would be a positive step in communicating needed information. Such groups could provide a focal point for the global development and evaluation efforts. Standard advice on consumer education programs, safety, and technical feasibility would be provided.

In this international context, problems relating to industrial applications and public education could be addressed in a manner similar to the solving of the toxicological and wholesomeness questions during the 1970s. The formation of an international project under the auspices of FAO/WHO/IAEA could address concerns on an international basis and develop generic guidelines, information packages, etc., of benefit to all countries in their efforts to assess the food irradiation technology relative to other alternative processes.

Finally, it must be noted that some form of protection is needed against insect and bacterial contaminations that are natural occurrences in grains, fruits, spices, and products of nature. Some process, whether it be fumigation, irradiation or any other alternative, is required in order to provide a safe product to consumers. As implied by Mossel,[6] though it is laudable that agencies proceed with caution in adopting the food irradiation technology, the benefits of eliminating actual foodborne health problems, which must be made known to consumers and examined by academicians and regulatory agencies, would outweigh any presumed minor risks.

SUGGESTED RECOMMENDATIONS

1. Adopt Codex standard and code of operation relative to food irradiation in order to eliminate confusion caused by different countries adopting different dose levels and different foods.
2. Develop educational material on all aspects of food irradiation which explains the process in layman terms so that consumers can properly separate the myth from the fact and draw their own conclusions as to the validity of food irradiation.
3. Encourage further research on the public health problems that would be best eliminated by the use of the irradiation technology.

REFERENCES

1. Mentzer-Morrison, R. and Roberts, T. 1985. Food irradiation: New perspectives on a controversial technology. USDA December 1985.
2. Masefield, J. and Dietz, G. R. 1983. Food irradiation: The evaluation of commercial opportunities. *CRC Critical Reviews in Food Science and Nutrition* 19(3):259–272.
3. Kume, T. 1985. Economic viability and commercial experience with Shihoro potato irradiator, Proceedings ASEAN Workshop on Food Irradiation, ASEAN Food Handling Bureau, Bangkok, pp. 78–89.
4. Loaharanu, P. 1985. Food irradiation: an alternate technology. Proceedings ASEAN Workshop on Food Irradiation, ASEAN Food Handling Bureau, Bangkok, pp. 2–14.
5. Codex Alimentarius Commission. 1984. Codex General Standard for Irradiated Foods and Recommended international code of practice for the operation of radiation facilities used for the treatment of foods. CAC Vol. XV, Edition 1, Rome.
6. Mossel, D. 1985. Irradiation: An effective mode of processing food for safety. Proceedings of an International Symposium on Food Irradiation Processing. IAEA-SM-271/80, pp. 251–279. International Atomic Energy Agency, Vienna.

38. Irradiation Processing—Regulatory Procedures

Gerad McCowin

U.S. Food and Drug Administration

NATURE OF THE PROBLEM

Irradiation as a process for the treatment of food has been the subject of investigations and concerns for 40 years. In all of this time, the primary questions addressed have been (1) Is the food treated with radiation safe? (2) How is safety to be evaluated and the treatment to be cleared by the U.S. Food and Drug Administration (FDA)?

Irradiation of food is known to cause a variety of effects in food, much the same as the effects caused by more traditional processing of food, such as cooking. These include changes in chemical composition that could affect organoleptic properties or nutritional quality. For most of its history, radiation has also been popularly viewed with suspicion and fear. It is not surprising then that questions have been raised about whether the chemical changes in food caused by irradiation create any potential toxicity of the treated food.

Over the past few years, the FDA has made some changes in the way it has traditionally evaluated irradiation. These changes have resulted in the FDA being able to identify some uses of irradiation that it believes are safe. Additionally, FDA has identified points that need to be addressed to demonstrate the safety of irradiation of food.[1]

HISTORICAL PERSPECTIVE

The 1958 Food Additives Amendment to the Federal Food, Drug and Cosmetic Act included the use of radiation for the treatment of food within the scope of the definition of food additive, i.e.,

> (S) The term "food additive" means any substance the intended use of which results . . . in its becoming a component or otherwise affecting the characteristics of any food . . . including any source of radiation intended for any such use.

It also established provisions for the premarket clearance of food additives by the FDA.

The food adulteration section of the Act was also amended to provide that "a food shall be deemed to be adulterated—if it has been intentionally subjected to radiation, unless the use of the radiation was in conformity with a regulation or exemption in effect pursuant to Section 409."

The two basic premises of the Food Additives Amendment are that food additives should be permitted to be used under those conditions where their use is safe and that industry should bear the primary burden of demonstrating conditions of safe use for a food additive.

Since the inclusion of a source of radiation under the definition of food additives has led to some confusion, it may be helpful to discuss this first. The bill for food additives introduced by Congressman Miller of Nebraska in January 1956 did not include a reference to radiation in the definition of a food additive. Briefly, it defined the term food additive to mean "any substance or treatment used, directly or indirectly, in or on food. . . ." During the congressional hearings for the bill, however, Congressman Miller explained further what was intended:

> The use of the word "treatment," and I use the word "treatment" where some of the other bills do not, in this definition is also significant. It was included so that new and revolutionary treatment of food, the use of neutrons, ions, and so forth, that is being used in the treatment of meat, and things that may be used in the future, . . . ought to be included in the definition. If the ionic treatment, or the neutron treatment, or the isotope treatment of food shows that there is some need for investigation by the Department of Health, Education and Welfare, it may be done.

The FDA had been working closely with the Departments of Defense and Agriculture on the nutritional and toxicological aspects of irradiated foods during the early 1950s. Thus, in hearing testimony before Congress in 1956, FDA Commissioner George Larrick said:

> Experiments in the preservation of foods by ionizing radiation from X-rays, radioisotopes, and radiation from atomic piles have now advanced to a point where they offer a distinct possibility that the processes will be adapted to commercial use. These methods, as well as the use of radioisotopes as quality control measures, should not be permitted until it is shown that food products will be safe.
>
> We therefore recommend that the pretesting requirements and procedures of the legislation be made clearly applicable not only to radioactive substances that might be introduced into food, either

> deliberately or unavoidably, but also to any changes in food, or new substances formed in food, by subjecting it to radiation from internal and external sources.
>
> We would welcome committee consultation with representatives of the Atomic Energy Commission as well as our department, in this connection. Staff members of the Department and of the Atomic Energy Commission are prepared to cooperate in working out an appropriate amendment for assuring the safety of food affected by radiation.

Other than this, however, there was limited discussion relative to food irradiation. There were some suggestions that there should be a distinction made between intentional and incidental food additives. It was further suggested that irradiation should be considered an incidental additive. However, the FDA generally opposed any attempts to make a distinction between intentional and incidental additives on the grounds that it would establish two different procedures for demonstrating safety.

The Food Additives Amendment, as eventually enacted, did not differentiate between intentional and incidental food additives and thereby provided the FDA with great flexibility in dealing with the various issues it faces concerning food additives. The broader definition and inclusion of radiation are consistent with two important facts. First, it was generally understood that irradiation of food is a process that incidentally affects the food. Secondly, and most importantly, it was generally agreed that the use of this process of irradiation should be subjected to premarket evaluation and clearance to ensure safety.

The amendment also provides that the FDA will issue regulations that prescribe the conditions under which a food additive, including a source of radiation, shall be safely used. Although the amendments do not specifically define what is meant by the term "safe," the legislative history does contain some discussion of what was intended. For example, the concept of safety was explained in the Senate report on the Food Additives Amendment of 1958, as follows:

> The concept of safety used in this legislation involves the question of whether a substance is hazardous to the health of a man or an animal. Safety requires proof of a reasonable certainty that no harm will result from the proposed use of an additive. It does not—and cannot—require proof beyond any possible doubt that no harm will result under any conceivable circumstance.

The amendment did identify the various factors that the FDA would consider in determining whether a proposed use of a food additive is safe: (1) the probable consumption of the additive and of any substance formed in or on food because of the use of the additive, (2) the cumulative effect of such additive in the diet of man or animals, taking into account any chemically or pharmacologically related substance or substances in such diet, and (3) safety factors that, in the opinion of the experts qualified by scientific training and experience to evaluate the safety of food additives, are generally recognized as appropriate in animal experimentation.

POTENTIAL PROBLEMS AND SOLUTIONS

As implied earlier, the FDA has always been confronted with the problem of identifying the types of tests and data that would be necessary to demonstrate safety, i.e., to establish a reasonable certainty of no harm from the use of radiation sources in the treatment of food. It has traditionally relied upon the results of animal feeding studies for demonstrating safety of food additives, and required such animal studies as support for the early petitions for various uses of irradiation. At the time of these petitions, the radiation chemistry of food had essentially not yet been investigated, so there was much concern over possibly unknown substances being formed in the food during irradiation. The initial approach that the FDA attempted was to require a core of animal studies on a variety of foods to provide a matrix. This matrix would then be used to predict safety of other types of food.

However, the conduct of these studies has led to confounding results that made decisions on the safety of irradiation of food difficult, if not impossible. In particular, chronic feeding studies were performed in which up to 35% of the diet was replaced with irradiated foods such as beef, chicken, potatoes, onions, and papaya. This approach obviously could lead to nutritional difficulties for the test animals. In a number of instances, studies had to be terminated early because of premature mortality. However, during this period, the FDA did approve sources of radiation for limited use in the treatment of potatoes, wheat, and flour. An early approval of irradiation of bacon was withdrawn when it was determined that the supporting animal studies were not completely adequate, i.e., some of the conclusions were not supported by the data. Although there were several petitions before the FDA during the 1960s and 1970s, little progress has been made toward approval of additional uses, largely because of the difficulty in interpreting the results of the available animal studies.

Consequently, in 1979, the FDA initiated an in-house review of its approach toward evaluating the safety of irradiated foods. Over the next year, the Bureau of Foods Irradiated Food Committee (BFIFC) considered the available radiation chemistry information and other information relevant to irradiated foods. They made three general recommendations, corresponding to three categories of foods:

1. Food irradiated at doses not exceeding 100 krad is wholesome and safe for food consumption.
2. Foods, such as spices, comprising no more than 0.01% of the daily diet and irradiated at 5 Mrad or less will also be considered to be safe, without toxicological testing, for human consumption.
3. Foods irradiated at doses above 100 krad and comprising more than 0.01% of the diet are assumed to contain a sufficient quantity of unique radiolysis products to warrant toxicological evaluation, e.g.,
   a. A battery of nonmammalian mutagenicity tests
   b. Two 90-day feeding studies—one with a rodent species (*in utero* exposed) and the other with a nonrodent species

As a followup to this Committee's efforts, the FDA established a task group of toxicologists to evaluate all existing toxicology data for irradiated foods. The purpose of this review was to identify those studies that would support safety for the third category of foods listed above. Their review, which was completed in 1982, covered 409 studies from which they targeted 69 studies for detailed evaluation. Only five of these were judged totally adequate for evaluating safety; the remaining 64 were determined to be deficient for one reason or another. Extensive toxicological studies have also been conducted more recently with pork and poultry.

Other concerns remain to be addressed:

1. *Microbiological.* One of the concerns is that the irradiation of foods at intermediate doses not intended to destroy *Clostridium botulinum* might result in the extensive destruction of the bacteria and other microorganisms that would otherwise provide signs of spoilage and prevent consumption in the event of temperature abuse of the product.

2. *Nutritional.* It is well-known that irradiation, as with all methods for the processing of food, will have an effect on the nutritive value of foods. The BFIFC report concluded that such effects would be minimal for foods treated at levels below 100 krad. Additional data are needed for levels above 100 krad.

3. *Labeling*. The proposal to permit the irradiation of foods at levels up to 100 krad generated many comments. Most of these were from consumers who generally declared (1) that they did not want food to be irradiated and (2) that if irradiation of food is approved, the FDA should require label declarations on the use of irradiation. (The final order of April 18, 1986 requires that irradiated food bear the statement: "Treated with Radiation" or "Treated by Irradiation." The regulations also require the use of a logo. The concept behind the logo is that it may become a recommended symbol that can subsequently be used in place of label statements.)[2]

In response to the final order approving the use of irradiation for the treatment of food at 100 krad, the FDA received over 200 objections to that regulation along with some requests for a hearing and a stay of effectivess. The FDA is in the process of evaluating these objections and hopes to respond to them later in the year. A regulation on the use of irradiation for control of trichina in pork was published in the Federal Register of July 22, 1985. The FDA received approximately 50 objections to that rule, including several requests for a stay of effectiveness of the regulation and for an administrative hearing. It expected to respond to these later in 1986.

These regulations are seen by the FDA as the first steps toward the utilization of irradiation for the general treatment of food. Approvals for treatment above 100 krad depend on the submission of petitions supported by the results of studies that address the points mentioned above. The FDA currently has petitions covering the use of irradiation for the pasteurization treatment of poultry and seafood and the sterilization treatment of food in general. Most of the data being submitted in support are from studies conducted during the 1950s and 1960s. The most notable exceptions are the results of an extensive study done with irradiated poultry by Raltech for USDA and of the studies done with irradiated pork and poultry by CIVO in The Netherlands.

As discussed earlier, the BFIFC report outlined the types of data considered necessary to demonstrate the safety of irradiated foods. It also described the basis for considering foods of similar composition as a generic class for regulatory purposes. These recommendations and considerations can be combined with the earlier concept of a data matrix to support broad uses of irradiation, where applicable. As noted earlier, studies are also necessary to assess microbiological, nutritional, and organoleptical concerns. These also could be conducted on a variety of foods that will produce a matrix suitable for extrapolation to other types of food.

REFERENCES

1. A.P. Brunetti, et al. 1980. "Recommendations for Evaluating the Safety of Irradiated Foods." U.S. Food and Drug Administration, Washington DC.
2. U.S. Food and Drug Administration. April 18, 1986. "Irradiation in the Production, Processing and Handling of Food: Final Rule," 51 FR 13376.

BIBLIOGRAPHY

Advisory Committee on Irradiated and Naval Foods, "Report on the Safety and Wholesomeness of Irradiated Foods," U.K. Department of Health and Social Security.

39. Aseptic Processing of Particulates – Technical Advances and Industrial Applications

Daryl Lund

University of Wisconsin

NATURE OF THE PROBLEM

In the 1950s, it was predicted that by the 1970s, aseptic processing would be commonplace. In 1957, Feliciotti and Esselen[1] showed the advantages of high-temperature, short-time processing for retention of vitamin B_1 (thiamin) by examining the effect of high-temperature, short-time and low-temperature, long-time processing on microorganisms and nutrients. Subsequently, it was shown that many quality factors, including colors and flavors, show the same temperature dependence as vitamins. Thus, optimizing for vitamin retention also resulted in optimizing quality factors. Implementation of this concept required design of appropriate equipment for processing foods at high-temperature, short-time, resulting eventually in ultrahigh temperature (UHT) processing of food products.

Problems that inhibited industrial adoption of aseptic processing were initially associated with material handling and packaging. Line speeds for aseptic packing have been inherently slower than those for conventional in-container processing of food products. Even with FDA approval of hydrogen peroxide as a surface sterilant in 1981, line speed per packaging machine did not significantly increase. However, in the following five years, there have been numerous applications of aseptic processing to liquid food products including juices and starch-based, viscous foods (e.g., puddings). These developments were driven primarily by the lower cost of flexible packaging materials compared to metal containers and by successful marketing of aseptic products to U.S. consumers. Although line speed is a significant impediment to adoption of aseptic technology, marketing advantages appear to be driving the technology toward adoption.

Other problems, however, need to be addressed in order to ensure an improvement in food quality as a result of aseptic processing technology. Improved quality refers not only to enhanced chemical and physical attributes of the food product, but also to increased consumer safety. These problems include:

- Need for additional data on the kinetics of inactivation of microorganisms or their spores in the UHT temperature range (up to 155C)
- Lack of data on the kinetics of destruction of enzymes, nutrients, and quality factors at UHT temperatures
- Lack of data on the viscous properties of food fluids and local heat transfer coefficients in UHT processing equipment
- Acceptance by the regulatory agencies of the accumulated lethality in the process from both the heating and cooling stages, as well as from the holding stage
- Understanding and predicting the residence time distribution of non-Newtonian fluids as they are being heated or cooled

HISTORICAL PERSPECTIVE

Inactivation Kinetics

Calculation of an aseptic process to ensure commercial sterility (i.e., absence of pathogenic microorganisms that would grow under the conditions of packaging and storage) is based on the thermal death time (TDT) model developed by Ball[2] for in-container processing. This procedure requires information on the time-temperature profile of the coldest point in the product and kinetic parameters to describe the time-temperature dependence for inactivation of a microorganism or its spore.

Currently, thermal processes at UHT temperature are designed based on extrapolation of TDT data obtained at lower temperatures (100–125C), because it has been extremely difficult to generate TDT data at temperatures up to 150C. Since there has been some question of the appropriateness of extrapolating TDT data outside the maximum temperature investigated, the FDA has restricted the calculated lethality in aseptic processing to that obtained in the hold tube, to ensure satisfactory thermal treatment. Lethality that occurs during heating and cooling of the product is not counted toward the total lethality of the process. As a result, products generally are grossly overprocessed by a factor of more than ten.[3]

The problem with determining kinetic parameters for thermal inactivation of spores of pathogenic microorganisms (e.g., *Clostridium botulinum*) is that with very short heating times at temperatures around

150C it is technically extremely difficult to heat, hold, and cool the product with a known time-temperature profile. For example, if the D value of spores of *Clostridium botulinum* is 0.2 minutes at 120C and the z value is 10C, then the D value at 150C is 0.0002 minutes (0.012 seconds). The resulting thermal death time (i.e., time to reduce the population by a factor of 10^{12}) is 0.144 seconds. Thus, the problem is one of trying to handle a spore suspension while heating it from room temperature to 150C, holding it, and then cooling it to room temperature again, when a process of 0.144 seconds at 150C has such a devastating effect on the microbial population.

Recently, David and Merson[4] described a computer-controlled reactor designed to study thermal inactivation at UHT temperatures. The reactor was based on the design of the resistometer.[5] David and Merson[4] studied the thermal inactivation of *Bacillus stearothermophilus* in the temperature range of 130 to 155C. Furthermore, they compared data from their reactor to data generated by the capillary tube method. Their results showed that the thermal death time curve could best be described by two z values, 9C for temperatures below 142.5C and 17C for temperatures above 142.5C. Using an Arrhenius activation energy model, they could adequately describe the entire range of data with a constant activation energy of 77.5 kcal/mole. Their data show that the assumption of a linear TDT curve is not justified and that the Arrhenius model fits the thermal inactivation data better than the TDT model over a broad temperature range. In their studies, they reported that the *Bacillus stearothermophilus* strain had a D value of 680 seconds at 120C. Since *Clostridium botulinum* spores have a D value that is approximately 12 seconds at this temperature, it is questionable whether this equipment can be used to study thermal inactivation of *Clostridium botulinum* spores.

Temperature Dependence of Quality Factors

Many studies have been reported in the literature on thermal destruction of enzymes, nutrients, and quality factors in foods as a function of thermal process parameters.[6] Unfortunately, most of these studies have maximum temperatures of 130C. Data from these studies have been used to design thermal processes for maximizing retention of nutrients and quality factors.[7] These data have also been used to justify improved product quality associated with UHT aseptic processing by extrapolating to the UHT range.

It is especially important to note that enzymes have been a source of product quality deterioration in some aseptically processed products. At lower temperatures (less than 130C), the thermal process that will inactivate the spores of microorganisms will also adequately inactivate the heat-resistant enzymes. However, at UHT temperatures, the thermal process for inactivating the spores may not inactivate heat-resistant enzymes that can cause product quality loss (e.g., amylases, which cause a decrease in viscosity of starch-based foods). An examination of thermal death time data for heat-resistant enzymes and for spores of pathogenic organisms shows that there is a crossover point in the temperature range of 120 to 130C. Above this temperature, the enzyme is more heat resistant than the spores of the microorganism. Although this difference has been shown in some cases, it is essential to conduct more research on thermal inactivation of enzymes, nutrients, and quality factors in the UHT range. This difference could have potential impact on the availability of nutrients from aseptically processed products, since inactivation of heat-resistant enzymes and thermal destruction of nutrients exhibit similar temperature dependences. Thus a process that would be based on inactivation of a heat-resistant enzyme may not produce significant benefit for retention of nutrients by processing at a UHT temperature above 130C.

Rheological Factors

In aseptic processing, a liquid or a liquid containing suspended particles is pumped continuously through a heat exchanger, a hold tube, and a cooling heat exchanger to a filler head where product is deposited aseptically into sterile packages. Knowledge of the rheological behavior of liquid food products is essential for process design and evaluation, quality control, and consumer acceptability. Although there has been extensive research on the rheological properties of liquid foods,[8] there is still a lack of correlations for viscous properties as a function of temperature for most liquid foods. Most of these liquid foods are pseudoplastics (i.e., shear thinning fluids in which the viscosity decreases as the rate of shear increases). Viscosity of these fluids is temperature dependent, and design of the heat transfer equipment is dependent on viscosity. In addition, as these fluids move through the heat exchanger (for either heating or cooling), there is a change in the surface heat transfer coefficient, which controls heat transfer from the wall to the fluid. There is little information on the temperature dependence of surface heat transfer coefficients.

Heat Transfer

For fluids containing suspended particulates, the surface heat transfer coefficient adds another dimension to the problem that has rarely been considered in design of thermal processes. At the surface of the particle, there exists a finite surface heat transfer coefficient for the transfer of heat from the fluid to the particle. In many thermal process calculations, it is assumed that the particle is sterile in the interior so that the thermal process is only designed for inactivation of spores on the surface of the particle. For in-container processing, the temperature sensing device for collection of heat penetration data is placed in the center of the particle; thus the surface heat transfer coefficient is taken into account in the heat penetration measurement. However, for aseptic processing, the particle moves with the fluid, and the surface heat transfer coefficient is not taken into account for calculating lethality at the center of the particles. DeRuyter and Brunet[9] considered the processing of nonsterile particles in a liquid under such aseptic processing conditions. They assumed in their model that the surface heat transfer coefficient for transfer from the liquid to the particle in the scraped-surface heat exchanger was infinite (e.g., there was no surface resistance to heat transfer, and the surface temperature of the particle was the same as the temperature of the surrounding liquid). They showed that for a one-inch particle the lethality delivered to the liquid could be up to 100 times more than the lethality delivered to the center of the particle. Obviously, under these circumstances, the liquid was grossly overprocessed compared to the center of the particle. This nonhomogeneity of delivery of thermal process to the liquid and to the particles remains a major impediment to adoption of aseptic processing technology for large nonsterile particles in the liquid.

The Jupiter System by APV Company was designed to overcome the disparity in thermal process delivered to sauce and particles.[10] This batch-continuous system processed the sauce in one system and the particles in a batch heating device. The sauce and particles were then mixed and pumped to the filler head after cooling. Unfortunately, the system was not widely adopted in industry.

For the FDA to accept the inclusion of lethality delivered during heat-up and cool-down operations, it is essential that the convective heat transfer coefficient to the fluid and, in the case of fluids with particles, to the particle must be incorporated into the calculations. Lenz and Lund[11] incorporated surface heat transfer resistance to particles into the calculations of lethality delivered to forced convection heated foods in containers. Using the Monte Carlo procedure, they showed that the 95% confidence interval for the lethality ranged from 20 to 60% of the median value depending on process conditions. Thus to ensure an adequate process that would include the lethality accumulated during heating and cooling, it is necessary to determine the surface heat transfer coefficients during those stages of the process and to use them in the thermal process calculations.

Residence Time Distribution

Currently, the design of aseptic processes for inactivation of spores of microorganisms is based on residence time and temperature in the hold tube between the heating section and the cooling section. Furthermore, the residence time of the liquid in the hold tube is based on the maximum velocity of the liquid. For a tube of circular cross section, the maximum velocity will be the center of the tube. In the limiting case of a Newtonian fluid, the maximum velocity is twice the mean velocity. For pseudoplastic fluids, the maximum velocity will be one to two times the mean velocity. For processors who have characterized the flow behavior of their product, they may base the maximum velocity (i.e., minimum residence time) on the actual flow behavior. However, in many cases the processor will calculate the maximum velocity as twice the mean velocity in order to provide a safety margin in the thermal process. This calculation could have detrimental effects on product quality because the product is therefore overprocessed. This overprocessing results in loss of nutrients and loss of quality factors such as color, texture, and flavor.

Teixeira and Manson[12] described a thermal process control system for aseptic processing which accounted for the non-Newtonian behavior of food fluids. Their control system was based on an Eolkin plastometer for continuous measurement of apparent viscosity in the liquid food. Thus, with interaction to computer control, either a drop in temperature or a change in viscous behavior of the liquid could activate the flow diversion valve.

Residence time distribution of product in the heat exchange system used in aseptic processing is important for two reasons. On the one hand, the minimum residence time will dictate the minimum lethality achieved in the thermal process and therefore must be known. On the other hand, product that is in the heat exchanger longer than the minimum required for lethality will be overprocessed, resulting in potential loss of product quality (color, flavor, texture, and nutrients). Thus, knowledge of the residence time distribution is essential. When a liquid is flowing in a tube or across a surface, the velocity of the liquid at the wall is near zero, and the velocity in the fluid increases away from the wall. The difference results in a velocity distribution in the liquid, ultimately resulting in a residence time distribution in the fluid. Milton and Zahradnik[13] investigated residence time distribution in a non-Newtonian fluid in a swept-surface heat exchanger. Their results showed that the residence time distribution was highly dependent on the rpm of the scraper blade (dasher speed), on the viscous behavior of the fluid, and on the mass flow rate. Recently, Cuevas, Cheryan, and

Porter[14] showed that current models for residence time distribution in scraped-surface heat exchangers do not adequately represent residence time distributions for non-Newtonian food fluids. They concluded that a residence time distribution would be required on each food fluid that is processed in a scraped surface heat exchanger.

POTENTIAL SOLUTIONS

Each of the problems identified with aseptic processing could be solved by conducting appropriate research. Data on the kinetics of inactivation of microorganisms or their spores in the UHT temperature range can be generated using the integrated lethality approach of Hayakawa et al.[15] From the data of David and Merson,[4] it would appear that the Arrhenius activation energy model is more suited to describe the temperature dependence of inactivation of spores of microorganisms over the UHT range than the TDT model. Similarly, data on kinetics of inactivation of enzymes, nutrients, and quality factors can be obtained. Data on viscous properties of food fluids over the UHT temperature range of interest, as a function of fluid composition, can be obtained through appropriate statistical experimental design.

Experiments to measure local heat transfer coefficients in UHT processing equipment can be conducted as a function of viscous characteristics of the fluid, temperature range, size of particle, and flow velocity of the system. Using these data, it is possible to design aseptic processing systems that include the lethality accumulated during heating and cooling, which would result in an improvement in product quality characteristics. All data should be subjected to peer review and should be generated at several research laboratories in order to confirm results. For studying residence time distribution of non-Newtonian fluids, techniques using radiolabeled materials or other chemically identifiable substances need to be employed. From these distributions, models for residence time distribution as a function of the non-Newtonian character of the liquid and the thermal profile can be generated. These can be used in predicting thermal processes delivered to fluids and particles.

SUGGESTED RECOMMENDATIONS

Aseptic processing has the potential for delivering an improved quality product to the consuming public. These products will have extended shelf life comparable to in-container processed products. The existing problems that can be addressed by the public sector for increased adoption of this process technology have been identified. These problems are researchable, and it is imperative to launch a concerted effort to apply appropriate research techniques to develop the knowledge base. Concomitant with the development of this knowledge is the requirement that private industry develop appropriate machinery that will adequately handle product at rates that are economical and competitive with current in-container processing.

It is also imperative in addressing the research priorities for aseptic processing that several public institutions initiate research projects in each of these areas. This approach should be viewed as a confirmation of research rather than as a duplication of effort. This confirmation will lend increased credibility to the conclusions of the research when they are presented to the regulatory agency for potential adoption. Furthermore, it is appropriate to explore combined processes for preservation of these products. For example, irradiation combined with continuous thermal processing as a means of extending the shelf life of these food products requires additional attention. Each of these research programs will required financial support from federal agencies and private industries. This is an excellent opportunity for public and private sector financing of research to increase the quality of foods delivered to the consumer.

REFERENCES

1. Feliciotti, E. and Esselen, W.B. 1957. Thermal destruction rates of thiamin in pureed meats and vegetables. *Food Technol.* 11:77–84.
2. Ball, C.O. 1923. Determining by methods of calculations, the time necessary to process canned foods. *Bull. Nat. Res. Coun.* 7(37):9–72.
3. Hallstrom, B. 1977. Heat preservation involving liquid food in continuous-flow pasteurization and UHT. In: Hoyem, T. and Kvale, O. (Eds.). "Physical, Chemical and Biological Changes in Food Caused by Thermal Processing." Applied Sci. Publ., Ltd., Essex, United Kingdom, pp. 31–41.
4. David, J.R. and Merson, R.L. 1985. Kinetic parameters for thermal processing at high temperatures: Heat resistance of *Bacillus stearothermophilus.* Presented at Int'l Cong. Engr. Food. IV. Edmonton, Alberta, Canada.
5. Pflug, I.J. and Esselen, W.B. 1953. Development and application of an apparatus for study of thermal resistance of bacterial spores and thiamin at temperatures above 250°F. *Food Technol.* 7:237.
6. Lund, D.B. 1975. Effects of heat processing on nutrients. In: Harris, R.S. and Karmas, E. (Eds.). "Nutritional Evaluation of Food Processing." AVI Publ. Co., Inc., Westport, CT. Pp. 205–240.
7. Lund, D.B. 1977. Design of thermal processes for maximizing nutrient retention. *Food Technol.* 31(2):71–78.
8. Rao, M.A. 1977. Rheology of liquid foods—A review. *J. Texture Studies* 8:135–168.

9. deRuyter, P.W. and Brunet, R. 1973. Estimation of process conditions for continuous sterilization of foods containing particulates. *Food Technol.* 27(7):44–51.
10. Hersom, A.C. and Shore, D.T. 1981. Aseptic processing of foods comprising sauce and solids. *Food Technol.* 35(5):53–62.
11. Lenz, M.K. and Lund, D.B. 1978. The Lethality-Fourier Number Method. Heating rate variations and lethality confidence intervals for forced-convection heated foods in containers. *J. Food Proc. Engr.* 2:227–271.
12. Teixeira, A.A. and Manson, J.E. 1983. Thermal process control for aseptic processing systems. *Food Technol.* 37(4):128–133.
13. Milton, J. and Zahradnik, J.W. 1973. Residence time distribution in a Votator pilot plant using a non-Newtonian fluid. *Trans. Amer. Soc. Agr. Engr.* 16(6):1186–1189.
14. Cuevas, R., Cheryan, M. and Porter, M.L. 1982. Heat transfer and thermal process design in scraped-surface heat exchangers. *Amer. Inst. Chem. Engr. Sym. Ser.* 78(218):49–57.
15. Hayakawa, K.-I., Matsuda, N., Kamaki, K. and Matsunawa, K. 1980. Computerized estimation of reaction kinetic parameters for thermal inactivation of microorganisms. *Lebens. Wiss. und Technol.* 14:70–78.

40. Aseptic Processing of Particulates — Approval Procedures

D.I. Chandarana
A. Gavin
D.T. Bernard

National Food Processors Association

NATURE OF THE PROBLEM

Aseptic processing of homogeneous low-acid products has been a commercial reality in the United States for over 40 years. However, the technology has undergone rapid expansion, and many manufacturers have shown renewed interest since the 1981 approval for use of hydrogen peroxide as a sterilant for packaging material. Current interest in aseptic technology within many companies seems to center on the development of low-acid aseptic products that contain discrete particles. Examples of such heterogeneous foods would include soups and stews with meat and vegetable chunks.

As with other foods sold unrefrigerated in hermetically sealed containers, these low-acid heterogeneous products must receive a process that will render them commercially sterile. For aseptic products, this sterilization can be accomplished by the application of a high-temperature, short-time process, which has the potential to yield a very high quality product with minimal degradation of nutrients. Processes for homogeneous products such as juices or milk products are based on specific time-temperature relationships. Processes for heterogeneous or particulate products become more complex due to the nature of the product. Such processes may be very conservative in order to satisfy regulatory requirements with minimal effort. Conservative processes may unnecessarily penalize the nutrient and quality attributes of the product. Conversely, if the processes are calculated incorrectly and prove to be insufficient, spoilage may result, and toxin formation by *Clostridium botulinum* is a possibility.

It is, therefore, of the greatest importance that the factors that significantly affect the aseptic thermal processing of particulate-containing foods be identified and quantitated to allow the calculation of safe and optimum processes for these products.

The objective of this chapter is to analyze various process establishment procedures for aseptic processing of heterogeneous foods. The main problem is knowing the amount of heat that penetrates to the center of a food particle while that particle is continuously moving through an aseptic system. The solution would include a model for predicting heat penetration into the particulate matter, a model for predicting the residence time of the fastest moving particle, and a method to validate the above models.

HISTORICAL PERSPECTIVE

Aseptic processing offers many potential advantages, including:

- Superior product quality, because as temperature is increased in a thermal process, the bactericidal effectiveness increases at a greater rate than do chemical changes (nutrient destruction)
- Higher efficiency, because processing is continuous
- More uniformity, because heat transfer to the food is faster
- Lower cost, because applicable containers are potentially less expensive than metal cans

There are also some possible disadvantages, such as:

- High initial capital investment
- Substantial cost of training personnel
- More complex instrumentation and controls

System Description

Heterogenous foods potentially could be aseptically processed in a variety of systems. Current system designs include the use of a positive displacement pump to move product through heating scraped-surface heat exchangers, a hold tube, and cooling scraped-surface heat exchanger and a lobe-type pump to produce a "back pressure" throughout the system. Other possible system designs are being investigated.

Process Establishment

Traditionally, process establishment for low-acid foods involves the marriage of the heating characteristics of the product with the bacterial resistance of certain target microorganisms. A mathematical model is used to develop thermal processes, which allows maximum exploration of process alternatives while minimizing the amount of process confirmation using bacterial inoculated pack studies. The Ball formula method essentially serves this purpose for traditional in-can retort processing.

The conventional approach for aseptic processing of homogeneous products considers only the thermal lethality accumulated in the product as it passes through the hold tube. A mathematical model is used to determine the minimum residence time and the minimum product temperature necessary to produce a desired lethality. The flow rate, physical dimensions of the hold tube, and the rheological properties of the fluid are used to calculate a residence time for the fastest fluid particle. The minimum temperature is verified by physically measuring the temperature at the end of the hold tube. Since no heat is applied to the hold tube, this temperature represents the minimum for the product flowing through the hold tube.

When considering processes for particulate-containing foods, however, the situation becomes much more complex due to the increase in the number of variables that must be considered. The process design must be such that the lethality imparted to the center of the largest, fastest moving particle is adequate to achieve commercial sterility within that particle, while at the same time minimizing the overprocessing of the fluid carrier medium.

POTENTIAL SOLUTIONS

Several possible approaches could be considered when designing aseptic particulate processes. Examples are selected for discussion.

Lethality Only in Hold Tube

An approach that considers only the lethality incurred in the hold tube is one possibility. Within the system, the carrier medium serves to transport the particle through the system while providing the vehicle for transferring heat to the particle. This approach assumes the center of the particle would be brought to a minimum processing temperature in the heat exchanger, transported through the hold tube, and then cooled. One drawback associated with this approach is that a means would have to be developed to assure that the slowest heating region of the particle has reached processing temperature prior to entering the hold tube. Such assurance must rely on research not only on factors affecting heat penetration and heat transfer rates for individual particles, but on factors that affect particle residence time within the scraped-surface heat exchangers and temperature distribution profiles as well.

Since the carrier medium is used to heat the particle, the thermal process received by the carrier medium is substantially greater than that received by the center of the particle. This differential will increase as particle size increases. The amount of heat used to bring the particle center to processing temperature may degrade the carrier medium and particle surfaces. Therefore, processes that consider only the lethality in the hold tube may prove to be impractical for some heat-sensitive products.

Particle Heat Up Credit and Hold Tube Lethality

A second possible approach was introduced by Dail.[1] This approach considers only the heat received by the particle as it moves through the hold tube. The assumption made is that the temperature of the slowest heating region of the particle as it enters the hold tube is the same as when it entered the heat exchanger part of the system. The thermal contribution of the heat exchangers, therefore, is ignored. A second assumption is that the temperature of the carrier medium in the hold tube is constant and equals the minimum temperature at the end of the hold tube. A mathematical model is used to calculate the lethality at the center of the particle given the above conditions. This approach should be very conservative and would result in a more drastic "overprocessing" of the carrier medium than the first approach discussed.

Both the first and second approaches involve a determination of the residence time for the fastest particle as it moves through the hold tube. The rate of particle movement in relation to the carrier medium and the influence of both solid-to-liquid ratio and particle density on the flow behavior of the particulate slurry are crucial factors.

Combined Heat Exchanger and Hold Tube Lethality

A third possible approach, and by far the most complex, involves using the combined thermal contribution of the heat exchangers plus that accumulated in the hold tube. The contribution during cooling would not be considered, since particle integrity cannot be controlled, and a fractured particle may cool at a faster rate than predicted. In order for this approach to work, minimum temperature profiles would have to be determined for the carrier medium and combined with fastest particle residence times in both the heat exchangers and the hold tube.

KEY CONSIDERATIONS

All three possible approaches involve a determination of the heat transfer condition at the carrier medium/particle surface interface. These heat transfer conditions will be influenced greatly by the relative velocity of the particle compared to the surrounding medium. The relative velocity of a particle may be minimal within a hold tube but could be radically different when considering the same particle moving through a scraped surface heat exchanger. The rheological properties of the carrier medium at the critical interface between particle and medium will also need to be defined in order to reliably predict the amount of heat transferred to the particle surface and, via conduction, into the slowest heating region of the particle.

Whatever approach is selected, a mechanism would have to be established to verify that the model accurately predicts conditions within the aseptic system. Process simulators could be used that would allow direct measurement of particle temperatures under controlled conditions. Coupled with bacterial inoculated pack studies under controlled and commercial conditions, these mechanisms would provide the strongest evidence that the predicted process reliably produces the intended lethality and can be controlled within predefined operating conditions.

Of the three approaches outlined, a model that allows the processor to take into account all thermal lethality accumulated in the heat exchangers and hold tube would allow the calculation of an optimal process in terms of nutritional and other quality attributes in the finished product. While this approach is more complex, if all of the factors necessary for the establishment of safe processes are identified, few problems should be encountered in obtaining regulatory acceptance.

SUGGESTED RECOMMENDATIONS

- Develop mathematical models for predicting heat transfer through suspended particles and residence time distributions.
- Determine microbial resistance and enzyme inactivation at ultrahigh temperatures.
- Develop process verification techniques.

REFERENCE

1. Dail, R. 1985. Calculation of required hold time of aseptically processed low-acid foods containing particulates utilizing the Ball Method. *J. Food Sci.* 50:1703–1706.

41. Novel Processes—Ultra High Pressure Processing

Daniel F. Farkas

University of Delaware

NATURE OF THE PROBLEM

The effects of high hydrostatic pressure on both foods and microbes in food are receiving increased attention. Recent work by Macfarlane[1] and Elgasim[2] indicates that pressures in the range of 1000 to 1500 atmospheres will accelerate the tenderization of pre-rigor beef. These pressures combined with temperatures of 60C were found to hasten the tenderization of postrigor meat. Recent work by Kurth[3] indicates that high pressure can activate in vivo proteolytic enzymes that may act on connective tissue. Pressures in the 2000 to 4000 atmospheres range will inactivate vegetative forms of microbes at 25C. Elevated temperature of about to 100C is needed to potentiate the effect of high pressure on spores. Metrick et al.[4] showed that *Salmonella sp.* in chicken could be inactivated at 23C by pressure treatments in the range of 2000 to 3300 atmospheres.

Commercial high-pressure metal working equipment capable of treating one cubic foot of material at 2000 atmospheres is available off the shelf. This equipment has the potential for cost-effective pressure treatment of large piece form foods (beef, poultry, fish) at or near room temperature to achieve tenderization, shelf life extension, and other useful applications.

Pressure is uniform throughout the food, and treatment is unique in that the effects do not follow a concentration gradient nor change as a function of time. Liquid foods can be pumped to treatment pressures, held, and then decompressed aseptically for filling as with other aseptic processes.

Hydrostatic pressure exerts its effect on biological systems by volume change. At 3000 atmospheres, water can be compressed about 10%. Presumably, hydrostatic pressure is a physical treatment that should not cause extensive chemical changes in food systems. However, native proteins appear to become denatured after exposure pressures in excess of 1000 atmospheres. Studies have shown that denaturation is reversible, depending on such variables as exposure pressure, time, temperature, and pH.

The interaction of these variables and their effects on the creation of unique reaction products, while not likely, should at least be studied in parallel with conventional heat treatments at atmospheric pressure. To date, very little is known about the effects of hydrostatic pressures above 1000 atmospheres on food with respect to:

- Flavor changes
- Nutritional changes, particularly effects on protein digestibility
- Structural changes
- Package-food interactions

HISTORICAL PERSPECTIVE

Extensive research findings are available on the effects of hydrostatic pressure on isolated enzyme, protein, and microbial systems. Much work on intact animals and cells has been done in conjunction with deep sea marine biology studies. Ocean depths can create a pressure of 1000 atmospheres at the bottom of the deepest trenches (35,000 feet), according to Sleigh and Macdonald.[5]

Numerous articles can be found in the biophysical literature. Much of the work is an attempt to understand the effect of pressure on the rate of protein, DNA, RNA, and membrane structure change as a function of pH, ionic or chemical composition of the suspending media, and temperature. Pressures in excess of 6000 atmospheres have been studied, and sample sizes are on the order of a few cubic centimeters. Most studies used well-defined biological samples and suspending media.

Workers in Australia (CSIRO) and later at Oregon State University have studied extensively the effect of pressure on meat biochemistry. The objective of this work, which is summarized by Macfarlane,[1] was to develop a commercial method for tenderizing beef. Nutritional evaluation of pressure-treated beef in terms of protein digestibility showed pressure-treated samples to be similar to nontreated controls.

Since the application of high pressure is very new, there are few studies addressing potential food safety problems. Two areas may be identified that may be of food safety interest.

POTENTIAL PROBLEMS AND SOLUTIONS

Microbiological

In some respects, the use of high pressure to reduce microbial levels in foods is similar to mild heat or irradiation treatment. These treatments can selectively reduce the number of certain groups of microbes while others, particularly spore formers, are little affected. If these treatments are combined with refrigeration and possibly modified atmosphere packaging, it may be possible to extend product shelf life from several days to several weeks. Since pressure treatment has only a slight effect on food enzymes, some heat may be required for certain foods to inactivate in vivo enzymes to prevent off flavors and changes in color, nutritional value, and structure. The overall microbiological safety of foods pressure-treated to extend refrigerated shelf life should be evaluated. The influence of oxygen tension, water activity, pH, and temperature on the specific lethality for different microorganisms should be investigated. Guidelines for exposure times and other processing parameters need to be established.

Packaging

A second area of food safety associated with high-pressure treatments is package-product interactions. Little information is available on the effect of high pressure on food packaging materials and on the interaction of food and polymeric materials during and after pressure treatments to 4000 atmospheres. Work on the transport of extractables in packaging material is needed to determine the effect of pressure on the rate of extraction as a function of pressure and temperature.

REFERENCES

1. Macfarlane, J.J. 1985. High Pressure Technology and Meat Quality. Chapter 6 in: Ralston, Lawrie, Ed. Developments in Meat Science—3, Elsevier Applied Science, NY, pp. 155–184.
2. Elgasim, E.A. and Kennick, W.H. 1980. Effect of pressurization of prerigor beef muscles on protein quality. *J. Food Sci.* 45:1122–1124.
3. Kurth, L.B. 1986. Effect of pressure-heat treatments on cathepsin Bl activity. *J. Food Sci.* 51(3):663–664.
4. Metrick, C.M., Farkas, D.F., and Hoover, D. 1986. Effect of pressure on survival of *Salmonella enteritis*. Abstract Symposium on High-Pressure Biology, Toronto, July 8–10, 1986.

5. Sleigh, M.A. and Macdonald, A.G. (Eds.). 1972. The effect of pressure on organisms. University Press, Cambridge.

42. Biotechnology and Food Protection*

Nanette Newell

Biotechnology Consultant

INTRODUCTION

Imagine animal feed that does not have to be supplemented with amino acids and vitamins, nonnutritive sweeteners that really taste like sugar, low-calorie beer, natural and inexpensive flavors and fragrances, and simple diagnostic kits to detect pathogenic organisms. Biotechnology is bringing us all this and more.

New biotechnology has been touted as the most important industrial advance since semiconductors, and the predictions of sales of biotechnology products by the year 2000 range from tens to hundreds of billions of dollars. One reason biotechnology has caught the interest of so many people is its broad applicability across many industrial sectors, from pharmaceuticals to chemicals to electronics. The food and food processing industries have found that biotechnology can be used for a variety of products and processes, including food additives, improved processability of crops, improved processing methods, and detection of contaminants in food.

There is a wealth of potential biotechnology products in the food processing industry as seen in Table 1. Some of these products, such as amino acids and vitamins, are already made biologically, and new biotechnology is being used to improve their production processes and lower costs. New products, such as the nonnutritive sweeteners and natural flavors and fragrances, are more attractive commercial targets now that biotechnology can be used to produce them cost-effectively for the first time. Additionally, enzymes improved through biotechnology will find their way more and more into a variety of manufacturing stages, thus improving processing economics.

The food processor will benefit from biotechnology in a multitude of

*This paper was presented in general session to underscore the interdisciplinary nature of food protection. The reactions of a panel reflecting other disciplines represented in the Conference are not included here.

Table 1. Biotechnology Products for the Food Industry

Product	Use
Amino Acids	
Glutamic acid	Food additive
Aspartic acid	Aspartame
Phenylalanine	Aspartame
Methionine	Animal feed
Lysine	Animal feed
Tryptophan	Animal feed
Vitamins	
B_2	Human and animal diet supplements
B_{12}	
C	
E	
Enzymes	
Alpha-amylase	High-fructose corn syrup
Glucoamylase	High-fructose corn syrup
Glucose isomerase	High-fructose corn syrup
Amyloglucosidase	Light beer
Pullulanase	Light beer
Proteases	Food treatment
Rennin	Cheese
Low-Calorie Products	
Aspartame	Nonnutritive sweeteners
Thaumatin	Nonnutritive sweeteners
Monellin	Nonnutritive sweeteners
Stevioside	Nonnutritive sweeteners
Modified triglycerides and fatty acids	Cooking oil and food additives
Microbial Polysaccharides	Thickeners and gelling agents
Flavors, Fragrances, and Colorants	
Improved Organisms	
Yeast	Alcoholic beverages and bread
Lactic acid bacteria	Milk fermentation
Crop plants	Better nutritional and processing characteristics
Single-Cell Protein	Animal and human food supplement
Food Testing	
Monoclonal Antibody Kits	*Salmonella*, aflatoxin, *botulinum*

ways. The benefits could include improved product margins, improved processing economics, or unique product advantages. In addition to the use of new enzymes, much of the improvement in processing will come from newly engineered characteristics of crops. For instance, the processed tomato industry could significantly reduce overall processing costs by using tomatoes with higher than average solids content, since a major cost of processing tomatoes is water removal.

Biotechnology in the food industry will be important, but how quickly and to what extent is difficult to predict. The actual timing for the research and development for these products and for their incorporation into retail food and feed will undoubtedly not move as quickly as the technology will allow. Compared to the pharmaceutical industry, which has pushed the limits of biotechnology, the food industry has moved at a snail's pace in adopting biotechnology. Whereas the pharmaceutical industry is technology driven and spends more than 10% of its sales on research and development, the food industry is more marketing and cost driven and spends less than 1% of sales on research and development. Thus, the food industry can be expected to move slower in its research and development, both because of the cost of the research and because the industry is less willing to take risks with new technology.

With a few notable exceptions, the introduction of biotechnology into the technical repertoire of the food industry will present no new food safety issues. The products of biotechnology should be treated like any new food or food additive, regardless of the process by which they are made. To date, regulatory and industry groups have agreed on this approach. For instance, issues of toxicity of food or food additives are the same whether the process by which they were made used either biotechnology or some more conventional process. Biotechnology will, however, change some aspects of the food industry.

NEW FOODS

The most obvious change that biotechnology will bring to the food industry is new foods. We have all heard of the grand schemes to devise the perfect crop, one that is easily harvested, tastes great, is nutritionally excellent, and is not fattening. This particular crop is a long way off, considering our ability to manipulate genetic information, but aspects of it are being developed now. For instance, much research is being done on the quality of seed storage proteins. Eventually, we will be able to manipulate the seed proteins of crops, such as corn and wheat, so that they contain all the essential amino acids. Another possibility is that the fatty acid chain length could be manipulated in oilseed crops, such as soybeans and rapeseed, to produce fatty acids that are shorter and consequently easier to digest.

New foods from biotechnology can also come in the form of microbial products. Biotechnology can be used to tailor microbial polysaccharides for use as thickeners and gelling agents. And single-cell protein, which is just a simple extract of a large-scale microbial growth, can be used as animal and feed supplements. A new single-cell protein is being made now by a joint venture between ICI and Rank Hovis McDougall in England. They are growing a particular fungus that, when isolated and processed, gives a stringy texture. This single-cell protein has been processed into food bearing good resemblances to chicken, beef, and pork. The Queen of England is rumored to have liked it when it was served to her!

Because of biotechnology's ability to mass produce compounds that are normally found in nature in very small quantities, some of these compounds for the first time are amenable to large-scale production. Examples include nonnutritive sweeteners found in plants, such as thaumatin and stevioside, and flavors and fragrances, such as vanilla and rose oil. These compounds are the products of complex metabolic pathways and, therefore, are probably more likely to be made in the short term by culturing the particular plant cells rather than by cloning the pathways into production microorganisms.

Biotechnology's promise for new foods is nearly limitless. But there are other contributions biotechnology is making to the food industry.

FOOD PROCESSING

Biotechnology can contribute to the processing of foods in two ways. The first is through the development of new crops with improved processing characteristics. As mentioned earlier, if processing tomatoes could be made with higher solids, or less water, the processing industry could function more economically. It has been estimated that for every one-percent increase in tomato solids, the tomato processing industry would save $80 million per year. These savings would come from lowered raw product volumes, lowered raw material transportation costs, lowered processing energy costs, and increased productivity without facilities expansion.

The other mechanism by which biotechnology can contribute to food processing is through the development of new and improved enzymes for the processing of food products. The food processing industry is currently the largest consumer of industrial enzymes, making up about 40% of a $400 million market. For instance, glucose isomerase is used to make high-fructose corn syrup, and rennin is used to make cheese. However, only a handful of enzymes are now used in the food industry. In many cases, they present enough disadvantages to outweigh their major advantage: to carry out complex chemical reactions with a high degree of specificity. Enzymes used industrially must be able to withstand the rigors of production processes, e.g., heat. With advances in biotechnology, the use of enzymes in the food processing industry should increase.

PRODUCT TESTING

One of the themes that emerged from the Second National Conference for Food Protection is the need for rapid and inexpensive screening methods for contaminating microorganisms in food. This is one area where biotechnology will have an enormous impact. Diagnostic kits are being developed using extremely specific monoclonal antibodies to detect contaminating microorganisms. These test kits are inexpensive, easy to use, and results are generally available in less than one hour. There are currently kits on the market that test for *Salmonella* and aflatoxin. The development of new kits is not a difficult process, and market demand should prompt the development of many useful kits.

REGULATION

The development of new biotechnology has caused a dramatic response from the regulatory agencies. The FDA's position is that it will regulate the product and not the process by which it is made. Thus, for food additives, the regulatory process is the same whether or not the additive is produced by a recombinant DNA organism. The FDA's statements on biotechnology regulation have concentrated on this type of example—production of compounds from genetically engineered microorganisms.

For foods themselves, the FDA's position is less clear. Food does not normally go through a regulatory process. Now, with the ability to alter the composition of food, the regulatory situation changes. For instance, the FDA would regulate a genetically engineered plant that makes an increased amount of a particular amino acid or vitamin by viewing the increased quantity as an "added substance." It should be noted, though, that the food contains no substances that are not normally found in food. On the other hand, if the plant has been engineered to contain a pathway for the production of a sweetener, that food now contains a compound that has never been in food. According to the FDA, this food also contains an "added substance." Even though the FDA puts both these examples in the same regulatory category, they present very different risks to those who ingest them. Another example of regulatory uncertainty is the modification of the fatty acid chain length in oilseed crops. If a naturally occurring oil with a particular fatty acid chain length already has GRAS status, will an oil that comes from a plant that has been engineered to contain that same fatty acid also be GRAS? Or will that oil need to go through a GRAS application? These examples show that the FDA still needs to clarify its position with respect to genetically engineered foods.

EDUCATION

With any new technology, education is of paramount importance. New technology tends to frighten people because they do not understand it and because there have been occasions where technology did not prove to be as safe as first thought. We are seeing this distrust of technology emerge again with the advent of biotechnology, and there certainly is good reason to fear this technology if one does not understand it. If all one hears is that the technology gives us the ability to change life at will, then one does have a right to be concerned. But with education, the public can be taken through the steps of placing a new gene into an organism and shown that this change is not dangerous and that, in fact, we are not really "creating new life-forms" like something out of a science fiction book. It is critical that we do this because the public has to be able to make an informed decision on the risks vs the benefits of this new technology. In this country, the public as a whole has the ability to slow or halt the development of a technology, as exemplified by the nuclear power industry.

It is terribly unfortunate that our public education system does not place more importance on teaching the basics of science. The consequence is that all of us must make the effort to educate the public by visiting schools; talking to civic groups; teaching continuing education classes; encouraging industry to make films, produce booklets, and fund museum exhibits; and encouraging the press to be honest and unbiased in its reporting. The benefits of biotechnology are far too important to have them taken away by an ill-informed public.

CONCLUSION

Biotechnology's greatest contribution to the food industry will be new foods and new processing methods. Biotechnology will also be used to produce new food additives, which will be "natural" because they will be biologically produced. However, the major contribution of biotechnology to food safety will be the introduction of rapid and inexpensive diagnostic kits to detect the presence of harmful microorganisms in food. And finally, we must all be educators so that the benefits of biotechnology are not delayed or lost because of irrational fears about a new technology.

43. New Foods, Processing, and Packaging Committee Recommendations

On Genetic Engineering

RECOMMENDATIONS

1. Prepare a position paper on research needs for genetically engineered food to be widely distributed to government agencies, Congress, consumer groups, industry, and academia. The paper should cover, among other things, DNA probe technology for toxicological tests, potential deleterious effects of gene regulations in plants or other foods, and potential effects on nutrient composition.

2. Prepare a position paper for wide distribution on regulatory considerations for genetically engineered foods. Specific topics would include the need for criteria and protocols consistent with federal and standardized international regulations to ensure the timely review of genetically engineered foods and food ingredients.

3. Develop an educational liaison activity of the Conference for Food Protection to provide scientific lectures on topics including genetic engineering principles, risk-benefit analysis, and nutritional considerations.

On Food-Package Interactions

RECOMMENDATIONS

1. Establish a government, industry, and academia consortium whose goal is to secure and administer sufficient funds to support the following critical needs:

    - Extraction methodology to include wide temperature ranges, pressure ranges, radiation and heat transfer techniques (conduction, convection, and radiation)
    - Vastly improved solvents and food simulants which will

reflect the effects of food components on plastics, thus altering their extractability under a wide range of conditions

- Fundamental research to establish a database for prediction of diffusion and solubility of plastic packaging construction materials.

2. Establish a central clearinghouse and database on diffusion and solubility of plastic package construction materials.

On Irradiation Processing

RECOMMENDATIONS

1. Encourage further research and assessment of the public health problems that would be best eliminated using irradiation technology, such as irradiation of poultry for the control of *Salmonella*.

2. Promote the adoption of the General Codex Standard for Irradiated Foods and the Recommended International Code of Practice for Operating Radiation Facilities for the Treatment of Food promulgated by the Codex Alimentarius Commission.

3. Assist the International Atomic Energy Agency (IAEA) in the expansion of the database on radiation sterilization of food with doses above 10 kGy for establishing a framework for extrapolating wholesomeness data within generic classes of foods.

On Aseptic Processing of Particulates

RECOMMENDATIONS

1. Promote the convening of a workshop to identify research needs and develop priorities and implementation for aseptic processing of particulates. Topics would include:

   a. Research and verification of heat transfer to particles
   b. Methodology for measuring residence time distribution of particles under practical operating conditions
   c. Equipment to produce and control uniformity of size of food particles
   d. A model for lethality delivered to particles that includes heating and holding

   e. Kinetic parameters for effects of UHT on microorganisms or their spores, on enzymes, on nutrients, and on other quality factors

2. Encourage the Education and Training Committee to address effective and comprehensive training of operators and supervisors of aseptic systems.

On Novel Processes

RECOMMENDATION

1. Promote development of a framework to be used to evaluate the safety and regulatory aspects of novel processes and products.

44. Conference Program Committee Conclusions and Recommendations

The structure of the Conference is a matter of concern. A full and equal partnership between industry and government is necessary and desirable at the committee level. At the council level, however, a mechanism must be developed to allow state regulatory ratification and acceptance of items coming from any committee.

RECOMMENDATIONS

1. A state regulatory ratification section will be established within the government council whereby each of the 50 states in the United States will have one vote. Any item that impacts on U.S. regulatory policy in the states must be accepted through this section. A method of certification of state delegates, as well as council revision, will be referred to the Constitution and By Laws Subcommittee. This revision will be formulated and sent to the membership for ratification so that states may be requested to furnish delegates to the 1988 conference.

2. The Committee identified constitutional issues related to items such as committee and council structure, membership fees, and voting which are in need of revision. These will be referred to a Constitutions and By Laws Subcommittee which will be created to review the entire Constitution and By Laws and to formulate recommendations for the Executive Committee to consider.

3. The Committee ratified the proposal of a change in the Conference's articles of incorporation which, in accordance with the requirements of the Constitution, had been mailed to members (in the Conference quarterly newsletter) 45 days in advance of the 1986 Conference. The proposal relates to the addition of three amendments to the articles of incorporation which the Internal Revenue Service has advised are necessary for the Conference to qualify for tax exempt status. These are:

   AMENDMENT 1. Said corporation is organized exclusively for charitable, religious, educational, and scientific purposes, including, for such purposes, the making of distributions to organizations that qualify as exempt organizations under section 501(c)(3) of the Internal Revenue Code, or corresponding section of any future federal tax code.

AMENDMENT 2. No part of the net earnings of the corporation shall inure to the benefit of, or be distributable to all its members, trustees, officers, or other private persons, except that the corporation shall be authorized and empowered to pay reasonable compensation for services rendered and to make payments and distributions in furtherance of the purposes set forth in Article Third hereof. No substantial part of the activities of the corporation shall be the carrying on of propaganda, or otherwise attempting to influence legislation, and the corporation shall not participate in, or intervene in (including the publishing or distribution of statements) any political campaign on behalf of any candidate for public office. Notwithstanding any other provision of these articles, the corporation shall not carry on any other activities not permitted to be carried on (a) by a corporation exempt from federal income tax under section 501(c)(3) of the Internal Revenue Code, or corresponding section of any future federal tax code, or (b) by a corporation, contributions to which are deductible under section 170(c)(2) of the Internal Revenue Code, or corresponding section of any future federal tax code.

AMENDMENT 3. Upon the dissolution of the corporation, assets shall be distributed for one or more exempt purposes within the meaning of section 501(c)(3) of the Internal Revenue Code, or corresponding section of any future federal tax code, or shall be distributed to the federal government, or to a state or local government, for a public purpose. Any such assets not so disposed of shall be disposed of by the Court of Common Pleas of the county in which the principal office of the corporation is then located, exclusively for such purposes or to such organization or organizations, as said Court shall determine, which are organized and operated exclusively for such purposes.

4. The Committee discussed the frequency of the general conference which currently is every two years. It was agreed by the Committee that no change in the frequency should be made and further recommended that more activities should take place during the off year and that the membership should be kept fully advised of actions taken through the Conference quarterly report.

5. The Committee recommends that the constitution be changed by adding representatives of the World Health Organization (WHO)

and the Pan American Health Organization (PAHO), to the list of organizations (EPA, USDA, HHS, and FAO) now represented on the Executive Committee as advisory members. The addition of these organizations would benefit the international scope of the Conference. The change would read as follows (new addition in brackets):

> *Section 3.* The Executive Committee shall consist of the three officers of each Conference Committee, one nonvoting representative each from the EPA, USDA, HHS, FAO, [WHO, and PAHO] who serve in an advisory capacity, and the executive officer.

If this recommendation is approved, the change would be solicited by a mail vote of all the members in the next membership mailing.

6. The Committee recommends that the Executive Committee give consideration to holding the next meeting either in St. Louis, Missouri in May or in Orlando, Florida in March.

7. The Committee recommends adopting a theme which includes the idea of establishing cooperation between diverse food organizations, bridging their diverse interests into a unified effort, or progress through partnership. Whatever the theme, it should be related to topics of immediate concern at the time of the Conference.

8. The Committee agreed that the Conference agenda should emphasize issues of immediacy rather than traditional, boilerplate problems. The Committee discussed and recognized that there were a number of significant issues of immediate concern that were not covered under the current agenda. Examples of such issues are: food tampering, *Listeria*, veterinary drug residues in foods, and preparation and acceptance of national standards for plan review for retail level food establishments. The Committee agreed that the issues were of such high concern that consideration should be given to the establishment of a series of interim task forces to review the issues and recommend appropriate action for implementation.

9. The Committee reviewed the recommendation for establishing additional committees, e.g., a Food Chemistry Committee, and agreed that at this time no additional committees should be established but rather that consideration should be given to include the specific matter of food chemistry under the work of the Toxicology Committee. [No action was taken on this recommendation.]

10. The Committee further agreed that the establishment of task groups

under the various committees might be highly desirable. This was especially true under the Standards and Regulations Committee where two task forces—one on food service, food vending, and food stores, and the other on food processing—might be established. Further, the Education and Training Committee might be divided into two work groups—one on consumer affairs and the second on training of industry and regulatory officials.

11. It was the unanimous view of the Committee that the Conference should operate as an action organization, existing not merely to identify problems and formulate recommendations, but to resolve issues through the implementation of recommendations, much as the Weights and Measures Conference and the Interstate Milk Shippers Conference do. Specific recommendations in this regard will be presented prior to the next Conference.

INDEX